WHAT THE PHILOSOPHY OF BIOLOGY IS

Nijhoff International Philosophy Series

VOLUME 32

General Editor: JAN T.J. SRZEDNICKI
Editor for volumes on Applying Philosophy: LYNNE M. BROUGHTON
Editor for volumes on Logic and Applying Logic: STANISLAW J. SURMA
Editor for volumes on Contributions to Philosophy: JAN T.J. SRZEDNICKI
Assistant to the General Editor: DAVID WOOD

What
the Philosophy of Biology
Is

Essays dedicated to David Hull

Edited by

MICHAEL RUSE

University of Guelph
Guelph, Ontario, Canada

Kluwer Academic Publishers

DORDRECHT / BOSTON / LONDON

Library of Congress Cataloging in Publication Data

What the philosophy of biology is : essays for David Hull / edited by
  Michael Ruse.
        p.   cm. -- (Nijhoff international philosophy series)
    "Publications of David Hull": p.
    Bibliography: p.
    Includes index.
    ISBN-13:978-94-010-7020-1   e-ISBN-13:978-94-009-1169-7
    DOI: 10.1007/978-94-009-1169-7

    1. Biology--Philosophy.   I. Ruse, Michael.   II. Hull, David L.
III. Series.
QH331.W466 1989
574'.01--dc19   88-24075
                                                         CIP

ISBN-13:978-94-010-7020-1

Published by Kluwer Academic Publishers,
P.O. Box 17, 3300 AA Dordrecht, The Netherlands.

Kluwer Academic Publishers incorporates
the publishing programmes of
D. Reidel, Martinus Nijhoff, Dr W. Junk and MTP Press.

Sold and distributed in the U.S.A. and Canada
by Kluwer Academic Publishers,
101 Philip Drive, Norwell, MA 02061, U.S.A.

In all other countries, sold and distributed
by Kluwer Academic Publishers Group,
P.O. Box 322, 3300 AH Dordrecht, The Netherlands.

David Hull

Table of contents

VIII

Preface

Philosophers of science frequently bemoan (or cheer) the fact that today, with the supposed collapse of logical empiricism, there are now no grand systems. However, although this may or may not be true, and if true may or may not be a cause for delight, no one should conclude that the philosophy of science has ground to a halt, its problems exhausted and its practioners dispirited. In fact, in this post-Kuhnian age the subject has never been more alive, as we work with enthusiasm on special topics, historical and conceptual. And no topic has grown and thrived quite like the philosophy of biology, which now has many students in the field producing high-quality articles and monographs.

The success of this subject is due above all to the work and influence of one man: David Hull. In his own writings and in the support he has given to others, he has shown true leadership, in the best Platonic sense. It is now twenty years since Hull first gave his seminal paper 'What the philosophy of biology is not', and to mark that point and to show our respect, gratitude and affection to its author, a number of us who owe much to Hull decided to produce a volume of essays on and around themes to which Hull has spoken. This is not a *Festschrift* in any conventional sense – that must wait another twenty years – but rather an attempt to show that Hull's work has born fruit. Hence, although these are certainly essays 'for' David Hull, in the best philosophical spirit they are often 'against' him also. Although we (and he) have failed if we are not yet ready to stand on our own feet, we have invited Hull himself to make a contribution, not so much to comment on us as to point the way forward.

As editor, I have served purely as coordinator. The contributors have read and commented on each others' papers, which have obviously been ordered alphabetically. The two exceptions are Hull, who naturally comes at the end, and my own piece which was written deliberately as an introduction. Any royalties from this volume will be given to a charity of David Hull's choice.

David Hull Through Two Decades

MICHAEL RUSE

Departments of History and Philosophy, University of Guelph, Guelph, Ontario, Canada, N1G 2W1

In conclusion, there are many things that philosophy of biology might be. A philosopher might uncover, explicate, and possibly solve problems in biological theory and methodology. He might even go on to communicate these results to other philosophers, to scientists, and especially to biologists. He might show what consequences biological phenomena and theories have for other sciences and for philosophy or to show what consequences other sciences and even philosophy have for biology. These are some of the things which philosophers of biology might do. With rare exception, they have not. What philosophy of biology is not? It must be admitted that thus far it is not very relevant to biology, nor biology to it.

I first met David Hull in the fall of 1968, at the first meeting of the Philosophy of Science Association, in Pittsburgh. I had just returned from England, writing a thesis on the philosophy of biology, and so (thanks to a major paper on taxonomy in the *British Journal for the Philosophy of Science* (Hull 1965)) I knew who Hull was and was eager to meet him. We became good friends – part of a small group that included Kenneth Schaffner and William Wimsatt – and we remain good friends today. Like everyone else (a point to which I will return in a moment) Hull and I have had an extensive correspondence and I have received lots of friendly (but critical) advice. We have often worked on the same problems, for instance the philosophical underpinnings to Darwinism and (recently) the implications of an evolutionary perspective for the development of knowledge. The one big exception has been the sociobiological controversy, where I rushed in and he feared to tread, a not entirely inappropriate metaphor which I will explain shortly. (I do not, of course, imply that we have put equal effort into problems. I am a mere dilettante

M. Ruse (editor), What the Philosophy of Biology is. pp. 1–15.
© 1989 *Kluwer Academic Publishers, Dordrecht*

compared to his professionalism when it comes to taxonomy.)[1]

Perhaps we have been too close, although more likely I have been too concerned with my own interests and with what I can get out of the relationship, but I have never previously really thought in depth about Hull's own Philosophy (with a capital 'P') of biology. After twenty years, this is a good time to stop and take stock, for I think a particular position has emerged, although (something which an evolutionist like Hull would fervently trust) I think it is still evolving. Let me begin with Hull's 1968 PSA paper (Hull 1969), the concluding paragraph of which is quoted at the beginning of this essay. Although it is not the first of Hull's publications, by its very nature it is the most revealing.

'What Philosophy of Biology is not'

This paper was an extremely critical review of the then-recent literature in the philosophy of biology, a small but not vanishingly small body, written by those interested in the subject for its own sake (like T.A. Goudge (1961)) and by those who wanted to make a point somewhere else (like Mario Bunge (1961), who used Darwinism to underline some themes about simplicity). It was addressed to both philosophers and biologists – significantly it was published both in *Synthese* (a philosophy journal) and the *Journal of the History of Biology* (in those days, I would think, with a rather higher biological readership than today) – and had naught for the comfort of anyone. Typical was the treatment of Bunge, who gets taken to task for every one of his statements, right down to his having confused butterflies with moths. I remember when the paper was delivered, Bunge protesting that anyone could think such a point significant. But that was the very matter at issue. It was! 'It is certain that he would not treat quantum theory in such a cavalier fashion. The differences between mesons and pions are important. The differences between moths and butterflies apparently are not' (p. 244).

Reading again through the paper, I detect a dual message. To the philosophers, on the one hand, Hull wanted to say: 'Look! Take biology seriously. You may learn from the experience. You may, for instance, find that there is that in biology which challenges some of your thinking about science – thinking which has been based, hitherto, exclusively on the physical sciences.' This point comes through particularly in Hull's critique of a little analysis of Darwin's theory, given by the philosopher Anthony Flew (1959), claiming that the theory of the *Origin* is hypothetico-deductive. Hull shows that whatever the theory may be, Flew has singularly failed to make his case. 'Everything which Flew says is true, important, and needs saying, with one exception. Neither of his conclusions follows deductively from the premises which he presents' (p. 253). In this context, it is worth remembering that Hull was the student of Michael Scriven, who had previously argued that evolutionary theory does not fit then-standard models of explanation

(Scriven 1959), and who gets one of the few notes of praise in the essay.

To the biologists, on the other hand, Hull wanted to say: 'Perhaps you chaps might turn to philosophy once in a while, for you will find help to untangle some of the knots into which you are perpetually tying yourselves.' Hull picked up on a controversy which was then attracting biological attention, about which came first, taxa of higher levels or taxa of lower levels. Did we, say, get vertebrates before we got particular species, or was it the other way around? 'The major difficulty with this controversy is the logical crudity with which it is frequently expressed' (p. 263). Not that philosophers can take much comfort from this confusion, for those who have discussed the issue have failed to do 'anything to clarify the situation' (p. 264).

This is all stirring stuff – it certainly stirred me at the time – although very negative. To his credit, however, Hull was heeding his own calls to action, for at about the time of writing this paper he did write (at least) two others, taking the message of philosophy to biologists (publishing in biology journals), showing that a lot of silliness could be avoided by drawing on what we (philosophers) have already done (Hull 1967, 1968). One of these papers, particularly, is still worthy of note, for Hull showed that Percy Bridgman's philosophy of operationalism is not quite the answer to life's problems that some have suggested it is, and that its popularity among taxonomists shows more of their ignorance than of a deep commitment to uncovering the truth. (Appropriately, Hull has recently written a repeat paper, with Karl Popper as its focus (Hull 1983b).)

With hindsight, though, one sees more than negativism. Hull was starting to grope, positively, towards some sort of view of biology which does recognize the subject as having a nature and integrity of its own. I suppose the best term for such a philosophy is 'organicist', and it is worth noting that although Hull did have methodological objections to Morton Beckner's positive treatment of that philosophy in his *The Biological Way of Thought*, Hull did also say that 'Beckner's book remains the single major contribution of a philosopher to biology in over a decade' (p. 267).

Partly, Hull's positive position was defined by the already-mentioned negativism towards then-existing work by philosophers. But also it was partly defined by Hull's curious treatment of Marjorie Grene's notorious paper 'Two evolutionary theories'. In this paper, published in the *British Journal for the Philosophy of Science* in 1959, Grene had argued for a saltationist theory of evolution, favoured by the German paleontologist O. Schindewolf (1950), over the orthodox neo-Darwinism of the American, G.G. Simpson (1953). Or, at least, this was what the biological community had taken her to be doing, and they had jumped on her accordingly. Although Hull spoke in his acknowledgements of having 'severely criticized' Grene, his treatment was in fact most moderate and one senses a strong sympathy for what Hull took Grene really to have been doing – trying to set up the conditions (in an almost Kantian way) for any adequate evolutionary theory whatsoever.

More importantly, although this is all mixed up with language and positions to which people would not now subscribe and is stirred with a double dose of hindsight. I sense a sympathy for the kind of holistic philosophy that Grene admired in Schindewolf. Not that Hull was a saltationist. This, by best biological standards, is wrong. But there was a liking for the idea that Schindewolf represents a position which says organisms have certain features which belong to organisms in their own right, and which cannot be reduced to lower levels of organization. (What makes this extrapolation so difficult is that, rightly, Schindewolf is read as an essentialist, and already Hull had staked out a position against this. The trick would be to get Schindewolfian non-reductive holism, without the essentialism. And without the saltationism, of course.)

There is another dimension to Hull's response to Grene, which I would argue is as important as the intellectual response. Although Hull 'severely criticizes' Grene, he had sent his paper to her and in discussion had given her work a remarkably fair treatment. It would have been hard for Grene to take umbrage, although as Hull noted she had responded with vigour to other critics. I think this attitude was very deliberate by Hull. He took very seriously the poor state of philosophy of biology and thought that he (having criticized) must do something. And the way to do something was by networking, by building alliances, and thus moving forward. I remind you of the correspondence that he and I have had over the years, and I note that I am one of many.

I want to push this point and argue that Hull's social behaviour is part of his philosophy. This is not such a ridiculous idea. Socrates' behaviour (his intense dialogues with his followers, his trial and death) were part of his philosophy, and the same seems to have been the case in a slightly different way of Wittgenstein. (Who can read anything by him without being affected by the aura of his tormented genius?) If you want a third example, Karl Popper surely provides it. In Hull's case, likewise, the man and the philosophy belong together. At a fairly uncontentious level, Hull's networking has surely led to choice of subjects. I suspect this is why he sat out the sociobiological controversy (despite sympathy with certain sociobiological ideas), because you had to alienate one group or the other, and perhaps in part this explains the delay in his much-awaited book on taxonomy. At a deeper level, the networking has become part of the philosophy, as I will explain later when I get to the work Hull is doing now. The ideas and the behaviour merge together. (In saying all of this, in no way do I want to question Hull's moral integrity or genuine kindliness. As I shall explain, if this were a critical paper, I would argue that his Humean sympathies undermined his Hobbesian theorizing.)

I suggest therefore, that in this early paper, we see the seeds of a positive organicist philosophy. How did they flower?

Anti-reductionism

Hull's first work in its own right, as opposed to criticizing philosophers or putting biologists right, was in philosophical taxonomy. But, although his interest continued in this (by 1974, he was conducting in-depth interviews with taxonomists about their positions), the major controversy engaging Hull for the first half of the 1970s was one about the nature of reduction. Specifically, there was a dispute over the proper relationship between the older, Mendelian genetics and the newer molecular genetics. The dispute reached a climax at an invigorating symposium of the 1974 PSA meetings, at Notre Dame, and matched Kenneth Schaffner (with me in support) against Hull, with Wimsatt somewhere on the side saying 'a plague on both your houses' (Schaffner 1976; Ruse 1976; Hull 1976b; Wimsatt 1976). So intense was Hull's own involvement, that the reduction question took up a very large portion of the text in the philosophy of biology he was then writing for the distinguished Prentice-Hall series *The Philosophy of Biological Science*. Reduction altogether squeezed out any discussion of taxonomy (to my amazement which continues to this day).

Basically, the dispute was straightforward. The leading logical-empiricist philosopher Ernest Nagel (1961) had argued that the relationship between an older theory T_1 and a newer theory T_2 is frequently one of deduction, where T_1 is shown to be a special case of T_2. (This is 'reduction', as opposed to 'replacement', where T_2 entirely wipes out T_1.) Nagel's student Schaffner (1967) picked up on this, noted that often what is deduced is a 'corrected' version of the older theory, T_1^*, but argued that as long as there was 'strong analogy' between T_1 and T_2^*, one could still properly speak of 'reduction'. Moreover, he argued that his analysis fits what has happened in genetics. Although one cannot deduce Mendelian genetics from molecular genetics, Mendelian genetics has been corrected into 'transmission' genetics (essentially by breaking up the classical gene into three sub-units), and this corrected genetics can be deduced, in principle at least, from molecular genetics.

Hull took off after this thesis with a fury. 'I find the logical empiricist analysis of reduction inadequate at best, wrong-headed at worst' (Hull 1974b, 12), adding that, 'the conclusion seems inescapable that the logical empiricist analysis of reduction is not very instructive in the case of genetics. For my own part, I found that it hindered rather than facilitated understanding the relationship between Mendelian and molecular genetics' (p. 44). His objections (I thought then and see now no reason to change) were three-fold. First, he did not think that an actual deduction had been effected between transmission and molecular genetics, nor was he *a priori* convinced that such could be done. Second, even if it were done, there would be no strong analogy between the deduced transmission genetics and the uncorrected Mendelian genetics. Third, Hull felt that the whole demand for a reduction was misleading and led you away from the true nature of science.

Looking back, as far as the first two objections are concerned, I am inclined to think that Schaffner and I gave as good as we got. We also got as good as we gave. Not all of Hull's arguments were watertight. However, as recent work on the Mendelian/molecular interface shows only too well, the relationship between old and new is a lot more complex than that dreamed of in the idyllic days of logical empiricism (Rosenberg 1985; Kitcher 1984). (What isn't?) But, as I noted then and still agree, it was the third charge that really stirred Hull. The very attempt to fit theories into a Procrustean bed of formalism struck Hull as wrongheaded, philosophically. It was not just that a Nagel-type reduction did not work. It *should* not work.

Why the opposition? I believe there were two main reasons, apart from the dislike of having a thesis devised for physics thrust down the biologists' throats.

First, there was the anti-reductionism of the holistic organicist. Francisco Ayala (1974) has usefully divided reductionist strategies into three. First, there is *ontological* reduction, where only material elements are supposed to exist. Second, there is *methodological* reduction, where the larger is explained in terms of the smaller. Third, there is *theory* reduction, where one theory is deduced from another. Hull's explicit target was the third category, theory reduction. But, he wanted to bite more deeply (as did all of the biologists, like Ernst Mayr (1969), whose work Hull respected greatly). Hull never aspired to ontological reductionism. At best this is dualism, at worst, vitalism. But he was after methodological reductionism, the dream (or fantasy, if you like) that organisms in their biological mode are no more than atoms, moving around together. Moreover, reference to atoms is all you need for full understanding. Obviously, if you get theory reduction at this point, any anti-reductionist stand at the methodological level is impossible. But, if you can block theory reduction, then the belief that organisms are 'something more' can be maintained.

If we interpret Hull's thinking in this way, then perhaps the following passage, which as I shall show fits uncomfortably if one interprets it historically, can properly be understood hierarchically.

> Most contemporary geneticists know both theories. They can operate success-fully within the conceptual framework of each and even leap nimbly back and forth between the two disciplines, but they cannot specify how they accomplish this feat of conceptual gymnastics. Whatever connections there might be, they are subliminal. In a word, those geneticists who work both in Mendelian and molecular genetics are schizophrenic. The transitions which they make from one conceptual schema to the other are not so much inferences as *gestalt* shifts... (Hull, 1973b, p. 626).

It is worth remembering that Hull was N.R. Hanson's student, so he had grown up with this kind of thought.

Second, leading Hull to his anti-reductionism, there was another major prong that Beckner finds characteristic of the organicist: historicity. By this is meant the taking seriously of the historical dimension (and not to be confused with 'historicism', the belief in sweeping historical laws). Not only is Hull a product of the department of *history* and philosophy of science at Indiana, by 1974 he had worked in detail on the Darwinian revolution (producing *Darwin and His Critics* in 1973 (Hull 1973a)). He had come to realize just how far real science is from the sanitized versions with which the logical empiricist works. Influenced also by Wimsatt, Hull appreciated that there really is no such thing as Mendelian genetics *per se*, just as there is no such thing as molecular genetics *per se*. At best, there are clusters of different views around two poles.

> On the logical empiricist analysis, scientific theories are individuated on the basis of substantive content. Two theories are two different theories because they differ with respect to one or more substantive claims. From this perspective, the molecular theory of Watson and Crick is different from the biological theory of Mendel, but Watson and Crick's theory is also different from the molecular theories of Jacob and Monod, Lederberg, and Kornberg as is Mendel's theory from those of de Vries, Correns, Morgan, Dobzhansky and Muller. But all these theories are not equally different. One would expect them to form two clusters, one roughly molecular, the other Mendelian. But they do not if these clusters are to be formed on the basis of the substantive claims made in these theories. As Wimsatt has pointed out, theories evolve in a process which he terms 'successional reduction' (Hull 1976b, 656).

What does this all add up to? The reductionist approach ignores the developmental nature of science.

> Ruse cannot see how the logical empiricist analysis of science could impede an adequate understanding of science. Here are but two ways. It ignores the temporal dimension to science and directs attention away from the chief means by which various stages of a scientific theory can be integrated to form a single theory (657).

What we have got here – or, rather, what we can infer from a footnote Hull made at this point – is the move on to the next two important aspects of his thought.

On the one hand, we are pushing towards an *evolutionary* conception of scientific change. The work of Toulmin (1972) is mentioned (I think approval was intended – elsewhere, Hull (1974a) was defending Toulmin against criticism).[2] On the other hand, Hull was wrestling with the way that he could characterize a scientific theory, or the group of people around a theory. As he pointed out, 'scientific theories are historical entities developing continuously in time' (Hull 1976b, 669, f.2). But, how is one to think of a group as an entity? The way to seek help is to switch for a moment from thinking about the group around a theory to thinking about the much better-known type of group, that which Hull had been worrying about in his ongoing concerns with systematics. I refer of course to the biological *species*. How does one characterize this and why does one think exemplifications (like *Drosophila melanogaster*) to be real or special? This is the species problem, and it is a problem with which the biologists Hull admired had been wrestling for years. Fortunately, help was at hand in a bold thesis by Hull's friend and fellow Darwin student, Michael Ghiselin (1974). Let us pick up first on this, for it will lead us back to the other of Hull's concerns.

Species as individuals

Obviously, at least most obviously, a species is a group, a class, a set. *Homo sapiens* is a class, and Michael Ruse and David Hull and Charles Darwin are members thereof. My dog Spencer is not. He has his own class: *Canis familiaris*. Hull himself affirmed this position (in his 1968 critique) – and yet, there were always tensions. If *Homo sapiens* is a class, and if it is a natural class – which it seems to be – then what is its essence? What are the defining characteristics, necessary and sufficient for membership? As Hull pointed out, in his first major paper (1965), species are just not the sorts of things which have essences. Take just about any feature you like to name, and the chances are that some perfectly good member of the species will not have it. Essentialism works just fine for mathematics. It may well work for physics and chemistry. (Recently, Hull (1985) affirmed that he has never denied essentialism *per se*.) It does not work at all for biology.

Until the mid-1970s, we all had to work with cluster or polytypic definitions (where you specify a list of features, possession of some of which are sufficient but no one of which is necessary for species membership). Then, Michael Ghiselin (1974) began championing a thesis which hitherto had seemed, not merely wrong, but just plain stupid. Species are not classes at all. They are individuals. And the relationship between organism and species is not one of member in set but of part to whole. As my hand is a part of me, so I am a part of *Homo sapiens*.

To Hull, needing protection against the threat of essentialism and needing a way of characterizing scientific movements, the species-as-individuals thesis fell as manna from heaven. He embraced it enthusiastically (Hull 1976a, 1978 – it is also accepted in Hull 1973c). Indeed, with his prosletizing skills, he became the Huxley to Ghiselin's Darwin. (This is an analogy that I am sure will please Ghiselin, and may appeal to Hull given what I shall have to say on his views about the need to sell one's ideas to one's fellows.)

Moreover, confirming the thesis I am proposing in this paper, we see Hull's organismic inclinations become more and more obvious. On the one hand, we have a crucial emphasis on *organization*. Why do we think that something is an individual? Because it is organized. That is why we think of a business or a country as an individual, even if not all of the parts are touching. But, we have organization in species. Therefore, it is right and proper to think of them as individuals.

On the other hand, what is it that marks an individual? That it has a unique *history*. An individual is born, lives and dies. It can never be repeated, but any amount of change is allowed so long as there is continuity. Likewise with species. They come into being and change through evolution and go extinct. They can never be repeated. The dodo is gone forever. And the important point about all of this – and we know now how important this is for Hull – is that the thesis that a species is an individual only makes sense when set against the context of biological theory. It does not matter what common sense says. Evolutionary biology treats species as individuals. That is enough – more than enough – for Hull.

The species-as-individuals thesis has attracted a great deal of attention, both from philosophers and from biologists. Some have rejected it, or looked rather warily at it, but a large number have embraced it enthusiastically. Above all else, it must be satisfying to Hull as a philosopher that his work – within ten years of his critical review paper – was being taken so seriously by practising scientists, who were using the s-a-i thesis in their actual science. (Paralleling this intellectual influence was the continuing social influence, which culminated recently in Hull's serving a president of the Society for Systematic Zoology – at the same time as he was president of the Philosophy of Science Association.)

However, to underline my theme about the organicist nature of Hull's thought, it is worth looking at precisely which scientists (philosophers) have embraced the s-a-i thesis, and which have rejected it. Without wanting to get into linguistic nit-picking about the nature of a 'true' Darwinian, those who have ultra-Darwinian yearnings, by which I mean they take adaptation to be the all-pervasive fact about organic nature and they think natural selection to be the beginning and end of causation, have tended to draw back from the s-a-i thesis. For them – or, rather, I should say 'for us', for I am somewhere to the right of Archdeacon Paley on adaptation – species just cannot have the desired integration to count as individual-hood (Ruse 1987). Moreover, with the gradualness that we see coming out of the selective process, we worry that a species could go on changing indefinitely and still be counted the same species.

However, those who reject ultra-Darwinism have readily and happily embraced the s-a-i thesis. This includes those philosophers with holist leanings like Elliott Sober (1980), and biologists of two kinds (which often overlap). On the one hand, we have the cladists, for whom the historical dimension of species is all important. Just as the Ghiselin/Hull thesis demands, they think a species lasts as long, but not a moment longer, as there is no speciation. (See Cracraft 1987 for full details.) On the other hand, we have the supporters of punctuated equilibria theory (Eldredge 1985; Eldredge and Cracraft 1980). Again you would expect sympathy for the s-a-i thesis, given the way that the theory supposes species to have fairly clean cut (or, rather fairly clean cut at each end) histories.

In fact, the story of this theory of punctuated equilibria has a rather revealing

history. The early version of the theory was presented as a straightforward consequence of Darwinism (Eldredge and Gould 1972). It is true that this was before the
s-a-i thesis (or, at least, before the thesis was recognized as such), but the tenor was
very much against the idea of the thesis. We were then assured that any group
effects in a species stem from individual properties. However, as the theory
evolved into something at least expanding, if not outrightly challenging ultra-
Darwinism, and as the significance of adaptation was played down, enthusiasm for
the s-a-i thesis grew by leaps and bounds. The thesis dovetailed nicely with the
increasingly *hierarchical* structure that was being built into punctuated equilibria
theory, with that which happens at the macro-level in no way reducible to that
which happens at the micro-level (Gould 1982). Species as individuals have their
own lives and forces and histories, and these are not necessarily the lives and forces
and histories of species at lower levels.

I suggest, therefore, that Hull has contributed not only to an organicist
philosophy of biology but also to an organicist biology. I confess that I am a little
surprised that, in recent years, Hull has not entered more enthusiastically into the
task of making his philosophy work as science. However, by and large he has
stayed on the sidelines publically endorsing neither cladism nor punctuated
equilibria theory. In part, I put this down to his general tactic of not taking sides,
but of being one to whom all can turn. I know also that as he has been working on
his large book which covers taxonomic disputes in some depth, he has felt that
more is to be gained intellectually by restraint. However, in greater part I put down
Hull's unwillingness to follow through his philosophy as science to the fact that he
has other concerns – concerns which are important to him *as a philosopher*. With
the species-as-individuals thesis safely launched, particularly with its implications
for a hierarchical view of natural processes, he has been able to turn back to the
problem of scientific change.

Evolutionary epistemology

The hope that biology can provide *the* true world picture, from which all understanding flows, is the ultimate organicist dream. With the coming of evolutionary
ideas, it has seemed to many that the dream might be realizable. Before Darwin,
even, Herbert Spencer (1857) was arguing (inasmuch as he ever argued, as opposed
to stating flatly) that everything is in a state of development, from the
homogeneous to the heterogeneous, and that one major law of progress leads to all
understanding. Grand systems like that of Spencer have long been out of favour,
but in recent years, a number of people have argued that scientific change is indeed
best understood on a biological model. Thus, we have had the development of
'evolutionary epistemology', chief exponents of which have been Donald Campbell
(1974), Stephen Toulmin (1972) and Karl Popper (1972). The first worked for

many years in Chicago, Hull's home town, the second still does (he and Hull are now colleagues), and Hull was a student of the third (imbibing, amongst other things, Popper's well-known dislike of essentialism).

We have seen already how, by the mid-1970s, Hull was moving towards an evolutionary conception of science (citing Toulmin favourably) and how this move to the individuality of species was certainly connected with this.[3] In recent years, Hull's aim has been to develop and articulate to the full such an adequate evolutionary epistemology. (I suspect that, apart from reasons already mentioned, the long gestation of Hull's much-promised book on taxonomy has come because his interest has shifted from taxonomy *per se*, to its use as a case study for his model of evolutionary epistemology.)

As several commentators have noted, there are two basic ways in which you can attempt this kind of epistemology (Bradie 1986). Either you take the biology literally as your starting point, and try to show how science and the rest of knowledge could be produced by a naturally selected organ like the human brain; or, you take the biology as a model for other kinds of change, like scientific change, with the implication that perhaps everything falls beneath some universal process. Obviously only the second, the Spencerian-type approach (and, for once, I am using the term 'Spencerian' merely descriptively and not prescriptively, if that is possible) is truly organicist. You can hold to the former and think there to be nothing very autonomous at all about biology. (I do!) It is the latter approach which is organicist, for it is it which thinks in terms of processes happening at different levels (hierarchically), causally connected no doubt, but without being reducible to or collapsible into each other. Expectedly, therefore, we find that it is this second kind of epistemology that excites Hull. The aim is 'to present a general analysis of evolution through selection processes which applies equally to biological, social and conceptual evolution' (Hull 1982, p. 275).

How is this to be done, or, at least, how are we to unpack the working of these processes in the social/conceptual realm, which is Hull's area of special interest? You will have noticed the reference to '*selection* processes' in the quote just above, and my sense is that he is still more of an enthusiast for selection than many who have taken up his ideas. Within the species – which in the conceptual world corresponds to a scientific movement like Darwinism – Hull relies remarkably heavily on sociobiological ideas, like reciprocal altruism (Hull 1988). Scientists want to get on and they want to push their ideas. Truth *per se* will get you nowhere. The only good idea is a well-known idea. The only successful scientist is the one who is pushing his/her own ideas/career. But, in order to succeed in science – particularly, to have your ideas succeed in science – you must have other scientists. To work with you, to take up your ideas, to offer criticism. No one is going to help you, quote you, critique you, if you never help back, quote back, critique back. So we get the integration and development of science.

Parenthetically, when I say that Hull relies on sociobiological ideas, you should now understand I mean that he has used them in the non-controversial sense where they are applied to ideas and behaviour (e.g. Hull 1978). There are no claims about genetic causation. The main point for Hull is that, having underlined the hierarchical way of thought, through his endorsement and articulation of the s-a-i thesis, he can now legitimately think about conceptual/social evolutionary processes which are not a mere epiphenomena of the workings of the genes.

In respects, this theory of Hull's is obviously much like other evolutionary epistemologies – Toulmin's in particular – and if this were a critical paper I would point out that Hull faces (as he acknowledges) many of the same problems as Toulmin. The added sophistication comes from the deeper knowledge of the evolution of sociality. It comes also – and here we pick up a point introduced at the beginning of this essay – from the fact that one cannot separate Hull-the-social-being from Hull-the-conceptual-being. Just as in his written philosophy, social beings and concepts blend and mix together, so one must see Hull himself as the living embodiment of the philosophy he preaches, and conversely one must see that his philosophy is made meaningful by the fact that this is *his* life. Hull criticizes, helps, quotes. And as I know from one exchange which was very painful to me – and where I was entirely in the wrong – he gets upset if you do not quote in return. The point is that we have to work together, not for some disinterested goal of the greater good, but because if we do not we shall each one of us individually suffer. But, as we do work together, we get the integration of our group into an individual – the conceptual/social equivalent of a biological species.

I have argued already that the ultra-Darwinian tends to draw back from the supposed individuality of the species. For the ultra-Darwinian, selection is generally seen as a process that puts friend against friend and brother against brother (certainly in the hymenoptera, mother against daughter). Of course, this is a causal perspective. Hull turns it on its head, arguing that the effects are sufficiently integrating to confer individuality on the group. I will not argue this point here, although I would note that Hull would probably receive support from an unexpected quarter. Edward O. Wilson (1984) shares Hull's holistic philosophy and is always inclined to see the integrating effect of selection. What I will do, however, since the social and the conceptual have been brought together, is suggest that the reason why Hull cannot see the fragmenting nature of selection is because he himself does not live or feel the supposed cause of his philosophy. All genes may be selfish. Of at least one Hull, this is not true.

With the scientific movement (or theory or paradigm) established, with its individuality given, Hull's epistemology now moves its focus up to the next level. We have movements or theories (supported by scientists), and these now battle for supremacy. Thus the level of selection is raised, but (as in the Mendelian/molecular genetical case) the evolution of science moves forward. As far as I understand, Hull is not a realist – at least, not in the Popperian sense. Or if he is a realist, the real world does not function as quite a constrainer or goal-object as it does for Popper.

In this respect, Hull's is a subjective or relativistic philosophy. It does not matter how good your idea is, if no one knows about it. What counts is winning. But, because there are winners, there is change and development. Thus, Hull is able to offer an alternative picture to the sterile unreal idealizations of the logical empiricists. The true model of reality is not physics. It is biology.

Conclusion

Such, as I see it, is the philosophy of David Hull. From a negative reaction to what then existed, I see two decades of working towards a fully articulated organismic world view. But, as I have said, Hull is an evolutionist, and I expect to see much more in the next two decades. So just think of this essay as a progress report rather than as a definitive assessment.

Notes

1. As we shall see, Hull has used sociobiological ideas, but not in such a way as to upset the critics.
2. Hull has drawn my attention to the fact that, in *Science* (December 14, 1973) he wrote a detailed, and quite critically searching review of Toulmin's book (Hull 1973c). Obviously, there was no love at first sight, although through writing the review Hull had clearly thought hard about Toulmin's views. There is an interesting conclusion to the review, which (some fifteen years later) takes on added meaning: 'I will have to content myself here with remarking that at best all Toulmin has done in this volume is to set out a research program (p. 504), which will stand or fall on how well he and his future co-workers carry it out' (p. 1123).
3. I take the connection between the s-a-i thesis and the evolutionary epistemology to be at least three-fold. First, the thesis underlines the hierarchical nature of evolutionary thought. Second, the thesis does not demand that the parts of a theory be similar (which they are obviously not) or that the organisms of a species be similar (which some might think they are). Hence, if one argues for conflict (or 'conflict') between parts of a theory, leading to selection (or 'selection'), one is not open to the objection that in the biological world the conflict is between similar entities. (See Hull 1974a where he makes this point.) Third, one can argue that a group make one distinct scientific movement (or supports one theory) even though individual scientists have different views. (See Hull 1985 for this point – it is also made in Hull 1973c.)

References

Ayala F. (1974). Introduction. In Ayala F., Dobzhansky T. (eds) *Studies in the philosophy of biology*. Berkeley: University of California Press.
Ayala F., Dobzhansky T. (1974). *Studies in the philosophy of biology: reduction and related problems*. Berkeley: University of California Press.
Beckner M. (1959). *The biological way of thought*. New York: Columbia University Press.

14

Bradie M. (1986). Assessing evolutionary epistemology. *Biology and Philosophy* 1: 401–460.

Buck R., Hull D.L. (1966). The logical structure of the Linnaean hierarchy. *Systematic Zoology* 15: 97–111.

Bunge M. (1961). The weight of simplicity in the construction and assaying of scientific theories. *Philosophy of Science* 28: 120–149.

Campbell D. (1974). Evolutionary epistemology. In Schilpp P.A. (ed.) *The philosophy of Karl Popper*; 1, 413–463. LaSalle (Ill.): Open Court.

Cracraft J. (1987). Species concepts and the ontology of evolution. *Biology and Philosophy* 2: 329–346.

Eldredge N. (1985). *Unfinished synthesis.* New York: Oxford University Press.

Eldredge N., Cracraft J. (1980). *Phylogenetic patterns and the evolutionary process.* New York: Columbia University Press.

Flew A.G.N. (1959). The structure of Darwinism. *New Biology* 28: 18–34.

Ghiselin M.T. (1974). A radical solution to the species problem. *Systematic Zoology* 23: 536–544.

Grene M. (1959). Two evolutionary theories. *British Journal for the Philosophy of Science* 9: 110–127, 185–193.

Goudge T.A. (1961). *The ascent of life.* Toronto: University of Toronto Press.

Gould S.J. (1982). Darwinism and the expansion of evolutionary theory. *Science* 216: 380–387.

Hull D.L. (1965). The effect of essentialism on taxonomy: two thousand years of stasis. *British Journal for the Philosophy of Science* 15: 314–326; 16: 1–18.

Hull D.L. (1967). Certainty and circularity in evolutionary taxonomy. *Evolution* 21: 174–189.

Hull D.L. (1968). The operational imperative: sense and nonsense in operationism. *Systematic Zoology* 17: 438–457.

Hull D.L. (1969). What philosophy of biology is not. *Journal of the History of Biology* 2: 241–268. Also *Synthese* 20: 157–184.

Hull D.L. (1970). Contemporary systematic philosophies. *Annual Review of Ecology and Systematics* 1: 19–54.

Hull D.L. (1972). Reduction in genetics – biology or philosophy? *Philosophy of Science* 39: 491–499.

Hull D.L. (1973a). *Darwin and his critics.* Cambridge (Mass.): Harvard University Press.

Hull D.L. (1973b). Reduction in genetics – doing the impossible. *Proceedings of the IVth International Congress of Logic Methodology and Philosophy of Science*; 619–635. Amsterdam: North Holland.

Hull D.L. (1973c). A populational approach to scientific change (Review of S. Toulmin's 'Human Understanding'). *Science* 182: 1121–1124.

Hull D.L. (1974a). Are the 'members' of biological species 'similar' to each other? *British Journal for the Philosophy of Science* 4: 332–334.

Hull D.L. (1974b). *Philosophy of Biological Science.* Englewood Cliffs: Prentice Hall.

Hull D.L. (1976a). Are species really individuals? *Systematic Zoology* 25: 174–191.

Hull D.L. (1976b). Informal aspects of theory reduction. In Cohen R.S., Michalos A. (eds) *PSA 1974*; 653–670. Dordrecht: Reidel.

Hull D.L. (1978a). Altruism in science: a sociobiological model of cooperative behaviour among scientists. *Animal Behaviour* 26: 685–697.

Hull D.L. (1978b). Sociobiology: scientific bandwagon or travelling medicine show? In Gregory M.S., Silvers A. , Sutch D. (eds) *Sociobiology and human nature.* San Francisco: Jossey-Bass.

Hull, D.L. (1978c). A matter of individuality. *Philosophy of Science* 45: 335–360.

Hull D.L. (1982). The naked meme. In Plotkin H.C. (ed.) *Learning, development and culture: essays in evolutionary epistemology*: 273–327. Chichester: Wiley.

Hull D.L. (1983a). Exemplars and scientific change. In Asquitu P., Nickles T. (eds) *PSA 1982*; 2: 479–503. East Lansing (Mich.): Philosophy of Science Association.

Hull D.L. (1983b). Karl Popper and Plato's metaphor. In Platnick, N., Funk V. (eds) *Advances in cladistics*; 2: 177–189. New York; Columbia University Press.

Hull D.L. (1985). Darwinism as an historical entity: a historiographic proposal. In Kohn D. (ed.) *The Darwinian heritage*; 773–812. Princeton: Princeton University Press.

Hull D.L. (1988). A mechanism and its metaphysics: an evolutionary account of the social and conceptual development of science. *Biology and Philosophy* 3: 123–155.

Kitcher P. (1984). 1953 and all that: a tale of two sciences. *Philosophical Review* 93: 335–373.

Mayr, E. (1969). Commentary – Part I. *Journal of the History of Biology* 2: 123–128.

Nagel E. (1961). *The structure of science*. New York: Harcourt, Brace and World.

Popper K.R. (1972). *Objective knowledge*. Oxford: Oxford University Press.

Rosenberg A. (1985). *The structure of biological science*. Cambridge: University of Cambridge Press.

Ruse M. (1976). Reduction in genetics. In Cohen R., Michalos A. (eds) *PSA 1976*. Dordrecht: Reidel.

Ruse M. (1987). Species: individuals, natural kinds, or what? *British Journal for the Philosophy of Science* 38: 225–242.

Schaffner K.F. (1967). Approaches to reduction. *Philosophy of Science* 34: 137–147.

Schaffner K. (1976). Reduction in biology: prospects and problems. In: Cohen R., Michalos A. (eds) *PSA 1976*. Dordrecht: Reidel.

Schindewolf O.H. (1950). *Grundfragen der Paläontologie*. Stuttgart: Schweizerbart.

Scriven M.J. (1959). Explanation and prediction in evolutionary theory. *Science* 130: 477–482.

Simpson G.G. (1953). *The major features of evolution*. New York: Columbia University Press.

Sober E. (1980). Evolution, population thinking, and essentialism. *Philosophy of Science* 47: 350–383.

Spencer H. (1857). Progress: its law and cause. *Westminster Review*.

Toulmin S. (1972). *Human understanding*. Oxford: Oxford University Press.

Wilson E.O. (1984). *Biophilia*. Cambridge (Mass.): Harvard University Press.

Wimsatt W. (1976). Reductive explanation: a functional account. In Cohen R., Michalos A. (eds) *PSA 1974*. Dordrecht: Reidel.

Rethinking the Propensity Interpretation:
A Peek Inside Pandora's Box[1]

JOHN BEATTY[a] and SUSAN FINSEN[b]

[a] *Department of Ecology and Behavioral Biology, University of Minnesota, Minneapolis, MN 55455, U.S.A.*
[b] *Department of Philosophy, California State University at San Bernadino, San Bernardino, CA 92407, U.S.A.*

Introduction

Over the past ten years, the propensity interpretation of fitness has attracted a number of proponents[2] and a few, persistent detractors.[3] Here, two previous supporters turn critics, to acknowledge and reframe some old problems, and to introduce some additional difficulties. We are not sure whether a radically revised interpretation of fitness is necessary. But it does seem to us that certain gross oversimplifications of the propensity interpretation deserve more serious attention.

We most certainly do not propose to return to the interpretation of fitness that the propensity interpretation was designed to replace. Whatever fitness is, it is not actual offspring contribution, although it was long misconceived as such.[4] The misconception had two sources. The first source was the operationalist fallacy of conflating properties with the manner in which they are measured, as if temperature, for instance, were best conceived as just the height of mercury in a glass column. Actual offspring contribution remains the most common and surest means of measuring fitness (especially considering the alternatives, like optimality models). But that does not mean that fitness is *just* offspring contribution. The second source of the misconception was the false assumption that it is a law of nature that the fittest always leave more offspring, which is too often thought to be one of Darwin's main insights, although Darwin was actually much more careful in this respect (see further). Current evolutionary theory allows differential descendant contribution to be a matter of chance as well as a matter of fitness, to the extent that, in small enough populations, the fitter type can leave less descendants on average, and dwindle in frequency relative to the less fit type.

The conception of fitness as mere offspring contribution has by now received so much criticism that its difficulties do not merit any further discussion here. The most common substitute, the propensity interpretation of fitness, also requires only a brief introduction. Part and parcel of the propensity interpretation is the notion that fitter organisms have greater *ability* to leave offspring, and not just (and not even necessarily) greater *success* in that regard. The propensity interpretation thus avoids the operationalist fallacy by identifying fitness with offspring contribution

M. Ruse (editor), What the Philosophy of Biology is. pp. 17–30.
© 1989 *Kluwer Academic Publishers, Dordrecht*

ability rather than with some record of that ability. The other important aspect of the propensity interpretation is the notion that fitter organisms only *probably* leave more offspring than the lesser fit. Darwin himself always wrote of those organisms whose particular abilities gave them 'the best chance' of surviving and reproducing (e.g., Darwin 1859, pp. 61, 81). Fitness, then, is probable offspring contribution, where the probable contribution of an organism depends on its abilities to survive and reproduce in the environment it inhabits. Organisms whose properties confer on them the same offspring contribution abilities in an environment may, by chance, leave different numbers of offspring.

Viable alternatives

But the propensity interpretation leaves much to be desired. In the first place, it is all too often paraded as *the* propensity/probabilistic alternative to the more deterministic and operationalistic conception of fitness. In the course of this paper, we will discuss a number of more subtle variants on 'the' propensity interpretation as it is usually elaborated. There are also a couple more radical variants worth pointing out. One is Thoday's (1953) notion of fitness as probability of representation over a given period of time. Another is Cooper's (1984) notion of fitness as probable duration of representation, or, in Cooper's own terms, 'expected time to extinction' (ETE). These two interpretations, and the rationales behind them, are somewhat similar. We will focus here on Cooper's proposal.[5] The ETE of a particular population, or, more importantly, of a particular genotypic or phenotypic subpopulation, at a particular time and in a particularly specified environment, is just the probability-weighted sum of possible time intervals that might elapse before the (sub)population in question goes extinct. That is,

$$ETE_{x,e} = \sum_{t=0}^{\infty} p_t t,$$

where x is a particular (sub)population, e a particular environment, and p_t the probability of extinction in the t-th time interval (e.g., the t-th generation or t-th season). Although Cooper does not mention it, ETE_x, in order to represent anything like the fitness of x, must assume no source of new members of x except via reproduction of previous members. For example, recurrent mutation and migration alone must not be sustaining x's duration. ETE_x is intended to represent the physical propensity of x-type organisms to endure via reproduction.

The first lesson to be drawn from this brief discussion of ETE is that there are probabilistic alternatives to 'the' propensity interpretation of fitness; ETE is one of many. The relative significances and roles of these various interpretations in evolutionary explanations beg for further clarification. Are these various alternatives in conflict, or can they be fit together into a coherent 'family' of propensities for survival and reproduction? We shall subsequently refer to this as the *multiple propensities problem*, and we hope that other biologists and philosphers can aid in offering a solution to it. In what follows, we hope to provide some suggestions for fruitful avenues of pursuit, as we uncover further difficulties for the idea that there is one privileged propensity interpretation of fitness.

The long term and the short term

To have any claim to the term 'fitness', a propensity must be shown to sustain the role nominally played by fitness in evolutionary explanations. However 'fitness' is defined, it must at least be positively correlated with actual evolutionary success, i.e., with persistence or increase in frequency.[6] Increased fitness should, in other words, make more likely (if not simply consist in) evolutionary success. An obvious starting point in attempting to solve the multiple propensities problem is to compare various survival and reproductive propensities to see how they fare in this regard.

How does ETE, for example, fare regarding this minimal requirement? The role that fitness qua ETE plays in evolutionary explanations is in the form of inferences, so common in the evolutionary literature, that lead from premises about the relatively greater fitness of a genotype or phenotype to its continued presence in nature. This is the use so often made of optimality models, namely, to estimate the relatively greater fitness of one among a set of alternative types, in order to explain its presence today. Optimality models would not serve this end very well if they were not being used to estimate the endurability of a type. Since ETE is an estimate of endurability, it seems ideally suited to play the role of fitness in such models. Thus, ETE easily satisfies the requirement that higher fitness should increase the probability of evolutionary success.

Unfortunately, 'the' propensity interpretation of fitness, defined as probable offspring contribution, does not so straightforwardly satisfy this important though rather minimal requirement. For it is well known that, past a certain point, increased numbers of offspring can threaten the evolutionary success of a type, for example, by placing too great a demand on available resources, or by minimizing the parental care that can be provided to each offspring, etc.[7] To be sure, fitness as probable offspring contribution is positively correlated with very short term evolutionary success – e.g., with representation early in the life cycle of the next generation. But increased fitness, so construed, may be the very *cause* of decreased evolutionary success in the longer term.

The usual response to this problem on the part of supporters of 'the' propensity interpretation is that, just as one can distinguish between actual short-term and long-term evolutionary success, one can also distinguish between short-term and long-term fitness: between probable descendant contribution in the short term and probable descendant contribution in the long term (e.g., Mills and Beatty 1979). The relevant sort of fitness to invoke in order to explain short-term evolutionary success is short-term fitness; long-term fitness is most relevant to long-term success. In the apparent counterexamples just discussed, the inappropriate sort of fitness was invoked. In those cases, decreased long-term fitness should have been invoked to explain decreased long-term evolutionary success.

This rather casual acknowledgement of the need to distinguish probable descendant contribution in the short and long terms underscores the depth of the multiple propensities problem, showing the potential infinity of definitions hidden in the term 'descendant contribution'. We can create as many distinct such definitions as there are potential future generations and stages of the life cycle (see also Kitcher 1987, pp. 86–87). And shortly we will show how the family of probable descendant contribution interpretations must be extended even further – indeed, much further![8]

But the difficulty that the short-term/long-term distinction presents for 'the' propensity interpretation is actually more complicated. In order to avoid the apparent problem of explaining low long-term evolutionary success in terms of high short-term fitness, we are supposed to invoke long-term fitness instead. But the fact is that, in some cases, such as those just considered, high short-term fitness indeed *explains* low long-term success. It may, for instance, just be the inability of parents to care sufficiently for large numbers of offspring that explains why organisms that leave too many offspring do not have many descendants in the long term. And inasmuch as high short-term fitness causally underlies low long-term fitness (and hence low long-term success) in such cases, reference to high short-term fitness constitutes a deeper explanation of the phenomenon than reference to long-term fitness. In short, the apparent counterexamples in which fitness is negatively correlated with evolutionary success are not so easily circumvented.

One response to this dilemma is to deny that short-term fitness is a legitimate notion, reserving the term 'fitness' for a long term propensity such as ETE. Thoday and Cooper argue for this approach, claiming that long-term ability to obtain representation has a more legitimate claim to the nomenclature 'fitness' than short-term ability. Since this approach would resolve the multiple propensity problem, it deserves consideration.

The sorts of cases we have been considering, in which high short-term fitness is compatible (indeed causes) low long-term success might be considered by proponents of long-term notions of fitness to be arguments ad absurdum against short-term notions of fitness. Let us consider such an example more closely. Suppose there are two genotypes, A and B, where A organisms are presently capable of leaving more offspring than B organisms, and are in fact doing so, even though we have grounds for inferring that A's fate will change. Suppose that we can extrapolate as follows from the successes of A and B to date:

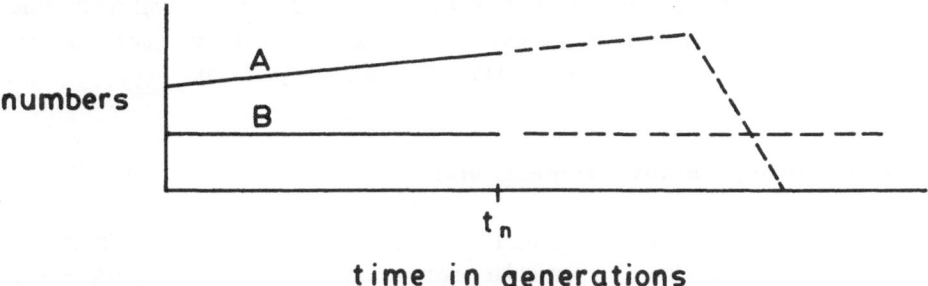

time in generations

At time t_n = now, a defender of the distinction between long-term and short-term fitness might say that A has higher short-term fitness than B, but lower long-term fitness. But others will insist that B is *simply* fitter than A. A's greater initial frequency and relatively higher offspring contribution ability *mask* its lower fitness, i.e., its lesser ability to persist. Far from granting that A has high short-term and low long-term fitness, and especially far from granting that A's short-term fitness plays a deeper role in explaining its evolutionary success or failure, such evolutionary biologists and philosophers will deny that the term 'fitness' even applies to short-term ability to gain representation. According to this line of reasoning, the only viable notion of fitness as probable descendant contribution is one where probable descendant contribution refers to a sufficiently large number of generations hence.

While it seems plausible to suggest that 'fitness' should refer to the long-term in the sort of case just considered, this approach leaves us without an adequate account of a wide variety of evolutionary changes that require appeal to fitness. Most prominent among these are many short-term, micro-evolutionary explanations that clearly require reference to fitness. We sometimes want to know what it is about some phenotype or genotype that accounts for its *sudden* increase or decrease in frequency. ETE has nothing to say about such fluctuations.

The problems of ETE as the privileged interpretation of fitness run even deeper. Consider that we can use short-term fitness (whether high or low) to explain long-term fitness and long-term success, but we cannot use long-term fitness to account for short-term fitness or short-term success. So, short-term fitness actually seems to be explanatorily more powerful.

The problem with ETE as 'the' interpretation of fitness is somewhat the same problem as that of the long-run frequency interpretation of probability as expressed by Keynes, namely, that 'in the long run, we're all dead'. There is much about what happens in the meantime that is important and interesting to explain. ETE fails as a guide to understanding all aspects of fitness for much the same reason that frequency interpretation of probability fails as a guide to life.

But if ETE fails as the definition of 'fitness', so does expected offspring contribution. What is needed is a more comprehensive analysis that takes into account both short- and long-term evolutionary success. As we shall argue, the problem of accommodating both short- and long-term evolutionary success in the definition of 'fitness' is one dimension of a much broader problem.

Supervenience and fitness components

The causal bases of fitness-qua-probable-descendant-contribution are multiple and multilayered. Possession of a particular pigment may underlie the camouflage that underlies the increased viability that contributes to the increased descendant-contribution ability of organisms of the appropriately pigmented type. A faulty meiotic mechanism may underlie the decreased fertility that detracts from the descendant-contribution ability of organisms of that type. Levels of viability, fertility, etc. are said to be causal 'components' of fitness; by the same token, camouflage (in this particular case) is a causal component of viability, and particular properties of the meiotic mechanism are causal components of fertility. Similarly, for those who respect the distinction, short-term fitness may be considered a causal component of long-term fitness.

Hand-in-hand with 'the' propensity interpretation of fitness goes the understanding that fitness cannot be universally expressed as any particular function of its components and subcomponents. Of course, the manner in which any organism achieves high fitness is a matter of its viability, fertility, etc., and ultimately a matter of the physical and chemical properties that underlie its viability, fertility, etc. levels. And yet different organisms achieve high fitness (and low fitness) in very different physical-chemical ways. What one member of a species accomplishes by increased viability, another may accomplish by increased fertility. What one accomplishes in terms of viability by way of possessing a particular pigment, another may accomplish by way of increased metabolic efficiency. And so on. So while each and every manifestation of high fitness can be explained in terms of underlying components and subcomponents of fitness, there seems to be no particular function of components and subcomponents in terms of which fitness can universally be expressed. Rosenberg (1985) speaks of the 'supervenience' of fitness upon its underlying causal components in his discussions of this aspect of fitness.

A criticism of 'the' propensity interpretation that we have already considered can now be restated and strengthened. Those who identify fitness with long-term ability to achieve representation might complain that supporters of the propensity interpretation, especially those who elaborate it in terms of probable *offspring* contribution, are mistakenly trying to construe fitness as a simple function of one of its *components*, namely short-term ability to achieve representation. In the process, supporters of the propensity interpretation overlook the fact that fitness is supervenient upon, and hence not expressable as a simple function of, its components.

What is, to us, a much stronger criticism of 'the' propensity interpretation – perhaps the very most compelling criticism of that interpretation as it is usually elaborated – takes the same form, namely, of pointing out that the propensity interpretation mistakes a component of fitness for fitness itself. To see what component is at issue here, it is first necessary to recall some important details of the propensity interpretation. We can think of descendant contribution as an event with different outcomes $o_1, o_2, ..., o_n$. Each outcome o_i represents a total contribution of i descendants. Now, given a particular organism x, there is a certain probability, p_1, that o_1 will obtain. And there are probabilities $p_2, p_3, ..., p_n$ that outcomes

$o_2, o_3, ..., o_n$ will obtain. These probabilities represent x's various propensities, based ultimately on physical-chemical properties of x, to leave various numbers of descendants. So, for example, the descendant contribution propensities of an imaginary organism x might be expressed in terms of the following probability distribution:

number of offspring	1	2	3	4	5	6	7	8	9
probability (x)	0.05	0.05	0.05	0.2	0.3	0.2	0.05	0.05	0.05

This distribution can be referred to as the offspring probability distribution of x. Alternatively, x can be a type – genotype or phenotype – of organism, in which case the various p_i represent average descendant-contribution probabilities.

One might think that such a distribution provides an adequate, albeit awkward, characterization of offspring contribution propensities. But beyond its unwieldiness, there is a problem in determining how to compare such distributions. For instance, is x fitter than, less fit, or equal in fitness to y and z, whose offspring probability distributions are as follows?

number of offspring	1	2	3	4	5	6	7	8	9	10
probability (y)					1.0					
probability (z)			0.5		0.3					0.2

In order to facilitate fitness comparisons, supporters of the propensity interpretation standardly move away from identifying fitness with entire probability distributions to identifying it with one particular statistic on the distribution, namely, the arithmetic mean, or 'expected value' (e.g., Brandon 1978, Mills and Beatty 1979). On such a construal, x, y, and z all have the same fitness (5.0), even though they have very different distributions. But of course there are many statistics on a probability distribution: the expected value, the mode, the median, the variance, the skew, etc. Why reduce fitness distributions to just one of these statistics? There are, after all, interesting differences between the descendant-contribution abilities of x, y, and z. Organism or type y will always leave 5 descendants, while z has a large chance of leaving a slightly smaller number, but also a small chance of leaving a much larger number. Is it legitimate to overlook these differences if our aim is to understand differences in the actual evolutionary successes of x, y, and z? As Rosenberg and Williams (1986) so pointedly put the question, is evolutionary expectedness no more and no less than mathematical expectedness?

No, in fact. Much work has recently gone into demonstrating that expected offspring contribution alone is sometimes poorly correlated with actual evolutionary success, even very short-term evolutionary success. For example, Gillespie (1974) has argued that organisms or types with the same expected offspring contributions can have very different actual evolutionary successes, depending on the variances of their offspring probability distributions. He argues in particular that of two organisms or types with the same expected offspring contribution, the one with smaller variance in its distribution will be evolutionarily more successful. An organism or type with a smaller expected value can even be more successful evolutionarily as long as its variance is also sufficiently smaller.[9]

A simulation example illustrates that variance in addition to expected value may be important in accounting for evolutionary success. This example (and all the others that follow) oversimplify natural processes in certain ways that do not allow us to generalize on the basis of them. But their purpose is just to serve as counterexamples to certain proposed definitions of fitness, not to serve as grounds for any new definition.[10] Let us consider a population composed only of two asexual, haploid genotypes A and B, each initially represented by 10 organisms (i.e., initially at equal, 50%, frequencies). Suppose now that they have the same expected offspring contribution, but that their offspring probability distributions differ in the following manner:

number of offspring	1	2	3
probability (A)		1.0	
probability (B)	0.5		0.5

The expected offspring contributions of A and B are both 2.0, but the variances in their fitness distributions differ greatly: the variance of A being 0.0, while the variance of B is 1.0. Now consider what their evolutionary successes would be if the variation in B's offspring contribution were distributed between generations only. That is, while every A in every generation leaves 2 offspring, all the B's of one generation leave either 1 or 3 offspring – in approximately half the generations, B's leave 1 offspring per capita, and in the other half they leave 3 (the order in which the variation of B's offspring contribution is distributed is irrelevant – let us suppose alternating generations of 3 and 1 offspring per capita). The evolutionary success of A and B can be charted as follows for 10 generations:

Generation	A	B
0	10	10
1	20	30
2	40	30
3	80	90
4	160	90
5	320	270
6	640	270
7	1,280	810
8	2,560	810
9	5,120	2,430
10	10,240	2,430

By the end of 10 generations, the frequency of B has decreased from 50% to 19%.

What Gillespie, and this example, show is, in effect, that expected offspring contribution is just one component of fitness, variance is another. Fitness is supervenient upon each of its components, and hence cannot be expressed as a simple function of any one of them in particular (like expected value), or any particular combination of them.

One can even construct examples where two asexual haploid genotypes with the same expected offspring contributions and the same variances in their fitness distributions nonetheless have very different evolutionary successes, as long as the skew (assymetry) of the fitness distribution of one exceeds that of the other. For instance, consider the types C and D whose fitness distributions are as follows:

number of offspring	0.83	1.0	2.0	2.17	3.0
probability (C)		.6	0.3		.1
probability (D)	.5			.5	

The expected values of both are 1.5. The variance of C is 0.45 and that of D is 0.449. C's distribution, however, is skewed, with its right tail extending farther beyond its mean than does its left tail. D's distribution is not skewed. Now if we begin again with 10 C's and 10 D's, and if we suppose again that the variation in each of their offspring contributions is distributed wholly between generations, then we can chart their evolutionary successes as follows:

C's leave 1 offspring for the first 6 generations, 2 offspring for the next 3 generations, and 3 offspring the last generation.
D's leave 0.83 offspring for the first 5 generations and 2.17 offspring for the next 5 generations.

Generation	C	D
0	10	10
1	10	8.3
2	10	6.8
3	10	5.7
4	10	4.7
5	10	3.9
6	10	8.5
7	20	18.5
8	40	40.2
9	80	87.3
10	240	189.5

C, whose distribution has the greater skew, also has the greater evolutionary success.

One can even show that of two types with the same expected offspring contributions, the type with considerably smaller variance will have considerably less evolutionary success than the type with larger variance, as long as the latter has sufficiently greater skew. Consider, for instance, E and F, whose fitness distributions are as follows:

number of offspring	2.5	3.0	5.0	7.5	10.0
probability (E)		.5	.3		.2
probability (F)	.5			.5	

Here the expected values of E and F are both 5.0. The variance of E is 7.0 and that of F, 6.25. Suppose we begin again with 10 E's and 10 F's, and suppose again that the variation in the offspring contribution abilities is distributed wholly between generations:

E's leave 3 offspring for the first 5 generations, 5 offspring for the next 3 generations, and 10 offspring for the next 2 generations.
F's leave 2.5 offspring for the first 5 generations, and 7.5 offspring for the next 5 generations.

Generation	E	F
0	10	10
1	30	25
2	90	63
3	270	156
4	810	391
5	2,430	977

6	12,150	7,324
7	60,750	54,931
8	303,750	411,987
9	3,037,500	3,089,904
10	30,375,000	23,174,316

Fortunately, there is one statistic, geometric mean, that can be used in place of the combined three we have considered thus far (arithmetic mean, or expected value, variance, and skew) to predict evolutionary success in some of the sorts of scenarios we have been discussing. As Crow and Kimura (1970) among others argue, when offspring contribution varies between generations, geometric mean is the most appropriate predictor of evolutionary success. Indeed, under those circumstances, the geometric mean of A (2.0) is higher than for B (1.7), C (1.4) is higher than D (1.3), and E (4.4) is higher than F (4.5).

The problem that the use of geometric means introduces is just that they alone are appropriate only when a type's offspring contribution varies solely *between* generations. When a type's offspring contribution varies solely *within* generations (as in the second case involving A and B above), expected value, or arithmetic mean, is the better predictor of evolutionary success. But offspring probability distributions alone give us no indication of how a type's variation in actual offspring contribution will be distributed. The distribution alone does not tell us whether actual variation will be distributed mainly within or mainly between generations, or which of an infinite variety of combinations thereof will obtain. From the offspring probability distribution of genotype B, for instance, we cannot tell which of the following cases will obtain:

Case 1. Approximately half the B's of each generation leave 1 offspring, and half leave 3.

Case 2. In approximately half of all generations, B's leave 3 offspring, in the other half B's leave 1.

Case 3. In approximately half of all generations, half of the B's leave 1 offspring and halve leave 3; in approximately one-quarter of the remaining generations, B's leave 3 offspring, and in the other quarter of generations, B's leave 1 offspring.

Case 4. In each generation, approximately one-quarter of the B's leave 1 offspring and one-quarter leave 3; the other half leave 3 offspring in half the generations, and 1 offspring in the other half.

The evolutionary successes of A and B differ from case to case in each of these four scenarios, as is easy to demonstrate. Thus, we cannot represent fitness (even short-term fitness) merely in terms of probability distributions of offspring contributions; we must also specify whether the variation in those distributions is distributed within or between generations. If we know how the variation is in fact distributed (within or between generations), then we might be able to argue for the appropriateness of the geometric mean or the arithmetic mean as representing fitness. But it is often very difficult to determine this (as if it were not difficult to determine the distribution itself!), and there is even the possibility that a type switches reproductive strategies over time. In short, we may sometimes have no access to the sort of information we need in order to decide what statistic on the fitness distribution to employ in order to explain a particular evolutionary phenomenon.

There are other, perhaps more difficult, problems arising from the need to choose between different statistical representations of fitness (we do not aim to be comprehensive here). Consider, for instance, that within one population, there may be one type whose offspring-contribution variation is distributed wholly within generations, and whose fitness is (let us say) best represented by the arithmetical mean of that distribution; while in the same population there is another type whose offspring-contribution variation is distributed wholly between generations, and whose fitness is accordingly better represented by the geometrical mean of that distribution. These two types may have the same fitness, and may in the long run of generations enjoy the same relative evolutionary success. But in the short term, from generation to generation, their evolutionary successes may differ drastically in ways that reflect the differences between the types. More specifically, the type whose offspring-contribution variation is distributed between generations will periodically lag behind the success of the other type for a few or even many generations, depending on the range of the former type's offspring-contribution variation. Now if we really care to understand very short-term (down to one generation?) differences in evolutionary success in terms of fitness differences, then the multiple propensities problem (in this case as represented by the geometric mean vs. the arithmetic mean calculation of fitness) is going to cause difficulties.

A different perspective on basically the same problem is to imagine the alternative sorts of stochastic effects that infringe upon the evolutionary successes of the two types under discussion. If the two types are present in very large numbers, then there should be only negligible stochastic effects on the evolutionary success of the type whose offspring-contribution variation is distributed wholly within generations. But over a short number of generations, there may still be great stochastic effects on the evolutionary success of the type whose offspring-contribution variation is distributed between generations. Again, because of the multiple propensities problem, we must sometimes infer different evolutionary successes on the basis of identical fitness values.

The problems involved in choosing one or another statistic to calculate fitness aside, though, it is crucial to recognize that such choices will arise. The various statistics on fitness distributions, such as the geometric or arithmetic mean, the variance and the skew, are aspects of an organism's or type's reproductive strategy, and as such are components of fitness – they contribute to evolutionary success in different ways, depending upon the environmental (broadly construed) circumstances. Once it is acknowledged that these statistics represent components of fitness, it becomes apparent that identifying fitness generally with any one statistic, or any particular function of statistics, is mistaken. We stand to learn a great deal about fitness and its role in evolution by studying descendant-contribution distributions, how they bear upon evolutionary success, and how they may be compared.

Conclusion

To decide a priori that fitness is a propensity to contribute descendants to the next generation, or to the tenth generation, or to the thousandth generation, is to beg important questions that are rightfully settled in terms of the sorts of evolutionary questions we pose for ourselves. To decide a priori that fitness is to be identified with one particular statistic (or one particular function of statistics) from a probability distribution of descendant contributions is to beg further important questions that are rightfully settled empirically, in ways that depend upon the organisms and environments we are investigating. The propensity interpretation of fitness, as usually articulated, is shortsighted in both of these respects.

Nothing in the above analysis suggests that fitness is not a propensity. But 'fitness' does seem to stand for a very broad family of propensities – a family that is difficult to describe in general terms. It would be unfair to suggest that, lacking any generally agreed-upon definition of 'fitness', we therefore lack *any* understanding of evolution in terms of fitness differences and natural selection. On the other hand, until we have an appropriate general definition of fitness, it is not altogether clear *how much* we understand about evolution in terms of fitness differences and natural selection. Our aim here is not just to be pessimistic about prospects in this regard; rather, our aim is to be realistic. General satisfaction with the propensity interpretation of fitness has created, we believe, a false sense of understanding. There is clearly room for improvement.

Notes

1. We are grateful to Peter Abrams and Robert Brandon for helpful discussions concerning the issues treated here.
2. See Brandon 1978, Mills and Beatty 1979, Burian 1983, Sober 1984, Kitcher 1987, and Mayr 1988. This list is meant to be exemplary, not complete.
3. See Rosenberg 1985, Rosenberg and Williams 1986, and Hodge 1987. Again, the list is exemplary, not complete.
4. See the examples discussed in Mills and Beatty 1979.
5. See also the excellent discussion of alternative notions of fitness in Endler 1986.
6. This is just one sense of 'evolutionary success'; another would be speciation rate. But

we will confine our discussion to the micro-evolutionary sense.

7. The literature on evolution of clutch size contains many such discussions. See, for example, Lack's classic 1947, also Lack 1954, 1966, 1968. For a useful, brief summary, see Pianka 1978, pp. 135–138.

8. We are leaving out of our discussion ways in which the family of propensities that constitute fitness might need to be extended still further to cover different units of selection.

9. Several discussions with Robert Brandon over the past several years have been very helpful in developing the views that follow. Brandon also encouraged us to use simulation examples to illustrate the issues in question; in fact, the first of the following examples is his. He plans to discuss similar topics in his book, in preparation, *Adaptaptation and Environment*.

10. We will not refer to any actual taxa here – of course, that is nothing to brag about. For literature dealing with real organisms, see the excellent review by Seger and Brockmann 1987.

References

Brandon R. (1978). Adaptation and evolutionary theory. *Studies in History and Philosophy of Science* 9: 181–206.

Burian R. (1983). Adaptation. In Grene M. (ed.) *Dimensions of Darwinism: Themes and counterthemes in twentieth century evolutionary theories*. Cambridge: Cambridge University Press.

Cooper W.S. (1984). Expected time to extinction and the concept of fundamental fitness. *Journal of Theoretical Biology* 107: 603–629.

Crow J., Kimura M. (1970). *An introduction to population genetic theory*. Minneapolis: Burgess.

Endler J.A. (1986). *Natural selection in the wild*. Princeton: Princeton University Press.

Darwin C. (1859). *On the origin of species*. Facsimile edition by Harvard University Press, Cambridge, 1959.

Gillespie J.H. (1973). Natural selection for within-generation variance in offspring number. *Genetics* 76: 601–606.

Hodge M.J.S. (1987). Natural selection as a causal, empirical, and probabilistic theory. In Krüger, L. et al. (eds.) *The probabilistic revolution*, Volume 2. Cambridge: MIT Press.

Kitcher P. (1987). Why not the best? In J. Dupré (ed.) *The latest on the best: essays on evolution and optimality*. Cambridge: MIT Press.

Lack D. (1947). The significance of clutch size. *Ibis* 89: 302–352.

Lack D. (1954). *The natural regulation of animal numbers*. Oxford: Oxford University Press.

Lack D. (1966). *Population studies of birds*. Oxford: Oxford University Press.

Lack D. (1968). *Ecological adaptations for breeding in birds*. London: Methuen.

Mills, S., Beatty J. (1979). The propensity interpretation of fitness. *Philosophy of Science* 46: 263–286.

Pianka E.R. (1978). *Evolutionary ecology*. New York: Harper and Row.

Rosenberg, A. (1985). *The structure of biological science*. Cambridge: Cambridge University Press.

Rosenberg A., Williams M.B. (1986). Discussion: fitness as primitive and propensity. *Philosophy of Science* 53: 412–418.

Seger, J., Brockmann J. (1987). What is bet-hedging? *Oxford Surveys in Evolutionary Biology* 4: 182–211.

Sober E. (1984). *The nature of selection: a philosophical inquiry*. Cambridge: MIT Press.

Thoday J.M. (1953). Components of fitness. *Symposia of the Society for Experimental Biology* 7: 96–113.

Species as Entities of Biological Theory

JOEL CRACRAFT

Department of Anatomy and Cell Biology, University of Illinois,
P.O. Box 6998, Chicago, IL 60680, U.S.A.

Species play a central role in evolutionary biology. They are generally the locus of most discussions of taxonomic and structural diversity. Species are said to be the active participants in a host of postulated processes. Thus, species speciate, go extinct, compete, interact, disperse, predate, or are selected.

The proposition that species should be treated as individuals (Ghiselin 1974; Hull 1976, 1977, 1978, 1980, 1981) and not classes has refocused attention on the role of species in theories of process inasmuch as one of the major justifications for the 'species as individuals' claim is that if they function in natural processes, then they must be individuals. Since species are universally judged to participate in processes, the argument goes, they must be conceived of as individuals (e.g., Rosenberg 1985: 208).[1]

Although biologists have debated the 'species problem' for a very long time, many issues of that debate have become hopelessly entangled with one another. The Rosetta stone of these discussions has been the consummate definition, which, once discovered, will enable biologists to finally specify what is and is not a species. Species definitions are used much like evangelicals use the Bible: as a putative guide to reality. Species *are* what the definition says they are; the entity before us satisfies the definition, therefore it must *be* a species. Using a definition of species in this way is not altogether objectionable, even if it is difficult for an investigator to realize how one's view of the world can be constrained by the content of a definition,[2] but it leaves unstated the reasons for having a species definition in the first place. Systematic and evolutionary biologists have identified two:

a) as a basis for enumerating and classifying diversity (discontinuities), and
b) as a means of individuating entities that are thought to be participants in, or the results of, various biological processes.

By and large these two uses of species definitions have remained independent in that most systematists who describe and catalogue diversity have been little concerned with using species in theories of biological process, and their notions of

M. Ruse (editor), What the Philosophy of Biology is. pp. 31–52.
© 1989 *Kluwer Academic Publishers, Dordrecht – Printed in the Netherlands.*

what a species is are empirical and essentially independent of theoretical considerations of what a species should be. In fact, the relationship that does exist between these two uses of species definitions has been largely asymmetric. Those concerned with biological theory have adopted definitions that were developed by systematists for classificatory reasons, and biological theory itself generally has not been employed to generate a definition that best conforms to the requirements of theory. The biological (polytypic) species concept is a case in point. Widely applied within evolutionary biology (Mayr 1942, 1963; White 1978), it was developed primarily as a response to the tendencies of many 19th and early 20th century biologists to describe most geographically differentiated populations as distinct species. Consequently, the polytypic species concept arose, in part, because of arguments over just how different isolated populations can be from another and still be considered the 'same' species. In this case, therefore, the definition of species was derived not so much from some theory of species change as it was from taxonomic motives to simplify patterns of geographic variation.[3]

Surprisingly, the ontological status of species as entities in biological theory has been little explored, both in the scientific and philosophical literature, especially given the long-standing debate over definitions and their postulated importance of species in numerous biological processes (but see Eldredge 1985a, b, 1986; Cracraft 1987, in press). This paper will consider some aspects of species ontology by focusing on the following interrelated questions:

1. Is the definition of species strictly an empirical issue, or must it be informed by considerations of theory?
2. What does it mean to say a species is a discrete, real entity? Is there justification for claiming that different species concepts imply the existence of different kinds of 'species'?
3. Do species, as discrete entities, participate in biological processes?

Species definitions

To what extent can a definition of species be independent of considerations of theory? That is, should the term 'species' be viewed as 'theoretical' or 'observational'?[4] If the former, then we might expect that our concept of species will function in some process theory that uses an entity termed the species, and thus its physical (or conceptual) characteristics and its behavior will be specified by reference to the predicates of that theory;[5] the theory, so to speak, becomes a blueprint for the individuation and study of its entities. If, however, 'species' is an observational term, we might expect to individuate them on the basis of theory-free, observational criteria and then, if appropriate, to use these entities to evaluate alternative theories about them.

Framing a discussion of species concepts in terms of an observational/theoretical dichomotomy seems useful because many of the conflicts generated by debates over species can be categorized along these lines. Some definitions of species, it is claimed, fail to take into account current conceptions of the speciation process and are typological attempts at describing nature; others, in contrast, have argued that some theoretical and methodological approaches to speciation are flawed because they fail to define species in a way that individuates those entities as they appear to exist in nature. The former argument adopts a position that theory takes priority over observation in conceptualizing what things we should be calling species, whereas the latter implies that process theories themselves might be in jeopardy if they fail to accommodate 'species' as they are 'observed'. If a particular species concept identifies 'fictitious' entities in nature, is it because of a failure of observation in the absence of theory, or a failure of theory to lead to correct observations, or a combination of both? All of these issues seem to have been important in discussions about species concepts.

Consider the following sets of species definitions:

1a) 'We may regard as a species (a) the smallest (most homogeneous) cluster that can be recognized upon some given criterion as being distinct from other clusters, or (b) a phenetic group of a given diversity somewhat below the subgenus category, whether or not it contains distinct subclusters ... meanings of the term 'species' have always been closer to the first alternative; it indicates a distinct kind, and by implication the smallest distinct kind' (Sneath and Sokal 1973: 365).

b) 'Species are the smallest groups that are consistently and persistently distinct, and distinguishable by ordinary means' (Cronquist 1978: 15).

c) '... species are simply the smallest detectable samples of self-perpetuating organisms that have unique sets of characters' (Nelson and Platnick 1981: 12).

d) '... a species can be defined as an irreducible cluster of organisms, within which there is a parental pattern of ancestry and descent, and which is diagnosably distinct from other such clusters' (Cracraft 1987: 341; see also 1983: 170).

2a) 'Species are groups of interbreeding natural populations that are reproductively isolated from other such groups' (Mayr 1970: 12).

b) 'A species is a lineage (or closely related set of lineages) which occupies an adaptive zone minimally different from that of any other lineage in its range and which evolves separately from all lineages outside its range' (Van Valen 1976: 233).

c) 'We can, therefore, regard as a species that most inclusive population of individual biparental organisms which share a common fertilization system' (Paterson 1985: 25).

d) 'An evolutionary species is a single lineage of ancestor-descendant populations which maintains its identity from other such lineages and which has its own evolutionary tendencies and historical fate' (Wiley 1981: 25).

If we grant that the categorization of terms as being theoretical or observational should be seen not as a rigid dichotomy but as a continuum, then we might expect to interpret species concepts in much the same way. The first set listed above consists of species definitions occupying the 'observational' end of the pole, the second set those formulated from a more 'theoretical' perspective. Although differing in some substantial ways, the four definitions of the first set are united by two common themes:

a) species are detectable, or diagnosable, and
b) they are the smallest such clusters of individual organisms.[6]

This general definition has undefined terms but is sufficiently precise so that, once the terms 'diagnosable', 'distinguishable', and 'smallest' are explicated further, it can be applied to natural situations. On the face of it, this view of species appears to be 'theory neutral', at least in the sense that no specific theory of speciation (or evolution) has been invoked to justify or derive the definition, and indeed it has been claimed that this definition developed from the practical efforts of biologists to systematize character variation based on what they observe in nature (e.g., Cronquist 1978; Nelson and Platnick 1981). In making this argument for observability, the assertion is not made that these entities are devoid of interest for theoretical discussions of processes of speciation or evolution in general; rather, it is usually implied such species are agnostic with respect to the different theories that might be invoked to explain them. Theories may come and go, but the entities themselves, as observables, persist and are there to be studied by biologists.

The definitions collected together in the second set are more distantly removed from species-as-observables in that the concepts obtain some of their justification from particular views of the processes of speciation and/or evolution. A particularly explicit statement to this effect is that of Wiley (1981: 22): '... the concept of species as evolutionary entities takes precedence over species as taxonomic entities and ... the formalism of taxonomy must be subservient to the demands of evolution.'[7] From such a view, the argument could be made that species are theoretical entities. For one thing, species are generally unobservable in practice[8] – even proponents of definitions in the first set would not deny this – thus, it might be argued that species, like electrons, are what some theory says they are. After all, if species are not observable, what other than a theoretical conceptualization, or inference, could justify assuming their existence (Nagel 1961: 146–152, raises this same general argument in his discussion of what constitutes criteria for reality)? Second, each definition of the second set can be interpreted as being derived from a particular view of the evolutionary process. Proponents of the biological species concept, for example, see reproductive disjunction as the critical endpoint of the process of speciation and define species accordingly. Likewise, Paterson (1985) emphasizes the origin of shared fertilization systems as the outcome of speciation and his recognition concept of species defines species in those terms. In reflecting on this link between a theory of process and its implied species concept, Paterson (1985: 21) notes: 'Any view of species must be cast in genetical terms if it is to be useful in understanding the process of evolution...'.

Does one of these two approaches to definition have an inherent superiority over the other? Philosophers seem to argue both ways. Feyerabend (1978: 168), for example, sees little need for any distinction at all since we can always be led to abandon either theory or observation. Knowledge, he argues, arises from the interplay of both: observation cannot be entirely divorced from theory (see also Gaukroger 1978: 45).

A distinctly contrary view is that of Brady (1985). His argument specifically addresses the analysis of character variation, itself directly relevant to discussions of species: 'But surely a regularity of nature recoverable by independent workers is of some scientific interest. After all, the pattern recovered by a parsimony analysis *would seem to be inherent in the data*, and the discovery of it was not dependent upon evolutionary theory. Does this point not argue a privileged position for such a pattern?' (1985: 114; italics in original). Brady points out (p. 117), moreover, that to 'replace a description of an empirical condition with its explanatory hypothesis is self-defeating'. Then again, this is precisely the argument: are species to be viewed merely as 'empirical conditions' of the natural world?

It is obvious different species concepts allow us to enumerate the world's diversity in radically different ways. It is equally obvious that no species concept can guarantee error-free estimates of species identity. But given that alternative species concepts can produce disparate pictures of species identities, is there a basis for choosing among these concepts? At least three criteria seem relevant:

a) *Congruence of data.* The set of species identities that best explains or organizes their intrinsic attributes (shared similarities of morphology, behavior, biochemistry, or physiology, for example) would be preferred. Thus, the most parsimonious distribution of character-state data is sought.

b) *Information content is maximized.* Closely linked to the preceding, this criterion would discriminate among hypotheses of species identity by rejecting those which either ignore or confound phylogenetically relevant information, such as that pertaining to character-state change or spatial pattern. This criterion receives its justification from the fact that speciation is first and foremost an historical process, and we cannot hope to study this process if historical information is lost.

c) *Relationship to theory.* If a theory individuates its entities (in this case, species) in such a way that they become more useful for evaluating empirical and theoretical claims implied by this theory compared to other ways of individuation, then the former entities might be preferred (McMullin 1984; see also Hacking 1984). Physicists, for example, have substantially redefined the concept of 'electron' as theory has changed; earlier definitions would be inappropriate within the context of present-day theory.

It is possible these criteria might conflict. In particular, entities individuated by relationship to theory might not conform to those individuated on the basis of congruence of character-state data. This is not especially surprising since a theoretical justification might be invoked to prefer a species concept that does not order all available data most parsimoniously (in fact, this regularly happens in the application of various species concepts). The implications of this type of conflict are several:

a) either the entities proposed within the context of the theory have been misconstrued and individuated incorrectly, or
b) the theory itself is improperly or incompletely understood, or perhaps
c) we have misunderstood character-state data and the 'observational' entities implied by them.

The apparent contradiction over viewing species as observables or as theoretical entities is not sharp and perhaps can be resolved by two lines of argumentation, namely that the definition of species-as-observables is not without theoretical justification, and that those species concepts said to be most consonant with theory are actually not.

The 'observational' definition of species employs two aspects of individual organisms to unite them into clusters called species: shared similarity in character data that can be used to distinguish them from other such clusters and reproductive continuity *within* those clusters. Diagnostic characters may be restricted to only one of the sexes or to one stage of the life cycle, therefore the criterion of reproductive continuity ensures that those individual organisms not having the diagnostic characters will be correctly associated with those that do. A third component of this type of definition is relational: these clusters are the smallest capable of being diagnosed. They cannot be further subdivided into diagnosable units, which also implies that two or more diagnosable units should not be united into a single species.

The above discussion speaks for species-as-observables: species are manifestations of pattern analysis. This does not mean that this view of species could not be given a theoretical justification. In fact, the species-as-observable view just articulated is more easily justified by our current understanding of evolutionary theory than are the 'theoretical' definitions noted earlier. Evolutionary theory encompasses a loose set of propositions with regard to the origin of taxa (species):

a) populations become isolated (usually spatially),
b) through various processes novelties arise in the genomes of individual organisms,
c) these novelties may or may not manifest themselves in ontogeny, and finally
d) some of these novelties, at the genome level or as expressed in ontogenies, become fixed in populations.

It is these latter populations of organisms – the ones with new diagnostic characters – that are called species under the species-as-observable view just discussed. They are a direct expectation of a general theory of evolution, as presently conceived. Thus, if so desired, it would be possible to justify this concept of species by recourse to evolutionary theory.

What about the 'theoretical' definitions? In general, it would appear their relationship to evolutionary theory is perhaps more tenuous. The biological species concept (Mayr 1963, 1970), for example, is defined in terms of reproductive disjunction (isolation). Reproductive isolation, however, is not an intrinsic attribute of populations; it is instead a relational concept. Populations do not evolve reproductive isolation. Moreover, there is nothing in the propositions just listed that would lead us to predict reproductive isolation as a necessary outcome of any particular instance of taxonomic differentiation, and indeed, in many groups of organisms well differentiated populations are often not reproductively isolated. Reproductive isolation is best interpreted as an epiphenomenon or effect of differentiation.[9] In adopting reproductive isolation as the key component of a species definition, one is requiring its accompanying theory of species origins to be construed much more narrowly than our knowledge of evolutionary mechanisms would seem to require.

Defining species in terms of a shared fertilization system (Paterson 1985) presents a different set of difficulties. A shared fertilization system manifests intrinsic attributes of organisms, but it is possible for 'species' individuated on this basis to represent an assemblage of historically unrelated populations. The reason for this is simple: the attributes contributing to a common fertilization system are primitive and reflect a failure to differentiate (Rosen 1978, 1979). Thus, populations that are very strongly differentiated in characters other than those associated with reproduction might share a common fertilization system (e.g., such as species within many genera of orchids). Many of those populations, moreover, might be very distantly related genealogically. Such a species concept would not permit us to reconstruct the history of speciation accurately.[10]

Species as discrete real entities

In general, biologists have been little concerned with considerations of species reality because they have few, if any, doubts that the different kinds of entities they discern in nature are 'real things'. Most would think any discussion of species reality to be purely academic, with little intrinsic scientific merit. This would be a mistake, however. Neontologists and paleontologists repeatedly face practical problems concerned with species reality, and one's particular ontological stance – that is, whether one takes species to be discrete, real entities or not – has major theoretical and empirical implications for systematic and evolutionary biology. To take but one example: if the application of different species concepts results in different suites of species being individuated, to what extent are we justified in claiming that some of those 'species' are fictitious and not real units in nature? Because it would be nonsensical to suggest fictitious entities can play a role in natural processes, the answer to this question has important empirical consequences. As Gaukroger (1978: 41) notes: 'depending on how we construe 'reality', some discourses [explanatory structures] will not explain 'reality', or they will offer explanations in terms of entities which are not 'real''.

The issue of whether species are discrete real entities has a close relationship to the species-as-individual debate (this interrelationship is discussed in detail by Rieppel 1986; see also Brothers 1985). The ontological status of species as real entities has always been somewhat ambiguous within 20th century evolutionary thinking. The architects of the so-called 'evolutionary synthesis' – in particular, Dobzhansky (1973) and Mayr (1942) – accepted species as discrete, real entities when viewed 'nondimensionally', that is, at one instant in time (generally, the present). But when conceived as historical entities, species were thought to loose their discreteness: because evolutionary change within populations is a slow, gradual process, there are no boundaries between ancestral and descendant species, simply intergrading populations (for further discussion of this ontological conflict, see Eldredge 1985a, chapter 3; Cracraft 1987). For these same reasons, a denial of species reality has been a common component in the thinking of paleontologists (Cracraft 1979, 1987; Eldredge 1985a).

Conceiving of species reality from these two perspectives (the 'paradox of discrete entities': Cracraft 1987) may not seem at all unreasonable to an empirically-minded paleontologist or neontologist. Yet, it creates a severe problem for biological theory: species cannot be arbitrarily defined entities in space and time,[11] and at the same time be used as participants in biological processes. Species cannot be unreal and still do something.

There are two, mutually compatible escapes from the paradox of discrete entities, one conceptual and one empirical. The conceptual solution to the paradox is to conceive of species as individuals and to realize that the boundaries of *all* individuals are scale dependent. Thus, the degree to which individuals are seen as discrete entities depends upon the spatial and temporal scale of the observer. The empirical solution is to define species in terms of species-as-observables in which case species boundaries, whether in space or time, will be individuated by discrete character variation.

If species participate in processes, they must be discrete entities, and thus individuals (Ghiselin 1974; Hull 1976). Individuals are individuated on the basis of spatiotemporal restrictedness and continuity (Hull 1976): they have beginnings and endings and exhibit continuity in between. Species-as-individuals can be conceptualized as having discrete beginnings and endings when the time-scale of perception is appropriately large (say, 10^5 or 10^6 years). Perceived from the perspective of a narrower time-scale (e.g., 10^2-10^4 years), evolutionary change among species appears more gradual, and observations on this time-scale have led to the belief that species loose their discreteness as they are 'traced back in time'. It is essential not to loose sight of the scale problem, because species are typically judged to be nondiscrete when compared to the one thing humans perceive as being discrete, namely individual organisms. Yet, the latter are no more objectively discrete than are species; only the time-scales are different. Within the context of a time-scale measured in units of 10^{-2} to 10^{-4} seconds, for example, the beginnings and ends of individual organisms themselves become nondiscrete.

As a consequence of the species-as-individuals formulation it is no longer necessary to judge the ontological status of species as being tenuous on the grounds that their boundaries are not discrete. Thinking of species as individuals thus clarifies some of the *conceptual* problems existent in the neontological and paleontological literature. However, conceiving of species as individuals does not constitute a sufficient solution to the problem of species reality. Species – as things bearing proper names – could have the status of individuals yet still not have any objective reality. To be considered real things, species must fulfill other criteria. They should also satisfy the requirement of objectifiable, 'observable' entities, and here we confront the importance of species-as-observables:

> How is it possible to claim reality for the process of speciation before having first demonstrated the reality of species taxa? How is it possible to claim that one species descended from another ... before having first demonstrated the actual existence of those particular species taxa? Yet, in order to do this, operational criteria are necessary that permit the delineation of the species taxon and the application of its name to parts (or members) of it ... Thus, unless the reality of species taxa can be demonstrated on the basis of operational criteria, there is no possibility to claim reality for the process of speciation (Rieppel 1986: 298).

Defining species in terms of the smallest clusters of individual organisms that share diagnostic character variation seems to be the only way to objectify species in space and time.[12] Advocates of more theoretical definitions, such as the biological species concept, must look at species nondimensionally because the differentiae of these concepts – reproductive isolation, shared fertilization system – have limited or no applicability in space and time. The unity of individuals over time is said to be maintained by cohesive processes: 'the elements comprising an individual do so because of how they are organized and not because of any shared similarity' (Hull 1980: 313). Yet, with respect to species, this is a conceptual presupposition, for the only evidence suggesting that 'cohesive processes' might be operable is the temporal and spatial continuity of a particular pattern of character variation. The reality of one or more 'cohesive processes' is another question altogether.

Conceptualizing species as individuals and adopting a definition of species-as-observables constitutes a viable solution to the paradox of discrete entities, although this solution will probably not be accepted by some supporters of quasi-theoretical definitions, such as the biological species concept. Mayr (1982: 293), for example, objects to individuating species on the basis of a cladistic interpretation of character variation, because it 'omits any reference to reproductive isolation and is strongly typological in its exclusive reliance on a limited number of ancestral or derived characters'. There are, however, several difficulties with this line of argumentation:

a) viewing species-as-observables is patently not typological and such a claim, unless made purely for its pejorative effect, misunderstands the content of essentialism: species-as-observables are scientific hypotheses about character variation, not statements about essences, and

b) reproductive isolation is not a character, or intrinsic attribute, of any taxon, and as a consequence has no materal reality: species defined in these terms, therefore, have ambiguous ontological status in space and time (as admitted by Mayr 1982: 295, for example).

The twin problems of species definitions and species reality are closely related in that if a definition is said to derive from some theoretical expectation of how the world is structured, then it suggests the interpretation that entities not meeting the criteria of the definition will have no objective reality as species. This appears to be the reason why some supporters of the biological species concept either question or deny the reality of asexual species (e.g., Bock 1986: 33; Ghiselin 1987: 138). But something is wrong here. There may be reasons why asexual organisms do not form species in the same sense as sexual organisms, but the ontological question itself cannot be adjudicated by recourse to a definition.

Disputes over the reality of asexual species mask a more general problem. If it could be agreed that a particular species concept really, truly specifies what entities exist in nature, does this mean that entities identified as species by other concepts are not real? Unless those entities are identical to species identified by the canonical species concept, the answer would seem to be yes. Different species concepts organize the world differently and conflictingly, thus some of these species must be fictitious, at least in some sense of the word. But in what sense? In the end, a final judgment will probably rely upon variations of the three criteria discussed in the previous section: congruence of data, maximization of information content, and relationship to theory.

The role of species in biological processes

Throughout empirical and theoretical biology species are taken to be active participants in numerous biological processes: they are said to speciate, evolve, compete, predate, interact, disperse, be selected, or go extinct, among others. This section takes a closer look at the role of species in biological processes. As discussed earlier, some workers find justification for species individuality and reality in the premise that species are active participants in processes: they do something, therefore they must be individuals and real entities. On the empirical side of the ledger, considerable discussion has taken place in recent years regarding species as participants in processes, especially within the context of various 'hierarchical analyses' of evolutionary processes (e.g., Vrba and Eldredge 1984; Vrba and Gould 1986). All of this suggests that within evolutionary biology, much rides on our understanding of the role of species in processes.

At the outset we must clear the decks of a noxious semantic problem (also noted by Eldredge 1986: 361–362). Sometimes many of us do not use language precisely; we say one thing, but mean another. Such is the case with species-as-participants. Many biologists confound species-as-participants with organisms-as-participants. They say, for example, 'species A and species B compete' when they actually mean 'individual organisms of species A compete with individual organisms of species B'. Such misuse of the language complicates this discussion enormously, and it also tends to overemphasize the importance of species-as-participants within current biological thinking. It also leads to a confused understanding of the biological processes themselves. The following discussion assumes a hierarchical view of pattern and process and thus statements about process are taken at face value: if species are said to do something, it is taken to mean that the species, acting as a cohesive entity, is the actor and that the action cannot be ascribed to the additive actions of individual organisms. Only by maintaining this distinction can we accurately assess the role of species in biological processes.

What does it mean to participate in a process? Species have been viewed in three ways. As *actors*: they do something (e.g., speciate or compete), and thus are taken to be *active* participants. As *recipients*: they are acted upon (e.g., species are selected, predated), and thus in some sense are *passive* participants. As *effects*: they look as if they are actors or recipients, but instead only manifest the additive effects of entities (usually individual organisms) participating in processes at another (generally lower) hierarchical level. It would be a mistake to take these categorizations as hard and fast. A billiard ball rolling across a pool table can be interpreted as an actor (it is rolling) or as a recipient (it was pushed) or perhaps even as an effect (if one wanted to look at the motion from the standpoint of molecular interaction). Even though these roles may not always be distinct, they have been components of discussions about the participation of species and seem to have some usefulness.

There is now a substantial literature on species-as-participants. In recent years, much of that has arisen within three areas of discussion:

a) the debate over species-as-individuals,
b) the units of selection controversy (e.g., Brandon and Burian 1984), and
c) the hierarchical analysis of evolutionary theory (including micro- versus macro-evolution).

Many issues cut across these three areas, but none is more important than the notion that if species participate in processes they must somehow function as a discrete entity, as a cohesive whole. Yet, the way in which one might use the notion of cohesion to argue for species individuality (a metaphysical construct) or to argue for species-as-participant (an empirical question) could be quite different. We are concerned here primarily with the latter question.

Despite the extended discussion about species as cohesive units and their role in processes, there has been remarkably little dialogue about the concept of cohesion and its precise meaning relative to those processes which species are said to participate in. Most investigators identify gene flow as the mechanism maintaining cohesion of a species. But there are other ways for an entity to be cohesive, and the concept of cohesion is context dependent. Gene flow, for instance, might be a cohesive mechanism with respect to one process (speciation) but not to another (say, competition). In the evolutionary literature, however, relatively little attention has been paid to cohesion except within the context of gene flow.

In attempting to find an approach to the concept of cohesion that has empirical validity, we can look to the part/whole relationship of (individual) entities and ask whether a part of the entity or the whole entity itself is the participant in a particular process. Organisms, for example, are comprised of numerous parts, but it is the organism as a whole which participates in a process such as competition. Species have parts too, namely individual organisms, although there may be some reason for considering 'demes' or 'populations' as the parts of species (see below). An organism is the paradigm example of a cohesive entity. Its parts are physically tied together; it has well-defined functional integration of the various parts; it has a discrete history (ontogeny); and many properties can be easily identified that are emergent at the level of the individual organism rather than being merely additive effects of its parts.

The cohesiveness of species, as many have already noted (in particular see Hull, 1977, 1978; Vrba and Eldredge 1984; also see below), is much more ambiguous.

The fact that species can be individuated over space and time suggests the presence of mechanisms promoting some form of cohesion, namely phenotypic cohesion. Indeed, many factors – falling into three broad categories: population genetic, developmental constraints, and stabilizing selection – have been proposed to account for the phenotypic cohesiveness of species (e.g., see Van Valen 1982; Wake *et al*, 1983). But to what extent do these mechanisms result in the *functional integration* of the parts of species (i.e., individual organisms) and facilitate the development of emergent properties? Without functional integration over space and time and/or the existence of emergent properties, the role of species in biological processes will be necessarily limited. This is the critical issue.

We can explore whether species function in biological processes by considering the following questions: Are species interactors? Are species selected? Do species speciate?

Are species interactors? Does a species interact with its biotic and abiotic environment as a cohesive unit? Species are often thought to compete with one another, and a considerable body of theory on interspecific competition has been developed. As already noted, many biologists would ascribe such 'processes' to the additive effect of individual organisms, but some no doubt see species as cohesive, integrated wholes. It is difficult, however, to see how this latter position can be justified. Individual organisms are interactors because their parts are jointly functioning in near contemporaneous space/time to respond to or act upon the environment. Even allowing for a different time-scale, it is difficult to see how the parts of a species jointy function in any interactive process – dispersion through space would seemingly preclude this (Damuth 1985; Eldredge 1986; Mishler and Donoghue in press). One could argue in response to this that not all parts of an individual need participate in a process for that individual to be considered integrated (just thinking of individual organisms demonstrates that). However, parts of species – organisms – interact (e.g., compete) not as a manifestation of some causal mechanism promoting their cohesion but because they happen to possess similar properties (e.g., morphology, behavior, ecology) as a result of a common history. The fact that an interaction occurs is most parsimoniously explained by the properties of individual organisms, and species-level appearances are additive effects (see Williams 1966 for the classic exposition of this type of argument).

To the extent that this argument is valid – that is, it is not possible to demonstrate mechanisms promoting cohesion among the individuals of a species, and observations can be accounted for by reference to individuals, not species – then we would conclude species do not function as individuals within competition theory. A consequence of this, for example, is that one could not invoke interspecific competition as a mechanism structuring ecological communities, and the reasons for this are not only empirical, they are also ontological: species do not function as individuals within the context of competition theory. Similar arguments could be made about other roles of species, for example, as predators or dispersers.

Are species selected? Intellectual battles rage over the units of selection controversy (e.g., Hull 1980; Brandon and Burian 1984; Sober 1984), but we will briefly consider only a small portion of this debate, namely that which posits a role for species in a selective process. The name 'species selection' has been applied in two ways (Cracraft 1985a):

a) for a pattern of differential survivorship of species among monophyletic clades (species sorting) and
b) as a process whereby the relative survivorship of a species can be causally linked to the possession of an emergent, heritable species-property (Vrba 1984; Vrba and Eldredge 1984).

The latter process is an explicit analogue to natural selection among individual organisms based on their heritable emergent properties, and this way of formulating species selection as a process is slowly becoming standard terminology. As Vrba and Eldredge (1984: 147; see also Vrba and Gould 1986) note, sorting of species over time, through differential speciation and extinction, can come about for three reasons: species selection, natural selection at the level of individual organisms (e.g., Vrba 1983), or through a variety of nonselective mechanisms influencing rates (Cracraft 1985b). The only issue of concern here is whether species can be participants in a selective process in the strict sense.

Three criteria must be met for species selection to occur:

a) the property said to influence the rate of speciation or extinction must be emergent at the level of the species and not additive of the parts (organisms) of the species,
b) the property must be heritable, that is, passed on to its descendant species, and
c) the functional interaction between the emergent property and the environment must be causally related to speciation and/or extinction in such a way as to explain the observed spatiotemporal pattern of sorting.

These criteria are parallel to those required of instances of natural selection at the organism level, and proper assignment of causation at either level is dependent upon the stringency by which these criteria are applied. It is clear that earlier thinking on species selection was imprecise because of a failure to draw a distinction between true species selection, effects of lower level processes (Vrba 1983), and influences on speciation and extinction rates attributable to neither. Gilinksy (1986), for example, abandons this parallelism in his discussion of species selection and ascribes the latter to any variation in speciation and extinction rates that is not 'caused by fitness differences among organisms contained within species' (p. 257). This classification does not clarify matters much, for it unites under the rubric 'species selection' two very different phenomena having distinct causations:

a) species selection *sensu stricto* in which differences in rate can be ascribed to species-level properties, and

b) differences in rates that cannot be ascribed either to natural selection or species selection (e.g., Cracraft 1985b).

If we are to develop a general theory of speciation and extinction rate-control (='macroevolution'), it is essential that these causal distinctions not be conflated.[13]

Returning to the main theme of this paper, if species selection can be demonstrated, then an argument could be made that this is an instance of species functioning as discrete entities in a natural process. We could not ascribe causation to the level of individual organisms because the relevant properties causally correlated with differences in speciation or extinction rate are emergent at the level of species. The main difficulty for the thesis of species selection is finding examples that satisfy all three criteria, noted above. Our difficulties begin with attempting to find an unambiguous meaning for the concept of emergent character variation. The standard use of 'emergent' refers to character variation that cannot be attributed to 'sums of the parts':

> one can suggest that species characters that arise from *distribution and interaction among* organisms are emergent at the species level. Possible candidates include characteristic population size, spatial and genetic separation between populations, and the nature of a periphery (whether it is extensive and convoluted or not) [Vrba and Eldredge 1984: 154; the same list is given in Vrba and Gould 1986].

Yet, the emergent status of these characteristics can be questioned for various reasons. Properties such as population size or the length of species boundaries (Jablonski 1987) seem most properly attributable to sums-of-parts, much like body mass or height might be in individual organisms. An interpretation of a species' population/demographic structure or of the nature of its periphery, on the other hand, is more ambiguous. A particular pattern of demographic structure emerges from the behavior and ecology of individual organisms in conjunction with the physical characteristics of the environment. The nature of a species periphery also seems not easily reducible to sums-of-parts. By way of analogy, two organisms may have the same surface area and height but different shapes, shape being emergent at the level of the organism. Whether a species' periphery is convoluted or relatively smooth is not a function of the number of individual organisms, but rather of their individual distributions and, especially, the physical properties of the environment. Thus, the demographic or distributional characteristics of species thought to be emergent may not be determined by the properties of the organisms as much as by the structure of the external environment in which they are found. Do these, then, count as emergent properties at the species level? An answer to this question is not clear-cut, and in the long run may depend on particular examples and whether causal connections to speciation and extinction rates are best explained by considering these types of properties as emergent.

While granting that emergence might be postulated in any particular instance, applying the concept of heritability at the species level entails additional difficulties. What does it mean for a property to be heritable at the species level? At the organism level, heritability takes its meaning from a direct physical causality (genetics and development) for the correlation in character variation between parent and offspring. In terms of a direct analogue with the organism level, an ancestral species would pass its (emergent) species-level property on to its descendant species much as it might pass on intrinsic characters (as synapomorphies, which themselves arose as effects of lower level processes). Jablonski (1987) infers heritability of geographic range by showing a positive correlation in range size between putative sister species. At the organism level a causal basis for such correlations is found in developmental genetics, but if it could be demonstrated that the parent-offspring correlations have no basis in genetics and development, we would reject the hypothesis of heritability and seek another explanation (epigenetic effects, spurious correlation, etc.) At the species level, 'causes' of these correlations would presumably lie in similarities among sister-species in ecology, behavior, and other aspects of their biology. Thus, a correlation in and of itself does not constitute evidence for heritability, and it may be very difficult to justify an argument, ontologically and causally, that such correlations reflect physical properties of the species and are not just epiphenomena.

The third criterion of species selection is even more difficult to satisfy because it must be shown that postulated emergent properties have a direct causal influence on speciation rate. In some sense at least, all factors influencing extinction rates are effects, since extinction is nothing more than the accumulated deaths of individual organisms. Jablonski (1987) postulates a relationship between range size and temporal duration, on the one hand, and resistence to extinction, on the other. Although such associations may exist, range size, *as an emergent property under the influence of species selection*, cannot have an effect on extinction: extinction seems reducible to organism-level processes in all cases. In contrast, it might be argued that an emergent species-level property has an influence on speciation rate, because variation in the latter cannot be explained entirely by reference to organism-level processes.

Emergence aside, for any species-level property to influence speciation rate it is necessary that the species function as a cohesive whole such that there is a property-environmental 'interaction' causally related to speciation rate. It is difficult to envision species functioning as a cohesive unit relative to some environmental parameter. For precisely this reason Damuth (1985) rejected species selection: 'a collection of entities that do not share a selective environment can hardly be said to be adapting to any particular environment' (p. 1134). The argument seems to be that species are not integrated ecologically. Individuals of species A in one area may be functioning ecologically more similarly to individual organisms of sympatric species B than they are to their conspecifics in a different area. Responses to environments are relatively localized in space and time.[14] For this and other reasons already mentioned, the case for species, as entities, functioning in selective processes is tenuous, both on philosophical (ontological, causal) and empirical grounds.

Do species speciate? Perhaps no role for species, *qua* individuals, seems more secure than as the entities of speciation theory: *species* speciate. Yet, is this claim, like 'species compete,' merely a syntactic error or is there justification for the assumption that species function as cohesive, discrete units in some process called speciation? Put another way, we want to ask whether species speciate, or *whether they are speciated*; the distinction is of some importance.

Asking whether species speciate may seem like a radical question but from the early days of the so-called evolutionary synthesis, evolutionists have identified populations as the entities that differentiate. Widespread species are either subdivided by an extrinsic barrier, or individual organisms disperse over pre-existing barriers, and then these isolated populations differentiate into new species. This appears to place the entity of speciation back one level, to the population. Yet, if we consider populations to be the individuals of a theory of differentiation (speciation), and examine the content of this theory from the standpoint of the part/whole relationships of populations, it is apparent that not even populations, as discrete cohesive entities, themselves differentiate. Instead, change is generated within individuals, at the levels of the genome and developmental pathways. Sometimes, these changes become fixed in populations as diagnostic character variation through processes having individual organisms as their entities. If this view of evolutionary change is correct, then populations and species do not function as entities in speciation theory, rather they are the effects of lower-level processes. A further implication of this is that if populations are much like species in terms of their status as functioning entities, then doubts must be raised about populations as units of selection. Populations, like species, can exhibit differential sorting (species selection, if you will, as pattern), but is that sorting due to the effects of emergent character variation? Comparison with species would suggest that this will be unlikely, but only further analysis and examination of empirical cases will resolve this.

Conclusions

The themes addressed in this paper are interrelated, not only as empirical questions within biology but also as they touch on problems of an ontological nature. Many biologists and philosophers have identified species as being important to biological theory, and the endless discussions over the nature of species demonstrates the complexity of understanding just how and why species are important. This paper offers an unconventional view: species are discrete real entities through space and time, but seemingly they don't do much, or at least not as much as most biologists and philosophers generally think they do.

Species are much like developmental pathways within the ontogeny of individual organisms: both are effects rather than effectors. They are epiphenomena, developed or evolved – in the true historical sense of the term – from lower-level processes. Both are thus historical entities or 'by-products'.

Because species are historical entities, biologists have traditionally found it difficult to envision how they might be individuated in space and time. This has led to confusion over their ontological status and their roles in biological theory. Much of this confusion can be alleviated once species are seen as discrete entities that are products of processes, not participants. This clarifies, for example, much of the debate over species concepts. It clarifies much of the confusion that has risen around seeing species as 'units of evolution'. They are units produced by some evolutionary (historical) process but they do not themselves evolve.

References

Bock W.J. (1986). Species concepts, speciation, and macroevolution, in Iwatsuki, K, Raven, P.H., Bock W.J. (eds) *Modern aspects of species*; 31–57. Tokyo: University of Tokyo Press.

Brady, R.H. (1985). On the independence of systematics. *Cladistics* 1: 113–126.

Brandon R.J., Burian R.M. (eds): *Genes, organisms, populations*. Cambridge: MIT Press.

Brothers D.J. (1985). Species concepts, speciation, and higher taxa. *Transvaal Museum Monograph* 4: 35–42.

Cracraft J. (1979). Phylogenetic analysis, evolutionary analysis, and paleontology. In Cracraft J., Eldredge N. (eds) *Phylogenetic analysis and paleontology*; 7–39 New York: Columbia University Press.

Cracraft J. (1983). Species concepts and speciation analysis. *Current Ornithology* 1: 159–187.

Cracraft J. (1985a). Species selection, macroevolutionary analysis, and the 'hierarchical theory' of evolution. *Systematic Zoology* 34: 222–229.

Cracraft J. (1985b), 'Biological diversification and its causes,' *Annals Missouri Botanical Garden* 72: 794–822.

Cracraft J. (1987) 'Species concepts and the ontology of evolution,' *Biology and Philosophy* 2: 329–346.

Cracraft J. (in press), 'Speciation and its ontology: the empirical consequences of alternative species concepts for understanding the pattern and process of differentiation,' in D. Otte and J. Endler (eds), *Specification and its Consequences,* Sunderland, Ma.: Sinauer Assoc.

Cronquist A. (1978). 'Once again, what is a species?' *Beltsville Symposium Agricultural Research* 2: 3–20.

Damuth J. (1985). 'Selection among 'Species': a formulation in terms of natural functional units,' *Evolution* 39: 1132–1146.

Dobzhansky T. (1937). *Genetics and the Origin of Species*, Columbia University Press, New York.

Eldredge N. (1985a). *Unfinished Synthesis*, Oxford University Press, New York.

Eldredge N. (1985b). 'The ontology of species,' *Transvaal Museum Monograph* 4: 17–20.

Eldredge N. (1986). 'Information, economics, and evolution,' *Annual Review of Ecology and Systematics* 17: 351–369.

Feyerabend P. (1975). *Against Method*, Verso, London.

Gaukroger S. (1978). *Explanatory Structures*, Humanities press, Atlantic Highlands, N.J.

Ghiselin M.T. (1974). 'A radical solution to the species problem,' *Systematic Zoology* 23: 536–544.

Ghiselin M.T. (1987). 'Species concepts, individuality, and objectivity,' *Biology and Philosophy* 4: 127–143.

Gilinsky N.L. (1986). 'Species selection as a causal process,' *Evolutionary Biology* 20: 249–273.

Hacking I. (1984). 'Experimentation and scientific realism,' in J. Leplin (ed.), *Scientific Realism*, Univ. California Press, Los Angeles, pp. 154–172.

Hull D.L. (1976). 'Are species really individuals?' *Systematic Zoology* 25: 174–191.

Hull D.L. (1977). 'The ontological status of species as evolutionary units,' in R. Butts and J. Hintikka (eds) *Foundational Problems in the Special Sciences*, D. Reidel, Dordrecht-Holland, pp. 91–102.

Hull D.L. (1978). 'A matter of individuality,' *Philosophy of Science* 45: 335–360.

Hull D.L. (1980). 'Individuality and selection,' *Annual Review Ecology and Systematics* 11: 311–332.

Hull D.L. (1981). 'Units of evolution: a metaphysical essay,' in U.L. Jensen and R. Harre (eds), *The Philosophy of Evolution*, Harvester Press, Brighton, Sussex, pp. 23–44.

Jablonski D. (1987). 'Heritability at the species level: analysis of geographic ranges of Cretaceous mollusks,' *Science* 238: 360–363.

Mayr E. (1942). *Systematics and the origin of species*, Columbia University Press, New York.

Mayr E. (1963). *Animal Species and Evolution*, Harvard University Press, Cambridge, Ma.

Mayr E. (1970). *Populations, Species and Evolution*, Harvard University Press, Cambridge.

Mayr E. (1982). *The Growth of Biological Thought*, Harvard University Press, Cambridge, Ma.

McMullin E. (1984). 'A case for scientific realism,' in J. Leplin (ed.), *Scientific Realism*, University California Press, Los Angeles, pp. 8–40.

Mishler B.D., R.N. Brandon (1987). 'Individuality, pluralism, and the phylogenetic species concept,' *Biology and Philosophy* 2: 397–414.

Nagel E. (1961). *The Structure of Science*, Harcourt, Brace & World, New York.

Nelson G.J., N.I. Platnick (1981). *Systematics and Biogeography: Cladistics and Vicariance*, Columbia University Press, New York.

Paterson H.E.H. (1985). 'The recognition concept of species,' *Transvaal Museum Monograph* 4: 21–29.

Rieppel O. (1987). 'Species are individuals. A review and critique of the argument,' *Evolutionary Biology* 20: 283–317.

Rosen D.E. (1978). 'Vicariant patterns and historical explanation in biogeography,' *Systematic Zoology* 27: 159–188.

Rosen D.E. (1979). 'Fishes from the uplands and intermontane basin of Guatemala: revisionary studies and comparative geography,' *Bulletin American Museum of Natural History* 162: 267–376.

Rosenberg A. (1985). *The Structure of Biological Science*, Cambridge University Press, New York.

Sneath P.H.A., and R.R. Sokal (1973). *Numerical Taxonomy*, W.H. Freeman, San Francisco.

Sober E. (1984). *The Nature of Selection*, Massachusetts University Press, Cambridge.

Van Valen L. (1976). 'Ecological species, multispecies, and oaks,' *Taxon* 25: 233–239.

Van Valen L. (1982). 'Integration of species: stasis and biogeography,' *Evolutionary Theory* 6: 99–112.

Vrba E. (1983). 'Macroevolutionary trends: new perspectives on the roles of adaptation and incidental effect,' *Science* 221: 387–389.

50

Vrba E. (1984) 'What is species selection?' *Systematic Zoology* 33: 318–328.

Vrba E.S., N. Eldredge (1984). 'Individuals, hierarchies and processes: towards a more complete evolutionary theory,' *Palaeobiology* 10: 146–171.

Vrba E.S., S.J. Gould (1986). 'The hierarchical expansion of sorting and selection: sorting and selection cannot be equated,' *Palaeobiology* 12: 217–228.

Wake D.B., G. Roth, M.H. Wake: 'On the problem of stasis in organismal evolution,' *Journal Theoretical Biology* 101: 211–224.

White M.J.D. (1978). *Modes of Speciation*, W.H. Freeman, San Francisco.

Wiley E.O. (1981). *Phylogenetics*, J. Wiley & Sons, New York.

Williams G. (1966). *Adaptation and Natural Selection*, Princeton University Press, Princeton.

Acknowledgement

I especially want to thank David Hull for his comments on this manuscript and for his continued good friendship and counsel. He has influenced my thinking in many ways, hopefully most of them productively so, and for that it is a pleasure to dedicate this paper to him. I am also grateful to Elliott Sober and Niles Eldredge for discussions and to the National Science Foundation, through grant BSR-8520005, for support of my research.

Notes

1. It could be argued, of course, that even if species did not participate in any process, they should still be viewed as individuals, not classes. While this may be of interest to philosophers, the empirical consequences of using species in theories is of more importance to biologists.

2. As noted elsewhere (Cracraft 1987: 338), by incorporating a particular definition of species, namely the biological species concept, into their collective consciousness, many biologists accept that speciation is the origin of reproductive isolation, or that reproductive isolation is the most important outcome of speciation. This is not a criticism of the biological species concept per se, for any definition will entail looking at speciation in a particular way. Nevertheless, it is worth appreciating that our understanding of the patterns and processes associated with speciation may sometimes be shaped more by the definition of species than by 'empirical' observations.

3. It might be argued that the biological/polytypic species concept derived from a notion of how speciation takes place vis-à-vis the origin of reproductive isolation. There is, however, nothing in speciation theory per se that suggests its entities should be reproductively isolated. Botanists, in general, and many zoologists do not see reproductive isolation as a necessary outcome of taxonomic diversification. It just so happens that the developers of the polytypic/biological species concept – in particular T. Dobzhansky, B. Rensch, and E. Mayr – all worked on groups in which sympatric differentiated populations are generally, but by no means always, reproductively isolated.

4. As will become evident, it is *not* implied here that there is a sharp distinction between these two types of terms.

5. It seems reasonable to ignore here the line of argument that all definitions are 'theory-laden' because they will always contain undefined terms. Although this is true, we are concerned here with a different question: whether a term for entities (such as species) that seem important to a theoretical discourse (e.g., the theory of speciation) need be defined in terms of how we think those entities behave within the context of the theory.

6. Omitted from further discussion is a third component, namely that species exhibit reproductive cohesion ('self-perpetuating' or 'parental pattern of ancestry and

descent'). First, *all* species concepts, whether they are explicit or not, must include some statement about cohesion in order that males, females, juvenile stages, and morphs not be allocated to separate species (Nelson and Platnick 1981; Cracraft 1983, 1987). Second, a clear distinction must be made between reproductive *cohesion* and reproductive *disjunction*. As just stated, the former is a component of all species concepts, but the latter is a component of only some (e.g., the biological species concept of Mayr 1963).

7. Yet in the next paragraph, Wiley (1981: 22) adopts a view not altogether different from that of species-as-observables: 'Because we do observe order we suppose that this ·order was produced by a phenomenon or process which has worked on organisms to produce this particular kind of order. It is not necessary to understand the phenomena or process to recover the ordering'. Whether Wiley means this to apply also to species recognition – which is a form of ordering character variation in the view of those definitions in the first set – is uncertain given his definition of species.

8. It could sometimes be claimed that it is possible to observe completely some species, particularly those with very low population sizes and restricted distribution (a species on the verge of extinction, for instance). Even though this proposition is perhaps arguably true, it is of little consequence for the argument being developed here.

9. This is not to say that the study of reproductive isolation is uninteresting or unimportant. Rather, the claim is made that it is not general and therefore should not be used as a criterion for species recognition: all reproductive isolation manifests differentiation, but not all differentiation is manifested as reproductive isolation. Differentiation is therefore more general.

10. The same criticism applies to the biological species concept for much the same reasons: the condition of not being reproductively isolated is primitive (Rosen 1978, 1979; Cracraft 1983). Uniting taxa that are not reproductively isolated will often mean that distantly related populations will be placed in the same species. This problem will not arise with the species-as-observables view because species are defined as the smallest units based on diagnosable characters.

11. We will dismiss here the notion that species should be viewed 'nondimensionally' (Mayr 1963; Bock 1986). Evolutionary theory, if it is to be a general explanatory structure, must deal exclusively with entities having reality in space and time.

12. We can justify this assertion by comparison with individual organisms. For the sake of argument, we can consider two twins as being absolutely identical. How do we individuate them? Perhaps the easiest answer would be that they do not occupy the same coordinates in space-time. If they were not observed together in space, then we could not individuate them. If they were observed at two different (short) periods of time, we could not differentiate them. If they were observed at two different (very long, longer than any known human life) periods of time, we would conclude that, short of supernatural intervention, they represent two different individuals.

    Now consider two absolutely identical species. Can their space-time coordinates be used to individuate them? The answer seems to be no. If observed anywhere in space – allopatrically or sympatrically – and at the same time, we would say one species is present. If observed at very different times, we would say one species is present because it is not inconceivable (although perhaps very improbable) that the same species could persist for long periods of time. Position in space-time may be important for individuating some physical entities, but it hardly seems likely that we would individuate individual organisms or species using anything but character variation (along with reproductive cohesion in the case of species; see note 6).

13. Sober (1984) uses species selection in much the same way as Gilinsky. Sober makes a good point when he notes (p. 365) that 'The crucial ingredient in species selection, properly so-called, is not the *selection of* species but *selection for* species-wide properties.' However, his use of 'species-wide property' also includes non-emergent character variation at the species level and thus encompasses 'effect macroevolution' (Vrba 1983) as well as external (environmental) effects on speciation or extinction rates

(Cracraft 1985b); this is illustrated by his example (p. 366) in which winglessness in grasshoppers is correlated with high species number, whereas the winged condition is associated with low species number. Winglessness is not an emergent property but a statistical property of all the individuals in the species (it is possible, for instance, that an occasional winged morph might arise in the population).

Sober (personal communication) does raise an important issue, however: if species selection is defined as pertaining only to emergent character variation, and the criteria for emergence are so stringent that they virtually exclude any real-world examples, then we effectively define species selection out of existence. He notes further that an argument could be made that all properties having physical reality could be decomposed into other, smaller parts, thus perhaps obviating the notion of emergence.

All of this may be true, but it still seems useful to make a distinction between emergent and nonemergent variation because of its implications for regarding species being selected as cohesive units. A good deal of the confusion in the literature is the result of confounding different causal processes to explain a single pattern, thus the semantic distinction over species selection seems critical. It may well be that either emergent species-level properties cannot be specified, or if they can, that they will prove to be irrelevant for selective processes affecting speciation rates.

14. Damuth (1985) attempts to solve this problem by arguing that the localized population (the 'avatar') of a species within a community might serve as the higher-level analogue of an individual organism and be selected. It would lead us too far afield to evaluate this proposition in detail. Suffice to say, a case has not been made that avatars themselves are really cohesive units, that they have emergent properties, that these properties are heritable, and that these properties, through an interaction between the avatar and the environment, will have some causal influence on their rates of reproduction or death. Damuth's discussion raises other ontological difficulties, namely whether, as he claims, 'communities' are themselves functioning as entities in various processes.

Eldredge (1986) and Mishler and Brandon (1988) also argue that species will not generally be integrated (or cohesive) over the entire extent of their range to function as a unit in a selection process.

Individuality, History and Laws of Nature in Biology

MICHAEL T. GHISELIN

California Academy of Sciences, Golden Gate Park, San Francisco, CA 94118, U.S.A. and Wissenschaftskolleg zu Berlin, Wallotstrasse 19, D-1000 Berlin 33, F.R.G.

David Hull and I approached the problem of the ontological status of species from very different directions, and a long time elapsed before our views on such matters converged. Early in my career as a comparative anatomist I found myself involved in the philosophy of systematic biology. Some authors had argued from a nominalistic position, and maintained that species, being classes, are not 'real'. I pointed out that this argument rested on false premises. To me it seemed obvious that species are individuals, not classes, and their names have to be defined ostensively (Ghiselin 1966, 1969). At first Hull, who approached taxonomic theory from the point of view of analytical philosophy, did not find this solution attractive. We exchanged a long series of letters, in which I presented arguments based upon analogies and counter-examples, but to no avail.

Hull changed his mind upon encountering of the claim of J.J.C. Smart that there are no laws of nature in biology, and that biology therefore isn't really a science, but rather something more like engineering. Smart was quite correct in saying that there are no laws for *Homo sapiens*, but it was Hull who drew the proper conclusions. Species are individuals, and there are no laws about individuals in any science whatsoever. In the biological and physical sciences alike, laws have to be generalizations about classes of individuals. This was perhaps Hull's most important contribution, but it has largely remained unexplored.

Smart and others have taken the putative lack of laws in biology to indicate that biology is inferior to the physical sciences. Such views are very widely held, and applied *a fortiori* to the social sciences. Witness the popularity of drawing basic distinctions between nomothetic and ideographic sciences, and between *Naturwissenschaften* and *Geisteswissenschaften*. Smart did observe that such physical sciences as astronomy deal with individuals. But he thought that their role here was peculiar. Physicists test their hypotheses by means of observations on individual stars and galaxies. Obviously he cannot have known much about biology, for this is precisely what many biologists are up to when they study individual organisms and individual species.

Hull and I have taken the position that the biological and the physical sciences are not fundamentally different, and that we need not have two different

M. Ruse (editor), What the Philosophy of Biology is. pp. 53–66.
© 1989 *Kluwer Academic Publishers, Dordrecht*

philosophies for them (Hull 1974; Ghiselin 1981, 1987; see also Rosenberg 1985). This implication of the individuality thesis makes it much easier to treat knowledge in general from a unitary and comprehensive point of view. But the whole intellectual world has tended to become polarized with respect to that very issue. Some people maintain that biology is somehow different from the physical sciences, others that biology is nothing more than an extension of them. The notion that there are no laws in biology can be used to support either position. This creates some difficulties for the position taken by Mayr (1982, 1987a, b), that there are no laws in biology – at least once one tries to be consistent and carry the arguments to their logical conclusion.

The difficulties arise when we decide how to explain the order that we observe among organized beings. There are several possibilities, not necessarily mutually exclusive. Those who maintain that there are no laws of nature at all – either in the biological or the physical sciences – can be left out of the present discussion. But we should remark that many arguments against the existence of biological laws of nature would equally apply to physical ones as well. It would be difficult to defend the thesis that organisms are totally exempt from the laws of physics and chemistry. We behave like good Newtonians all the time, and an organism is just like any other object when one drops it. Since physical laws of nature apply to all physical objects, an organism that did not conform to them would refute the laws in question. We may also ignore as a side-issue the question of whether there are non-physical objects such as minds to which the laws of physics do not apply, but we should stress that physical and biological laws alike apply to processes, such as thinking.

A more attractive possibility for many people is to attribute biological order to laws of nature, but to maintain that the laws in question are only those of physics and chemistry. This represents a kind of reductionism, and one that is quite popular. It can be refuted by showing that biological laws of nature do in fact exist, and this will be attempted later in the present essay.

Natural order can also be explained as the result of contingency. Often such contingency is labelled 'chance', but this is misleading. By contingent we merely mean that things could have been otherwise, not that they just happened, or that there was nothing that might be labelled a cause. Such contingency is involved whenever we invoke historical factors as causes. In the physical sciences, to the extent that they are historical, such contingencies are commonplace. The facts that the earth has just one satellite and Mars two, are not a matter of law, though of course a narrative history of the solar system could explain those facts as a consequence of law-governed events and processes. In biology the invocation of historical causes is routine. We suffer from backaches because our ancestors made the stupid mistake of shifting to bipedal locomotion instead of running around on all fours as nature intended. The generalizations of systematic biology are largely descriptions of what did in fact happen although things might have been otherwise.

Descriptions of species and larger chunks of the genealogical nexus are all of that character, and of course such units are individuals.

Therefore there is no question that some of the order in biology can be explained in terms of historical contingency. But if indeed there does exist anything else, beyond such causes and perhaps the physical laws, then what might that be? Of course one possibility is biological laws. Maybe there is some other possibility, but the only one that has ever been invoked, at least to my knowledge, is miracles. By definition, miracles are happenings that transgress the laws of nature. So if one denies that there are laws of nature in biology, either one has to say that the only laws are those of physics and chemistry, in which case one is a reductionist, or one has to invoke something other than history and law, in which case one is a vitalist of the most extreme sort.

There are difficulties with accepting either horn of this dilemma, and not the least of these is that biological laws of nature do in fact exist. One can of course choose to be inconsistent in one's views, and I have accused Mayr of just such inconsistency. In his historical and philosophical writings he asserts that there are no laws in biology, while in his scientific writings he often behaves like any good scientist, and argues for and against precisely such laws. Part of the problem, however, is that he has idiosyncratic notions about laws of nature. For one thing, he rejects as laws of nature everything that makes only statistical predictions. Considering how much of physics and chemistry would cease to be lawful in the sense that is commonly accepted by the philosophers of those subjects and by scientists in general, we can dismiss this notion out of hand. At the same time he rejects as laws of nature those universal generalizations that assert something to be altogether impossible. According to such a criterion, we would have to reject as a law of nature the impossibility of constructing a perpetual motion machine. Finally, Mayr claims that there is no difference between a law and a description. Again, that would mean that there are no laws in chemistry of physics, not just in biology. There would be no difference between asserting that 'For every action there is an equal and opposite reaction', on the one hand, and 'The earth lies between Venus and Mars' or 'Ernst Mayr speaks German' on the other.

Before going on to discuss what are the laws of nature in biology and what role they play, it may help to clarify the notion of a law. The aforementioned misunderstandings and confusion are by no means unprecedented. The philosophical literature has tended to be rather vague about the difference between laws and other kinds of generalizations. For example the difference between a law and a principle is not entirely clear. In the present discussion I will attempt to clarify the distinction, but will emphasize the point that laws can often be derived from, or at least illuminated by, the underlying principles. When dealing with the perfect gas laws, for example, we know that their applicability depends upon the number of molecules in the population. We can infer that some gases do not obey the perfect gas laws because the molecules combine or dissociate when the temperature or pressure gets changed. Nineteenth-century authors often spoke of 'empirical laws', evidently meaning ones that were evident in the data of experience but not explained in terms of principles.

Biologists often enunciate laws without saying that they are in fact laws. They merely assert that certain kinds of events occur according to certain rules. Sometimes they cast certain generalizations in terms of principles when they could cast them in terms of laws. Good examples of principles in biology that have been explicitly formulated as such are what Dohrn (1875) called *Funktionswechsel*, and the 'founder principle' of Mayr (1954). Perhaps biologists are less concerned with laws of nature than physical scientists are because they habitually attempt to go directly from the principles to their data, without bothering to think in terms of laws. This is certainly true of much of my own work on the division and combination of labor (Ghiselin 1974, 1978). If the issue were merely a matter of quantitative emphasis upon laws, history, and other features of the various kinds of science, rather than an assertion that there are qualitative differences and that certain features are lacking, Mayr and I would perhaps have very little of importance to argue about. The qualitative differences between living and non-living objects clearly imply that there will be important quantitative differences between the biological and the physical sciences, and perhaps qualitative ones as well – but not the ones that Mayr has suggested.

Sometimes when speaking of a law, people mean a statement of a law, sometimes they mean that which the statement asserts about the natural world. In the former case this is apt to be in an epistemological context, in the latter an ontological one. To say that there are no laws of nature in biology would mean one thing if what we meant was that nobody has formulated any. It would be quite a different proposition to say that there are no laws of nature, in the sense that the kind of regularities that can be thus formulated do no exist. A lot of discussion in the philosophy of biology arises because the distinction between ontology and epistemology gets blurred.

Ontologically speaking, the laws of nature are regularities among what goes on in the material universe. Accordingly they are formulated as general statements – assertions about more than one thing. But 'general' is insufficient, because what we really mean is that they are about classes of individuals. There is a certain sense of 'general' that applies to a statement about individuals. For example a statement about me is more general than a statement about a part of me such as my right hand. Likewise a statement about Greece is more general than one about Athens. Laws of nature are also necessary truths, not contingent ones. By this we mean physical necessity, not logical or metaphysical necessity. It is physically impossible for things to be otherwise, even if that were readily imagined and not a category mistake. (It is physically impossible but not logically impossible for me to jump over the Empire State Building, metaphysically impossible for anybody to eat the weight of that edifice.) Another way to put it is to say that laws are true of everything to which they apply. Laws apply irrespective of place and time. Since individuals are spatio-temporally restricted, laws are not about them, but rather about classes of such individuals, classes being spatio-temporally unrestricted.

So if we are to find laws of nature in biology, we must identify some regularities in the behavior of biological objects that are necessarily true of everything to which they apply, irrespective of time and place. Physico-chemical laws of course apply perfectly well to living systems. Their applicability at the organismal level and below can readily be seen with respect to size relationships. Living organisms like all bodies of matter retain or lose heat depending uopn their size and shape. This because volume increases as the cube of the diameter of a sphere, surface as the square, and similar relationships hold for bodies of other shapes. In dealing with such relationships we can, but need not, express the relationships in terms of extensive quantities – intensive ones (large, larger, and largest) will do. If we understand the underlying principles, we can see how an animal's anatomy might differ were its surface to serve as a radiator or a heat collector, analogously with the products of mechanical engineering. Note that such relationships in and of themselves serve only to tell us what will happen under certain conditions. To gain insight that will allow us to predict and explain heat conservation behavior, for example, we need to know such things as the economic situation. Small animals lose heat faster than large ones do, but the heat saved by growing large may be offset by less food being available.

The peculiarly biological laws are indeed approximately co-extensive with those of economics. In the broadest sense, that of general economy, economics means the study of how the availability and utilization of resources affects the structure and activities of organized beings. It deals with such matters as competition, scarcity, and above all resource allocation. Economics thus conceived is not a social science, because it deals with solitary creatures as well as social ones, nor is it a branch of biology, because it deals with machines as well as organisms. That bioeconomic laws and principles are truly bioeconomic, and not just extensions of physicochemical ones is clear from the fact that such terms as 'resource', 'scarcity', and 'competition' apply only to living creatures, their parts and their products. The laws of nature that govern organic evolution are cast in terms of equilibria. The same is true of economics and much of physical science. Fisher (1930) explicitly compared evolutionary theory to the perfect gas laws. As an interesting aside we may point out that many theorists have emphasized the importance of the irreversability of certain changes that take place in both biological and economic systems.

Much of evolutionary theory deals with changes in populations. Populations are reproductive wholes composed of organisms. The largest such populations are called species, and these may be composed of subpopulations, the smallest of which are local groups called demes. Given such definitions, all populations, including species are sexual. This is important because sex integrates species. Were they not integrated they could not do anything, such as evolve. For any entity to do something it has to be an individual, but although this condition is necessary, it is not sufficient. Individuals that are not populations can evolve and undergo other kinds of change – asexual lineages are a good example. I mention this for the sake of clarity – the following discussion of biological laws deals only with those laws that govern the evolution of populations.

Some of the most fundamental among biological laws govern the process of species-formation or speciation – a process that is not to be confused with species transformation. When speciation occurs a species breaks up into two or more populations that cannot interbreed and therefore must evolve as separate individuals rather than a single reproductive whole. Generally this involves some change in reproductive behavior or physiology, changes that ordinarily are genetic but could be phenotypic. One of the most important laws of nature in biology has been advocated most compellingly for many years by Mayr (1942, 1963 and elsewhere), and therefore deserves to be called Mayr's Law. It asserts that under ordinary conditions speciation will not occur unless the ancestral species gets broken up into populations that are separated by some extrinsic (usually geographical) barrier. Mayr's Law follows from certain basic principles. The reproductive proclivities that unite and integrate populations are so strong that they will be maintained spontaneously so long as the component organisms are capable of exchanging genetic material either directly or indirectly. There are certain circumstances, both well documented (polyploidy) and speculative (disruptive selection) under which speciation can occur in sympatry. These possibilities likewise derive from considerations of fundamental principles, and define what is meant by 'ordinary conditions'. (Recall what was said about the perfect gas laws.)

Although Mayr's Law has long been controversial (Darwin and Wallace argued essentially the same issues at great length), the issue has rarely if ever been either the validity of the law and rarely the legitimacy of the underlying principles. Rather, the controversy has been about what constitutes the ordinary conditions, and how often these are in fact realized in nature. There is an interesting analogy here concerning the formation of compounds by the noble gases – something that up until some twenty years ago was supposed not to happen. That noble gas compounds form with highly reactive elements such as fluorine makes a lot of sense in the light of fundamental principles reflected in the periodic table of the elements.

Another law about speciation was propounded as such by A.R. Wallace (1855), who said that every species has come into existence at a place and time coextensive with a previously existing species. But is this really a law? It states that species are spatio-temporally restricted as to origin, and suggests that the relationships among them are genetic. In other words, they are genealogical individuals, which cannot have multiple origins. This seems like just a restatement of evolutionary principles, principles which, to be sure, are fundamental to inference in paleontology, biogeography, and much else besides.

The Knight-Darwin Law was cast in picturesque language as asserting that 'Nature abhores perpetual self-fertilization' (Darwin 1858). It is true that lineages degenerate and tend to become extinct when organisms become obligatory selfers. But the law itself on the one hand has been refuted to the extent that such selfers can and do evolve and flourish. On the other hand it was vaguely stated, and the underlying principles were not known until the present century.

Much of modern biology is cast in terms of genetics, and this is not altogether fortunate. One unhappy consequence has been the habit of defining 'evolution' as if it were changes in gene frequency rather than descent with modificiation. So phenotypic selection and the role of behavior get overlooked, and extinction and speciation de-emphasized. Speciation can occur without genetic change if for example there has been purely phenotypic behavioral modification that keeps the populations separated.

Elementary treatments of population genetics often begin with a discussion of an equilibrium state, one in which an infinitely large population exists without any 'forces' tending to produce change. The use of a large population is necessary to simplify matters so as to rule out the consequence of fortuitous changes (drift etc.). They then go on to discuss how such forces might affect gene frequencies. The laws of nature that are involved are rarely if ever enunciated as such, and it will be an instructive exercise to draw attention to a few of these. The main forces that affect populational genetic equilibria – whether maintaining them in static conditions or changing them – are mutation pressure and selection pressure. Sampling effects also come into play, and these will be mentioned later.

Two of the most obvious laws of nature are as follows:

1. The rate of change is directly proportional to the mutation pressure.
2. The rate of change is directly proportional to the selection pressure.

When allele A mutates to allele B, how fast the proportion of A to B goes from 0:1 to 1:0 depends on how often that change occurs. In an equilibrium state, of course, it is assumed that the mutations go both ways, and the two cancel out. In the case of selection, we need only modify the considerations so that the pressure results from differential reproductive success, rather than differential origins. These laws of nature can be derived from basic principles by anybody who knows a little molecular biology. That they are legitimate laws of nature is clear from the fact that their predictions are universal and can be tested by experience.

If the second of these above laws be true, then we should be able to change populations from equilibrium in a particular direction and at a slower or faster rate by varying the selection pressure. If we found that change is slower when selection pressure is increased, we would have the same doubts about the theory as we would about the perfect gas laws if a balloon shrank when we heated it up. The first law helps explain why the rate of change drops off in small populations subjected to artificial selection.

Mutations occur because of imperfect replication of the genetical material. The Second Law of Thermodynamics tells us that replication has to be imperfect, and that change will tend in the direction of disorder. From these considerations we may derive yet another law: In the absence of selection pressure, evolution must occur. This law is very interesting, because it points out that one of the most important consequences of selection is that it prevents evolution. Stabilizing selection maintains populations in a condition of stasis. People who do not understand evolutionary biology are sometimes under the impression that civilized human beings are no longer subject to natural selection. Where they correct, it would mean that we are headed for extinction, which might be a good thing for the innocuous species.

The foregoing assumes that mutation and selection are the only causes of change in gene frequency. It is more than just logically possible that events ruled out by the above laws could occur as a result of other causes. We do not know for sure how many such causes exist, but an important example is 'sampling error'. Random fluctuations in gene frequencies are known to occur independently of selection and mutation pressures. Such fluctuations are most pronounced in very small populations. One important biological law of nature asserts that the frequency with which a particular mutation occurs in a given population is directly proportional to the size of that population. This in itself suffices to keep the equilibrium away from what it would be in a population of infinite size. What genes get passed on from generation to generation is partly a matter of chance, resulting in fluctuations in frequency that deviate from the equilibrium value or even in the loss of an allele from the population. The extreme condition would be in a population reduced to a single self-fertilizing, diploid hermaphrodite. In this case, the number of alleles at a given locus can be no more than two, and some of the consequences of a genetic bottleneck of this sort can be treated deterministically. Now, one might want to argue that under some conditions a population might be held in an equilibrium condition, in spite of mutation pressure and without selection pressure, simply through sampling error. But this would be like saying that if enough atoms just happened to move in a certain direction my entire body might move spontaneously upward, in apparent defiance of the laws of universal gravitation.

Many of the biological laws of nature treat population density as a fundamental parameter. The underlying principles from which these can be derived have largely to do with the fact that organisms are more likely to encounter each other, and hence to interact in a particular way, when there are more of them in a given area. Often these relationships have an economic character, insofar as it may be necessary to expend energy in overcoming the consequences of being a rare organism. Simple examples are to be found in the study of what are metaphorically referred to as 'reproductive strategies'. (I give just one example here – for more see Ghiselin 1974 and Charnov 1982.) Consider a species in which, as is ordinarily the case, the sex ratio is one male to one female. A low population density means a low probability of encountering a conspecific, perhaps not being able to reproduce – a bad thing from a selectionist point of view. But even if one does encounter a conspecific, the probability is only one in two that he or she will be of the other sex. Simultaneous hermaphrodites, which are male and female at the same time, have no such problem. Hence we get the low density model for hermaphroditism, which can be expressed as a law of nature: as effective population size goes down, the amount of simultaneous hermaphroditism goes up.

Such laws of nature generate definite empirical predictions and can readily be tested. One might try to argue that the predictions do not always fit the data perfectly. They only generate what are called 'trends', and therefore we are not dealing with legitimate laws of nature. But this claim would like denying lawfulness to Newtonian physics because the wind sometimes raises a few particles of dust. To find the laws of nature applying in their pristine form, we have to control all of the pertinent conditions. This is much harder to do with respect to field data than in the laboratory.

Nonetheless, such trends can be treated in an unscientific manner, by insulating them from criticism and by failing to convert them into the form of laws. A well-known example is 'Bergmann's Rule', which states that in colder climates animals get larger (Bergmann 1847). According to the Synthetic Theory, this is because of the need for heat conservation, and this was taken as evidence for adaptation by natural selection. As I have argued at some length (Ghiselin 1974), the real reason for the trends in question is not heat conservation, but the availability of food. The appropriate law of nature is not: as the climate gets colder, the animals get larger, but rather: as food becomes less of a limiting resource, the animals get larger. The facts that contradict Bergmann's rule had long been explained away on the basis of the rule being nothing more than a rule, most notably by Mayr (1956). So perhaps in a certain sense Mayr is right. For some biologists, some of the time, there may not be any laws of nature. So much the worse for them.

The Synthetic Theory also took group selection more or less for granted, and invoked species-level adaptations where we now know that none exist. That aspect of the Synthetic Theory began to fall apart with the rediscovery of the individualistic character of selection, a principle that Darwin understood very well, but one that was widely neglected until around twenty years ago. In very general terms, we recognize that adaptation occurs only under a range of conditions that is far less general than one might imagine. For an entity to be selected, it has to be integrated so that it functions as genetical unit (see Hull 1980). It follows that there can be adapted organisms, and adapted families, but not adapted species. One could translate this point into an explicitly-formulated law, but I won't bother to do so. Darwin derived a fair number of laws from the same basic principles. One of these was that natural selection cannot produce an adaptation in one species solely for the good of another species. In other words, there is no altruism toward other species.

This was born out by showing that parasites lack such altruistic adaptations toward their hosts. Mutualism, of course, is another matter, it being no exception because organisms of both species benefit. Another matter too is artificial selection – we breed domesticated animals and plants so that they will have features beneficial to us.

Truths about individuals, including historical truths, in principle cannot possibly contradict the laws of nature, for these are true of everything whatsoever to which they apply. Therefore we can use the data of natural history, including those having to do with the geological past, to infer the legitimacy of hypothesized laws of nature. Likewise, we can use the laws of nature to test hypothetical narrative explanations of what has happened to individuals in the past. The former procedure is hardly controversial, except with respect to its details. The whole of nomothetic science appeals to particular experiments and observations as a means of deciding what is and is not true. The second procedure is routinely applied in the whole range of historical sciences. But in some areas of biology it is somewhat more controversial, if by controversial one can mean that it has been the object of a conspiracy of silence.

Basically the procedure is nothing more than this: conjecture a possible sequence of events that might explain what happened. Then work out the consequences, and if any of these include something contrary to a law of nature, reject it in favor of some alternative hypothesis. If a scenario implies that an individual had to be in two or more different places at the same time, then something has to be wrong with that scenario. In geology, there exists a methodological rule about superposition. Strata of sedimentary rocks have to be laid down one on top of another, so given a series of these one assumes that the upper layers are younger than the lower ones. But what does one do about the possibility that certain strata have been overturned? One possibility is being able to find a piece of landscape where one can follow the strata from an overturned to an upright position through a series of intermediate positions. But one can also use laws of nature, for the processes of deposition often leave signs that cannot be interpreted otherwise than indicating which way is up. Take a beaker of water, add mud and stir. The largest particles will settle out first, and a layer with characteristic size distribution will be formed. Such layers formed in succession are termed 'varves', and the orientation to gravity is obvious upon inspection.

It was Charles Lyell who first explicitly formulated the procedure for applying the laws of nature to reconstruction of geological history. His greatest follower, Charles Darwin, was a great master of Lyellian geology, and among his greatest accomplishments was applying the same kind of reasoning to the history of life (Ghiselin 1969). Unfortunately, we have a long tradition in biology of ignoring that accomplishment, and pretending that nothing of the sort is even possible. At least this is the position of many schools of systematic biology, and one that has been taken by a substantial number of methodologists. When non-Darwinian authors such as the cladists discuss 'phylogenetic reconstruction' they mean arranging groups of organisms in purely formal schemes, not trying to reconstruct actual narratives of what went on in the history of life.

The Darwinian techniques pose certain difficulties. For one thing, one has to know that the laws of nature really are laws of nature, and that they are applicable to the materials in question. Consider how a paleontologist reconstructs the way of life of an organism from a fossil consisting only of its skeleton. The paleontologist can put flesh on the bones because the skeleton is a mechanical apparatus that works through interaction with muscles. From basic principles the size, strength, agility and other features of the organism can be reconstructed. And a paleontologist can do far more than just speculate about its feeding habits and other aspects of its ecology as well. Certain kinds of teeth would be effectual at tearing flesh, others at crushing and grinding hard materials such as grass. We can substantiate the legitimacy of such inference by trying it on extant creatures. But not everything about living organisms can be extrapolated to their ancestors – only what results from the laws of nature. The diagnostic features of taxonomic groups – ones that are universally present, hence useful in identification – are only contingent truths (Ghiselin 1972). For example, all the extant cephalopods are carnivorous, but that does not imply that all the fossil ones were. The ancestral stock of mollusks probably was not carnivorous, and the same may have been true of many early cephalopods. It is also possible that carnivorous ones evolved a different mode of feeding. So if we are going to infer the feeding habits of fossil cephalopods, we had better not rely upon taxonomic generalizations, but use real laws of nature, such as the physical ones.

The development of an embryo is a very orderly process, and one that might be largely governed by laws of nature. Pre-Darwinian biologists, including some who are actively publishing today, have attempted to treat development as if it were purely a nomothetic affair. Such persons are not very happy about the individuality thesis, because it creates serious difficulties for their research programs. It is curious that the 'structuralist' movement in contemporary morphology condemns Darwinism for ignoring developmental constraint as limiting the kind of evolutionary change that can occur. The possibility that we could analyse evolutionary change from precisely that point of view was the subject of a very important book by Darwin (1868), and many subsequent Darwinians and even some neo-Darwinians have advocated quite similar views (Rensch, Huxley, Severtzoff, Schmalhausen).

There probably are many laws of nature that govern the development of an embryo, but it is easy to confuse these with contingencies no different from any diagnostic character. One putative law of nature has been the biogenetic law, which states (with certain qualifications and hedges) that ontogeny recapitulates phylogeny. There are few naive recapitulationists these days, but the tendency for parallelism between ontogeny and phylogeny is routinely applied as a guide to the sequence of ancestors and descendants. The features of development that exist as a matter of contingent fact are retained to some extent, and these are taken as hypothetical explanations for the data of comparative anatomy. Such hypotheses can be excluded by showing that they contradict the laws of nature. In the development of an insect, for example, one might reasonably infer that a unicellular stage represents pre-metazoan conditions, and that an ability to walk preceded the ability to fly, as seen in the egg, larvae, and adult of a butterfly. However, there is also a pupa, which undergoes metamorphosis, but does not feed. We reject out of hand the interpretation of a long period of geological time in which the species got along without an energy source. This is obviously because it would run contrary to the most fundamental laws of physics. It is one of the most fundamental principles in historical biology that all the stages in a reconstructed lineage must have been viable. Other laws that can be applied are more purely biological. Vestigial organs, for instance, have to be treated as remnants of prior conditions, not as something in the process of originating. Whales are traditionally divided into two groups, the Odontoceti and the Mystacoceti, respectively with and without teeth as adults. Embryonic specimens of Mystacoceti have long been known to possess teeth, which never even erupt from the jaw. Natural selection could easily lead to the gradual reduction and virtual or complete loss of the teeth, but in this instance there is no way in which it could possibly effect a change in the opposite direction. Some processes are, and others are not, reversible, as one can readily appreciate from watching a film being run backward. Nobody ever unchews food and spits it out entire. Dollo's Law asserts that evolution in general is irreversible. As with the Biogenetic Law, we are really dealing with a whole series of laws and principles.

One of the most fundamental principles in biology and economics is the division of labor, which actually breaks down into a group of principles, such as 'The division of labor is limited by the extent of the market' (Adam Smith 1776) and 'Where activities interfere with each other, labor is divided; where they complement each other, it is combined' (Ghiselin 1974). The division of labor has often functioned as a pseudo-law, in both the biological and the social sciences. In biology this took the form of an assumption that hermaphroditism is primitive and evolution always proceeds in the direction of separate sexes. Given what we know about hermaphroditism, we can see that there is no such law. Which changes occur depend upon the kind of organism and its conditions of existence. We can generate predictions from the basic principles, but these predictions are not about the inevitable course of biological history. In the social sciences, progress was largely equated with the increasing division of labor. Herbert Spencer provided a long enumeration of facts that purported to illustrate this supposed truth, both in biological and social evolution. Such thinking in terms of pseudo-laws had a most unfortunate effect upon the social sciences. 'Evolution' came to be defined as such wrong-headed thinking, and ever since the vast majority of social scientists have continued to behave as if nothing of interest had been published on November 24, 1859.

Hull and I have also both been interested in applying the individuality thesis to objects other than biological species. In particular it has seemed attractive to attempt to discover the laws of nature that govern scientific investigation. Hull has emphasized the point that the individuals in the history of science – such as theories – can change indefinitely and remain the same particular thing. Equally, there are no laws of nature for particular theories or particular discoveries, any more than there are laws of nature for individual scientists. The laws of nature that govern science have to be laws of nature about classes of theories, and classes of discoveries. Hull's search for the laws in question began with sociobiology and the sociology of knowledge. My own work has emphasized such principles of economics as the division of labor. The two perspectives should complement each other, and lend each other mutual support. In any event the basic implications are the same. The social sciences are no different from the physical or the biological ones, in that all can have perfectly legitimate laws of nature. By implication we can have one philosophy for all of the natural sciences, including those that deal with societies.

Indeed, we can have one philosophy for the entirety of knowledge. From commentaries written by philosophers, I get the impression that virtually all of them have missed this, the most important point. The philosophy of biology should not be an uncritical application of what teachers tell students about philosophy to what teachers tell students about biology. Rather, it should be an effort to come to grips with, and solve, problems in both branches of knowledge. When philosophy and biology contradict each other, the thing to do is go back to the fundamental premises, find out what was wrong, and reconstruct. In the case of individuality, we have found that much confusion has existed, but that once this is cleared up it becomes much easier to separate what was sound from what was defective in our predecessors' views. It turns out that the house of philosophy is not a house of cards. The edifice is solid in many of its parts. It does not have to be razed to the ground, but only remodelled, refurnished, and set upon a new foundation.

References

Charnov E.L. (1982). *The theory of sex-allocation.* Princeton: Princeton University Press.
Darwin C. (1858). On the agency of bees in the fertilization of papilionaceous flowers, and on the crossing of kidney beans. *Ann. Nat. Hist.* (3)2: 459–465.
Darwin C. (1859). *On the origin of species by means of natural selection, or the preservation of favoured races in the struggle for life.* London: John Murray.
Darwin (1868). *The variation of animals and plants under domestication.* London: John Murray.
Dohrn A. (1875). *Der Ursprung der Wirbeltiere und das Prinzip des Funktionswechsels. Genealogische Skizzen.* Leipzig: Engelmann.

66

Fisher R.A. (1930). *The genetical theory of natural selection*. London: Oxford University Press.

Georgescu-Roegen N. (1971). *The entropy law and the economic process*. Cambridge: Harvard University Press.

Ghiselin M.T. (1966). On psychologism in the logic of taxonomic controversies. *Systematic Zoology* 15: 207–215.

Ghiselin M.T. (1969). *The triumph of the Darwinian method*. Berkeley: University of California Press.

Ghiselin M.T. (1972). Models in phylogeny. In Schopf T.J.M. (ed.) *Models in paleobiology*. San Francisco: Freeman and Cooper, 130–145.

Ghiselin M.T. (1974). *The economy of nature and the evolution of sex*. Berkeley: University of California Press.

Ghiselin M.T. (1978). The economy of the body. *American Economic Review* 68: 233–237.

Ghiselin M.T. (1981). Categories, life, and thinking. *Behavioral and Brain Sciences*. 4: 269–313.

Ghiselin M.T. (1987). Species concepts, individuality, and objectivity. *Biology and Philosophy* 2: 127–143.

Hull D.L. (1974). *Philosophy of biological science*. Englewood Cliffs: Prentice-Hall.

Hull D.L. (1980). Individuality and selection. *Annual Review of Ecology and Systematics* 11: 311–332.

Mayr E. (1942). *Systematics and the origin of species*. New York: Columbia University Press.

Mayr E. (1954). Change of genetic environment and evolution. In Huxley J., Hardy A.C., Ford E.B. (eds) *Evolution as a process*. London: Allen and Unwin.

Mayr E. (1956). Geographical character gradients and climatic adaptation. *Evolution* 10: 105–108.

Mayr E. (1963). *Animal species and evolution*. Cambridge: Harvard University Press.

Mayr E. (1982). *The growth of biological thought*. Cambridge: Harvard University Press.

Mayr E. (1987a). The ontological status of species: scientific progress and philosophical terminology. *Biology and Philosophy* 2: 145–166.

Mayr E. (1987b). Answers to these comments. *Biology and philosophy* 2: 212–220.

Rosenberg A. (1985). *The structure of biological science*. Cambridge: Cambridge University Press.

Smith A. (1766). *An inquiry into the nature and causes of the wealth of nations* (2 volumes). London: Strathan and Cadell.

Wallace A.R. (1855). On the law which has regulated the introduction of new species. *Annals and Magazine of Natural History* 16: 184–196.

Interaction and Evolution

MARJORIE GRENE

When I first returned to the United States in 1965, the most articulate philosophers working on biological topics were William Wimsatt and David Hull. Since then, thanks partly to their efforts, philosophy of biology has become a vast and growing field. Ignoring here Hull's undoubted contributions to the history of Darwinism, notably his *Darwin and His Critics* (Hull 1973), I want to concentrate on what I consider his most significant contribution to evolutionary theory: the distinction between replicators and interactors (Hull 1980, 1981). About philosophy of science, and perhaps philosophy as such, he and I could hardly disagree more, and some of my problems will be reflected even in my attempt to discuss this, as I believe, his most original and important argument.

The major source for Hull's distinction is of course his 1980 *Annual Reviews* paper, 'Individuality and Selection' (Hull 1980). But instead of going through it in detail, I propose to take as my text part of the 'Conclusion' of a related paper, 'The Units of Evolution' (Hull 1981), where both the distinction and some of its background are clearly and briefly indicated. Hull writes:

> The purpose of this paper has been to set out the general characteristics which replicators and interactors exhibit and the roles which they play in the evolutionary process. A process is a selection process because of the interplay between replication and interaction. The structure of replicators is differentially perpetuated because of the relative success of the interactors of which the replicators are part. In order to perform the functions they do, both replicators and interactors must be discrete individuals which come into existence and cease to exist. In this process they produce lineages which change indefinitely through time. Because lineages are integrated through descent, they are spatiotemporally localized and not classes of the sort that can function in laws of nature (pp. 40–41).

Rather unsystematically, let me stroll through this passage, mentioning, and discussing a little, some of the thoughts it suggests. To begin with, obviously, we are dealing here not with any old evolutionary process, but with evolution by natural selection: fine. Whatever may happen to newer style theories, or revivals of

M. Ruse (editor), What the Philosophy of Biology is. pp. 67–73.
© 1989 *Kluwer Academic Publishers, Dordrecht*

old ones, we have here an attempt to clarify something that was unclear in the classic Darwinian theory as well as in the modern synthesis. Perhaps, indeed, it is the population-genetic and then biochemical and then molecular foundation of the twentieth century theory that has made it necessary for Hull to introduce a distinction formerly neglected or in effect denied. So keen have geneticists of all stripes been to follow what Wimsatt has aptly entitled the bookkeepers of the evolutionary process (Wimsatt 1980) that they have forgotten the central role in selection of organisms and (and in) their environments. Hodge has persuasively argued that Darwin was always (not only in the late days of pangenesis) a generation theorist (Hodge 1985); but clearly, what produces descent with modification – or rather, what produces the modification in descent – is the struggle for existence: that is, the differential success of slightly differing organisms in suiting themselves to the conditions of life, in being or becoming, as we would say, differentially adapted to their environments. Environments change, too, of course, and in part (though obviously not, for example, in the case of such rare events as mass extinctions) the activities of organisms help to constitute one another's, and the environment's, environment (Lewontin 1983). Darwinism is, as I have called it elsewhere (Grene, in press) an exoetiological theory: evolution occurs if there are slight variations, generated randomly with respect to the needs of the organism, and environmental circumstances that render the heritable results of such variation relatively beneficial to its carrier. There is no internal drive toward evolution; everything depends on the available variation in its adaptive relation to the existing environment. Yet the ecological aspect of Darwinian evolution was often swept aside in geneticists' enthusiasm for their calculations and their chemistry. Many were the triumphant cries in 1959 to the effect that Darwin, poor old Victorian, had thought nature red in tooth and claw (Tennyson – before 1859, but never mind), while 'we now know' that natural selection, far from entailing a struggle for existence, just is differential gene frequency. That is the kind of formulation that leads, with some justification, to the charge of tautology: and perhaps that wouldn't matter, since tautologies can provide useful guidance in some circumstances. What matters is that half the process, and the dynamic half, is thereby ignored. Dawkins distinguishes, indeed, between replicators and vehicles (Dawkins 1982), but 'vehicle' suggests, as it is meant to do, that organisms and even the broader phenomena included as 'extended phenotypes', however far they reach, are only bearers of the Almighty Gene, or Small Hunk of a Genome. There is a lot more to evolution than that, and the more is succinctly baptized by Hull in his concept 'interactor'. Sometimes he also speaks of 'interactions', and that distinction, too, is important, since it is the interaction of interactors with replicators that produces the lineages traced in evolutionary trees.

In introducing the replicator/interactor distinction in 1980, Hull had started from two assumptions that have led, he believes, to stalemates in a number of evolutionary questions. The first (I shall allude to the second shortly) is 'the view that genes and organisms are 'individuals' while populations and species are 'classes'...'

(Hull 1980, p. 311). Never mind for the moment the notorious species=individuals debate (Alas, I shall return to that briefly too in conclusion). But it follows from Hull's argument about replicators and interactors that our picking out of organisms as somehow uniquely individuals has blinded us, not only to the nature of species as historical entities, but to an important sense in which organisms are not individuals either. I do not mean here, in a Dawkinsian spirit, that only genes are *really* individuals, but rather that organisms, which both laymen and field naturalists are inclined to single out as the interesting and persistent unique units of study, themselves need to be placed in a divided evolutionary context. We may relax the criterion for replicator status a little and allow organisms to 'replicate' – or to make more of themselves, as Eldredge puts it (Eldredge 1986); but that is not all they do – at least most of them spend most of their time in other activities: they eat and are eaten, move about and are pushed, communicate, play and are played with, and so on and on. So in the context of evolution, each organism is not one, but two: a more-maker, or if you must, a carrier of replicators, and an interactor. I sometimes wonder why Descartes had to separate mind and body so sharply precisely in order to show how they can work together in mechanics, morality and medicine. It seems that in evolutionary biology, too, it is necessary to separate sharply two processes: the replicative (or genealogical) and the interactive (or ecological) in order to show clearly how lineages are generated by the cooperation of the two (Eldredge and Salthe 1984; Eldredge 1986). In the clarification of this situation Hull's *Annual Review* paper has been immensely influential, if not foundational.

The second assumption that has misled us, Hull alleges, is 'our traditional way of organizing phenomena into a hierarchy of genes, cells, organisms, kinship groups, populations, species and ecosystems or communities' (Hull 1980, p. 311). Here, so far as I can tell, Hull himself has not been much interested in sorting out the consequences for this kind of hierarchical thinking that follow from his replicator/interactor distinction. But I find them fascinating. On the basis of Hull's analysis, it is plain that the conventional hierarchy he outlines is thoroughly confused. There is a genealogical hierarchy that runs from genes through organisms through demes through species to monophyletic taxa in general. There is an ecological hierarchy that runs from organisms through populations to communities to ecosystems, possibly to the whole biota. But they are *not* the *same* hierarchy. It follows, of course, that populations are ecological entities and species are genealogical entities: so whatever, in the present debate, species turn out to be, they are not (conceptually) populations. Admittedly, an individual tree that is the sole surviving 'part' (in Hull's terms) of the genealogical chunk that is the species may thus, in these circumstances, *be* the species, and so this part of this community in this case is also extensionally identical with a genealogical unit. But then we must separate the genealogical function of this entity from its ecological function, as we would with any individual organism, which is both a more-maker and an interactor. It has been difficult, if not impossible, to extend the biological species concept, beloved of the synthesis, through evolutionary time (Eldredge 1985; Cracraft 1987; Stebbins 1987 (personal communication)). The reason is that a populational concept will not fit smoothly into a genealogical context. A workable species concept, whatever the differences in its formulation among different workers (Cracraft 1987; Mishler and Brandon 1987), must be a phylogenetic species concept of some sort or other, since a species is a unit of phylogeny, neither more nor less. Granted, the above is a very crude stab at an argument – of course the interactions the BSC was concerned with were reproductive and hence with a special bearing on genealogy – and so on; but the basic point seems to me a valid one. There is, to say the least, something paradoxical about the fact that the species concept developed by some of the greatest evolutionists of this century fits everything except evolution.

So far I have been reflecting on the first two sentences of my chosen text. The next sentence, and after that the last three, evoke some further, perhaps even more far-fetched comments.

It is, Hull stresses both here and in the earlier paper, not replicators as such that are perpetuated: particular genes are no more immortal than are particular organisms. But it is the *structure* of genetic units that is 'differentially perpetuated'. In Hull (1980) it is suggested that something a little less than identity of structure would do. But identity does not come in grades. A is identical with B or it is not. Shunning the dread notion of 'similarity', which he appears to associate with the horrors of essentialism or typology, Hull has, I believe – as have many population thinkers – overlooked the necessity of recognizing, in many individuals or in many circumstances, identity of properties, that is, structures. This is not Leibnizian identity, where A as a given entity is B, the same entity. It is an identity of form, where one can recognize, say, the Hapsburg chin or the song of the nightingale. There is nothing 'essentialistic' in the scholastic sense in the acknowledgement of such identities. Indeed, without them there would be no discourse, no science, no nature, no evolution. Ruse has recently acknowledged the existence in our tradition of a morphological strand that has often conflicted with the more flux-oriented, particulate emphasis of the chief Darwinian style of explanation (Ruse 1988). It is that strand that little words like 'structure' come to remind us of. To ignore them, and to polemicize against them on principle (though much that has been written in the name of Form if certainly overblown if not nonsensical) is to succumb to the kind of pseudo-substitution characteristic of Simpson's concept of 'adaptive type': types are evil, but if I, GGS, call something a 'type' – and of course so long as it's adaptive – that's another story. Let's notice the cake before we eat it and not pretend we have been nourished without it (for more on this, see Grene, in press).

The last three sentences of our text, apart from continuing to solidify the distinction between replicators, interactors and lineages, introduce another ominous word: individual. 'Individuality and Selection' was clearly conceived by its author as an elaboration of and consequence of the famous Ghiselin-Hull thesis that species are 'individuals rather than classes'. Hull has sometimes stated precisely what he means by this distinction, for example:

> The only category distinction I discuss is between individuals and classes. By 'individual' I mean spatiotemporally localized cohesive and continuous entities (historical entities). By 'classes' I intend spatiotemporal unrestricted classes, the sorts of things which can function in traditionally-defined laws of nature. (Hull 1978, p. 336)

Now of course Hull has a perfect right to make this distinction if he chooses. But unfortunately biologists have taken him to be presenting 'the philosopher's' concept of individual and of class – although it is clear from the above quotation that by 'class' here he is picking out a subclass of classes, not 'class' as such. There exists, as it happens, a highly technical discipline called set theory which contemporary logicians refer to, at least to start with, when they speak of classes, and in that discipline 'class' certainly does not signify the particular kind of entity Hull chooses to call 'class' in this passage. Nor does 'individual' in logic, let alone in 'philosophy' in general, signify historical entity, and only historical entity. The logical muddle consequent on this misunderstanding has been horrendous and I certainly do not wish to add to it. Readers who will patiently attend to Kitcher's recent argument and to the literature he cites should be able to find some light in this darkness (Kitcher 1987). Species are historical entities: fine; that's clear. But not all individuals are historical entities and not all historical entities are individuals. Further, species are not spatio-temporally unrestricted, intensionally defined classes; fine – but neither are most classes (or sets); according to some authorities, none are. But again, far be it from me to try to clarify where so many others have tried already (chiefly in vain, it seems).

What interests me, however, in the closing lines of the text I started from as well as in the passage just quoted, is the connection Hull makes between the kinds of classes he is talking about and 'traditionally defined laws of nature'. Hull grew up (academically speaking) in the day of the so-called received view in philosophy of science, where the role of laws loomed large, and I suspect that for him what was truly revolutionary about species as they exist in nature, and especially in a Darwinian nature, was their resistance to lawhood. So what? One of the seminal insights in Mayr's *Growth of biological thought* (Mayr 1982) is the recognition that in general biologists are not interested in laws. And if nineteenth century biologists were different in this, the laws they tried to reach, and thought they reached, inductively, were nice good generalizations, almost like Aristotle's 'what happens always or for the most part'. They were not those cramped and cramping contrary-to-fact conditionals that, up to recently, philosophers of science thought they needed to worry about. But philosophers of science had all those worries because, starting as they did by separating the context of discovery from the context of

72

justification, they were attempting a logical reconstruction of science almost wholly divorced from the practice of science – and therefore from science. Hull of course (as he protests in the recent species-individuals debate (Hull 1987)) has always worked closely with biologists and biology; but at the same time he has come to his subject-matter from within the carapace of the myth of the H-D method or the D-N method, either version stressing both the unimportance of discovery and the importance of law-like statements as the starting point of justificatory deductions. This orthodoxy also carried with it the ideal of a unified science, and the reduction of other theories to more 'basic' ones that that ideal entailed. Hull was perplexed by the failure of such predicted reductions (Hull 1974), and he clearly came to realize also that his favorite subject-matter, species, eluded the grasp of the law-bound orthodoxy. My concluding comment is perhaps impertinent and I do not expect it to be effective; but I must add, finally, that the appropriate response in this situation is to try, as a few of us have been doing for some decades, to practice the philosophy of science within an historical context, though still philosophically (Grene 1987 a, b). So far as I have seen evidence of it, I fear that Hull, out of his disillusionment with his former philosophical ideal, has abandoned philosophical analysis wholly for 'sociology of science', where the study of the kinds of (successful) cognitive claims that make the natural sciences our paradigmatic instances of knowledge are thrust aside in favor of the who-beat-up-whom-and-why kind of story characteristic of this new 'field'. Why abandon an inadequate philosophy for none at all? If I am wrong in this hunch, my felicitations and apologies; if I am right, regrets.

References

Cracraft J. (1987). Species concepts and the ontology of evolution. *Biol. and Phil.* 3: 329–346.
Dawkins, Richard (1982). *The Extended Phenotype.* San Francisco: Freeman.
Eldredge N. (1985). *Unfinished synthesis.* New York: Oxford University Press.
Eldredge N. (1986). Information, economics and evolution. *Ann. Rev. Ecol. Syst.* 17: 351–369.
Eldredge N., Salthe S.N. (1984). Hierarchy and evolution. In Dawkins R., Ridley M. (eds) *Oxford surveys in evolutionary biology*; 184–208. Oxford: Oxford University Press.
Grene M. (1987a). Perception, interpretation and the sciences: toward a new philosophy of science. In McKnight C., Tschedroff M. (eds) *Philosophy in its variety*; 107–129. Belfast: Queens University.
Grene M. (1987b). Historical realism and contextual objectivity: a developing perspective in the philosophy of science. In Nersessian N.J. (ed.) *The process of science*; 69–81. Dordrecht: Martinus Nijhoff.
Grene M. (in press). Evolution and human nature. In *Proceedings of Penn. Conference on human nature today*.
Hodge M.J.S. (1985). Darwin as a lifelong generation theorist. In Kohn D. (ed.) *The Darwinian heritage*; 207–243. Princeton: Princeton University Press.
Hull D.L. (ed.) (1973). *Darwin and his critics.* Chicago: University of Chicago Press.
Hull D.L. (1974). *Philosophy of biological science.* Englewood Cliffs, NJ: Prentice-Hall.

Hull D.L. (1978). A matter of individuality. *Phil. Sci.* 45: 335–360.

Hull, D.L. (1980). Individuality and selection. *Ann. Rev. Ecol. Syst.* 11: 311–332.

Hull D.L. (1981). Units of evolution: a metaphysical essay. In Jensen U.S., Harré R. (eds) *The philosophy of evolution*; 23–44. Brighton: Harvester.

Hull D.L. (1987). Genealogical actors in ecological roles. *Biol. and Phil.* 2: 168–183.

Kitcher P. (1987). Ghostly whispers: Mayr, Ghiselin and the 'Philosophers' on the ontological status of species. *Biol. and Phil.* 2: 184–192.

Mayr E. (1982). *The growth of biological thought.* Cambridge, MA: Harvard University Press.

Mishler B.D., Brandon R.N. (1987). Individuality, pluralism, and the phylogenetic species concept. *Biol. and Phil.* 2: 397–414.

Ruse M. (1988). Booknotes. *Biol. and Phil.* 3: 117–121.

Wimsatt W.C. (1980). Reductionistic research strategies and their biases in the units of selection controversy. In Nickles T. (ed.) *Scientific discovery: case studies*; 213–259. Dordrecht: Reidel.

Picturing Weismannism:
A Case Study of Conceptual Evolution

JAMES R. GRIESEMER[1] and WILLIAM C. WIMSATT[2]

[1]*Department of Philosophy, University of California at Davis, Davis (CA), U.S.A.;*
[2]*Department of Philosophy, University of Chicago, Chicago (IL), U.S.A.*

Part I: Analyzing Conceptual Change

I. Introduction

The problem of analyzing conceptual change in science has been substantially refined since its introduction by logical empiricists, particularly Karl Popper, and its radical critique by Kuhn. Recently, philosophers of biology have added to the growing wealth of conceptions of change through their studies of evolutionary biology. Those who study evolutionary theory recognize it as a very broad framework in which to characterize dynamic processes. Moreover, there have been numerous attempts to generalize evolutionary theories beyond organic adaptation to explain the origin of life (Eigen et al. 1981), the origin of moral systems (Darwin 1871), the development of reasoning faculties (Campbell 1965, 1974), the origin and spread of culture (e.g., Boyd and Richerson 1985), as well as the dynamics of scientific communities and conceptual systems (e.g., Toulmin 1972; Hull 1975, 1978, 1980, 1982, 1983, 1985, 1988; Richards 1977; 1981; see Bradie 1986 for a recent review).

David Hull has contributed a number of central and influential views on the topic of conceptual change in science. Our aim in this chapter is to follow up and elaborate on some issues he has delimited, in the context of a case study of the evolution of diagrams of Weismannism. Hull has been a proponent of an evolutionary analysis of change in science, but he also argues that application of this approach demands more than many of its advocates seem to suppose: in order to use evolutionary concepts to analyze scientific change, not only do we need to understand what biologists mean by evolution, but we must use their conceptual tools correctly. This means that a substantial amount of conceptual analysis of biological concepts and methods must go into an evolutionary theory of conceptual change, and that the metaphorical use of evolutionary language must not be considered a substitute for analysis. Hull has contributed most to the analysis and use of two of these conceptual tools:

M. Ruse (editor), What the Philosophy of Biology is. pp. 75–137.

1. his approach to generalizing Darwinian principles and
2. his elaboration and advocacy of the type specimen method, which he has applied to the individuation of scientific communities and conceptual systems.

Hull's contributions concern the conditions under which entities may function as units of selection and criteria for individuating those entities. In extending his views to the conceptual realm, Hull is led to characterize evolving scientific communities in terms of the changing views of their member scientists. We argue that, in addition to this evolution-of-organisms approach, there is a second important prong of evolutionary analysis, typical in population genetics, which concerns the quantitative analysis of character change as a function of evolutionary forces. An important modern version of this trait-evolution approach is the quantitative genetic analysis of changes in phenotype distributions due to selection (Mills and Beatty (1979) give a formal account of the relation between organism fitness and trait fitness relevant to the distinction we draw here; Falconer (1981) summarizes the mathematical apparatus used by quantitative geneticists to describe trait change). Moreover, once the tasks of individuating conceptual units and characterizing their traits have reached a degree of success, it is important to combine the two in 'phylogeny reconstruction'. We suggest that consideration of the nature and distribution of conceptual characters among conceptual organisms is needed to elaborate Hull's fundamental contributions into a full evolutionary program. (Note that here 'conceptual organism' means a conceptual entity which has the properties of organisms: a life cycle including birth, growth, reproduction and death, not a biological organism which uses concepts.)

After a brief consideration of Hull's contributions, we will develop a model of conceptual 'organisms' or 'characters' amenable to evolutionary analysis. The complex entities (organisms or characters) which concern us are representations of contexts in which problems are framed and discussed in scientific texts. (Whether these are best viewed as organisms or characters will be discussed below.) Developing such a model is an important first step toward the development of quantitative models of character change among conceptual organisms (cf. Boyd and Richerson 1985 for similar arguments about cultural evolution). We will argue that in some important cases, diagrams appearing in published scientific works are conceptual organisms because they exhibit properties of heuristics. Modeling these heuristics as conceptual characters is essential to make use of sophisticated tools of quantitative evolutionary analysis.

Hull's analysis of the concept of historical individuals (Hull 1975) leads to a conception of the objects of evolutionary change which is readily generalized beyond the organic realm. His characterization of interactors (phenotype-like entities which are the objects of selection), replicators (gene or genotype-like entities which reproduce themselves and produce phenotypes whose selection and differential replications mediates their transmission) and lineages (descent trees of replicators which evolve) provided the generalized tools for understanding grounds on which any individual might belong to a population subject to evolutionary change (Hull 1980; see also Lloyd 1988 for enlightening discussion). They provide crucial conceptual background for understanding how 'Darwin's principles' (Lewontin 1970) are to be applied to the important entities of an evolutionary analysis.

Hull's introduction of the type specimen method into philosophical studies is his second major innovation, and it brings the disciplined use of population thinking to bear on historical analyses of scientific change. Evolution works on variable populations; without variation, there can be no evolution. Thinking of scientific communities or conceptual systems as populations which (potentially) vary is not as easy as it sounds (cf. Hull 1983; Mayr 1983). Operational problems of applying the type specimen method abound, but even after we succeed in thinking populationally, it is by no means clear that conceptual evolution will be shown to be a fact. What is clear is that conceptual change is a fact with as strong an evidential base as biological evolution. But if the appellation 'evolution' is to have more than metaphorical significance here, we must find the right kinds of objects and relations between them so that the structure of an evolutionary theory can be applied successfully to the process. And of course, given the differences between the biological and conceptual cases, this structure may itself require some modification.

Hull's vision of the best way to study conceptual change in science is to delimit social groups of interacting scientists on the basis of genealogical descent relations and then to analyze variation in the style, content and structure of ideas and arguments within and among social groups (see, e.g., Hull 1985). We suggest that the latter task is clarified conceptually and operationally by thinking of components of representations of ideas in texts as conceptual organisms or characters. On this view, population thinking is itself just one simple model of variation within delimited social contexts: social groups consist of collections of individual scientists who differ in their conceptual 'characteristics'. The individually held ideas of the members of the group form a distribution which might ideally be characterized in terms of 'statistics' of conceptual variation; conceptual means (or norms) and conceptual variances (i.e., measures of the degree of acceptance or similarity of ideas in a population). Hull's approach to an evolutionary analysis of conceptual change is thus to develop a set of general dynamic principles and operational procedures for individuating the entities which may manifest those principles, and then to use these tools in case studies. We suggest that quantitative character analyses will play an important role in studies of scientific change, both for purposes of 'phylogeny reconstruction' and for describing patterns of character change among conceptual organisms.

Hull's analysis of Darwinism is penetrating in the same way Darwin's analysis of organic evolution was: the general account is compelling because it explains in a qualitative way a broad spectrum of features of scientific communities and conceptual systems, such as why cooperation rather than agreement is the fundamental glue holding scientific communities together and why uninfluential precursors 'don't count', i.e., that precursors whose ideas did not play a role in the later genesis, rediscovery, or promulgation of the ideas should not be included in historical reconstructions, no matter how similar they may be. Most doubts about such evolutionary views do not, in fact, question their generality (except when generality is attacked as mere metaphor). Rather, critics question whether conceptual evolution (if that is taken to mean anything more than conceptual change) has indeed occurred, and also whether a theory of conceptual evolution can be made sufficiently precise to make this an interesting claim.

Doubts about whether conceptual change is evolutionary fall into two classes. First, as Hull admits, there are almost no data from the history of science easily suited to evolutionary treatment, and many argue that the reason is that conceptual evolution does not occur. Thus lack of data is taken to count against the truth of the theory. Hull's case study of Darwinism was motivated by the fact that the historical materials are unusually good (Hull 1985). But if the conceptual 'fossil record' is so much poorer than the organic, there would seem to be little hope for supportive evidence beyond a few fortuitous case studies. Second, social groups may not be the kinds of entities which can be subject to evolution. In particular, conceptual change seems to depend heavily on social groups which:

1. are vastly more ephemeral than organic groups,
2. engage in substantial inter-lineage borrowing that defeats attempts to individuate lines,
3. exhibit messy modes of transmission of concepts across generations, and
4. appear to be subject to inheritance of acquired variation.

The first two observations raise questions about whether scientific communities and conceptual systems are sufficiently like organic groups to have resulted from a Darwinian evolutionary process. The last two may suggest that a conceptual 'genetics' would be so different from anything resembling the genetics of biological organisms that conditions for evolution in a generalized Darwinian theory are unlikely to be met very often if ever.

In brief, Darwin's principles (see Lewontin 1970; Wimsatt 1980) require that for evolution by natural selection to occur, individuals (at a given level of organization) must vary in their properties; those properties must have effects on fitness (i.e., the individuals must make differential contributions to the next generation based on their differing properties); and fitness differences must be transmissible to future generations (i.e., the fitness differences must be heritable). The truth of claims of conceptual evolution must thus be evaluated in terms of a set of principles which appear difficult to apply to social groups and conceptual systems. Questioning what fitness means or how conceptual or cultural characters can be transmitted in a way that satisfies claims about heritability, or other claimed conceptual problems are often substituted for an argument that an evolutionary picture is incorrect. But this philosophical strategy of attacking an evolutionary view because it seems *a priori* impossible, or even because it is contrary to fact, has never been very effective against the scientific stance taken by organic evolutionists. Evolutionary models of conceptual change are just that, models. Not only are such models likely to be false, but their developers typically recognize them to be false (Wimsatt 1987). The simplification of the models which makes them false, also make them analyzeable and, it is hoped, will point the way to increasing their realism.

The precision issue, that a theory of conceptual evolution may be valid but so imprecise as to be useless, is the kind of charge which evolutionists should take seriously, and is the one which this chapter will address in a discussion of methodological problems involved in the individuation and evaluation of traits, characters, or characteristics of ideational structures which function as conceptual organisms. Darwin was concerned in *On the origin of species* to establish the fact of organic evolution and a theory of the mechanism(s) by which it occurs. Hull has argued that in order to begin the evolutionary analysis of conceptual change, we require population thinking to allow us to apply a generalized evolutionary framework. Hull has also made some attempt at characterizing a mechanism (Hull 1978, 1988), which must be, as Hull would recognize, but one of many, given the diversity of relevant conceptual units and of their modes of transmission (see, e.g., Boyd and Richerson 1985, chapter 1, for an elaboration of the latter). But individuating social groups and making the case that evolution is a plausible explanation of temporal change and spatial diversity is incomplete unless the particular features of those groups can be given adaptive explanations. Hull is well aware of this issue, and has distinguished between the problems of individuating social groups and describing character distributions. One point of individuating groups is so that the *relevant* character distributions can be studied, and it is to Hull's credit that he has not confounded these problems.

Hull urges population thinking in order to elucidate the nature of conceptual and social structure in science. In an important sense, sensitivity to questions of assessment and measurement of variation is the mark of population thinking in biology. Application of the type specimen method is predicated on having a meaningful sample of specimens from a population. We raise issues in this chapter about the collection of conceptual specimens and the discovery and assessment of variation among those specimens.

II. Conceptual characters and their distribution in science

In order to assess conceptual variation, we begin with a simple model of the structure of reasoning about problems presented in scientific texts. We assume that scientific publications present problems to be solved, attempts to solve them, and arguments that the attempts are successful (Griesemer 1984). The ways scientists choose to make such presentations depend in part on decisions they make about their audiences, what the latter expect and are likely to accept. These decisions influence the structure of texts in manifold ways because scientists bring their varying histories as trained professionals to bear on the problems of communicating their work in public.

Tracking the histories of interaction among scientists is critical to the individuation of social groups. The manifestations of those histories may surface in the form of variations in published scientific texts (as well as in other, less easily trackable places such as conferences and laboratories). To study this variation, we need a model of conceptual structures which may vary, and a strategy for comparing, measuring, and assessing the significance of the presence of variants in social groups. In the remainder of this section we will develop a model of conceptual maps, structures we expect to vary among scientists working in controversial domains, and consider the possibilities for operational measures of conceptual variation. Then, in part II of this chapter, we will make a case study of diagrams of Weismann's doctrine of the continuity of the germ-plasm as instances of actual conceptual structures which bear some of the elements of conceptual maps.

Scientific texts present solutions to problems. In order to do this, problems must be framed and arguments must be given that the results presented actually solve the problems framed, whether from experimental, observational or theoretical investigation. Problem-framing is an important issue for scientists because their shared background of exemplars, laws, theories, procedures and metaphysical commitments (Kuhn's disciplinary matrices; see Kuhn 1970) are never sufficiently explicit or clearly shared to frame problems in ways which heterogeneous audiences can be expected to accept. We use 'acceptance' here in a special sense to mean the cooperation of an audience in reading, and then citing and using the published work (see Hull 1988). The audience must accept, for the sake of argument, the problem as framed, or it will not be a problem at all; if the text is to be successful, some members of the audience must cite and/or use information in the text as well (cf. Latour 1987). In other words, acceptance is a component of cooperation rather than of agreement (Hull 1978, 1988). The audience may ultimately disagree with the author's arguments or results, but non-acceptance of the *framing* of a problem is in an important sense a refusal to cooperate.

No matter how we look at background assumptions, as premises, as starting conditions or as constraints, they function as a conceptual 'environment' or context of a problem stated in a text. This environment must be fixed so that the problem can be entertained by an audience *as* a problem, and its proposed solution judged *in that context* as to adequacy or correctness. Differently put, scientific texts must manage the openness with which an audience reads the text. Readers may come from varying paradigmatic backgrounds, and it is important to specify concepts relevant to getting an intended audience to accept the presented work as solving a problem (and even, to accept the problem as legitimate, important, interesting). The concepts and work taken to *count* as relevant by any given audience, in conjunction with factors such as the effectiveness of the arguments presented and the authority of the authors will determine whether the work presented is judged a solution to the problem presented.

Fixing the conceptual environment for a problem, in other words, means rendering a problem solvable by representing the world (including the disciplinary matrix deemed relevant) in such a way that the results presented solve the problem. To solve a problem, scientists have to do things (in their labs, in the field, in their heads). This activity sets a context in which they interpret what their actions mean. To get their work recognized as solving a problem, they have to limit the possibilities of acceptable interpretation so that the work presented will likely be counted a correct, adequate, or good solution. In other words, *scientists must close the conceptual world in order to make problems solvable* (Fujimura 1987; Gerson and Star 1986).

We can define a conceptual environment to be a model of a possible world in which the problem at hand is solved by the results presented. Fixing the conceptual environment is therefore following a reductionistic research strategy as described by Wimsatt (1980), but at the level of entire research projects rather than at the level of individual scientific problems, including problems of argumentation, presentation and communication as well as those usually deemed 'scientific' in a narrow internalist sense. In general, reductionistic research strategies divide an organized phenomenon into a system and its environment and use heuristic rules to interpret the behavior of the system in relation to its *specified* environment, which is often then detached and ignored. The latter is fixed in order to simplify modeling and to focus attention on what the author(s) wish the audience to take as problematic. A *conceptual map*, then, can be defined as an instance of a fixed conceptual environment relative to a problem which functions as a conceptual system.

The motivation for calling these instances 'maps' stems from a consideration of the context in which differences between texts are most easily recognized: scientific controversies. The conceptual environment of a problem presented in a text marks the 'lay of the conceptual land' in such a way that participants in a controversy can identify and trace the history of differences of framework or problem solving strategy. In other words, scientists present maps of 'how they got to be where they are' in the nebulous, heterogeneous conceptual scheme presumed to be shared by the widest expected audience. Such maps depict not only the history of the authors' work leading to their present considerations, but also re-representations of the work of others, including *their* environment-fixing and problem-solving activities (Griesemer 1983). This does not mean that scientists' depictions are correct or veridical, or even expressions of what the scientists actually believe. They are, simply, the data with which we are concerned.

There is one other image intended in calling instances of fixed conceptual environments maps. Maps are typically thought of as diagrams of relationships (e.g., road maps which show relative spatial relationships of landmarks). The graphical image of a road map is useful for framing a discussion of a problem central to assessing conceptual variation: how can we think about *measures* of conceptual variation? If we think of conceptual environments as road maps of concepts, the spatial metaphor can help to address questions of assessment.

III. Problems with finding appropriate conceptual units for study

In order to think about conceptual variation, we need to consider conceptual variants, i.e., structures which can (potentially) vary whether we conceive of them as conceptual units, as below, or as properties of scientists as in Hull's discussion of social groups. In philosophy, a host of conceptual structures are commonly discussed including concepts, propositions, arguments, and theories. But outside philosophy where canonical forms for conceptual structure have been developed, these structures are often hard to identify and individuate, and even harder to compare. Moreover, there are numerous conceptual structures of great importance in science which have not been given adequate philosophical consideration, such as citations, diagrams, charts, graphs, tables, maps, photographs and other visual displays of qualitative or quantitative information (but see Goodman 1976).

In philosophy of science, the problem of how to represent (and hence talk about) conceptual structures has been limited to considerations of theories, models, and explanations. In part, expression of this bias has been an essentialist goal: there should be a single representation of a theory which captures its essence such that all variation between holders of 'the theory' are reduced to accidental differences of notation or to translatable differences in the use of terms. Logical consistency among canonical forms becomes a criterion through which variation among scientists is expunged from philosophical consideration. On such a view, scientists simply cannot knowingly maintain logically inconsistent views because it would be irrational to do so. Essentialism about theories makes the problem of representation one of getting 'the' theory right, and any disagreements among philosophers will be due to their not all having captured the essence of the theory in their representations, rather than that they have discovered significant differences among conceptual structures. This is to define out of existence the possibility of continuous rational scientific change by denying the existence of its basis in individual variation among scientists. An evolutionary account from this starting point would look much more saltative, is likely to exaggerate the role of scientists expressing large conceptual variations which demand recognition as new, separate canonical forms (or even styles of reasoning, see Hacking 1982), and leads naturally to a Kuhnian polarization of types of scientific change into the dichotomous 'normal' versus 'revolutionary' science.

But if we are *interested* in conceptual differences among scientific products, this typological approach to representation becomes a hinderance. In order to focus on conceptual variation, we need a representational scheme which will allow us to distinguish between errors in philosophical assessment of concepts and actual conceptual variation within and among the scientific products studied. We need a canonical notation which can accommodate the variety of conceptual structures we expect to find in science.

Let us focus on one dimension of this problem to illustrate the issues. Consider an imaginary continuum of quantitative analyses of scientific texts, along which the entities subject to analysis vary in complexity. At the simple end of this continuum, there has been a well known progression of 'citation analyses' which try to capture the organization of scientific fields through study of a single kind of conceptual structure – the bibliographic citation. The method has progressed from simple citation counts to co-citation studies to content analysis to co-word analysis (e.g., Small and Griffith 1974; Small 1977; Sullivan, White and Barboni 1977; Edge 1979; Callon, et al. 1983; Callon, Law and Rip 1986).

A sequence of conceptions of the structures being studied orders this progression. In citation analysis, the basic datum is the citation of one published work by another. Co-citations are pairs of published works cited together. Content analysis recognizes that simple citation analyses lump together works being cited for different reasons, and so the basic datum is a work (or set of works) cited for a categorized reason, where the categories are imposed by the content-analyst. Co-word analysis takes pairs of significant words listed together in a keyword index as the basic datum in order to look at changing constellations of words in a database of keyword entries representing a scientific literature over time.

There are several virtues of these analysis which take very simple text components as 'data structures' (Gerson, in preparation). Citations and keywords are small in size (relative to the size of a whole published text) and abundant. If one is going to study variation, it is useful to be able to get large samples. Moreover, both citations and keyword lists can easily be counted and obtained in machine readable form, so that large samples can be studied quantitatively. Moreover, citation conventions are sufficiently standardized that there is very low probability of error in identifying and individuating citations compared with other sorts of conceptual structures. The trade-off for the manageability and reliability of this data is that it is far removed conceptually on our hypothetical continuum from the sorts of structures which interest historians and philosophers. The fact that much of the literature on citation studies is devoted to arguing that the units are good indicators suggests that there are substantial problems of interpretation in this regard (see, e.g., MacRoberts and MacRoberts 1986).

At the other end of the continuum would be the conceptual map structures outlined above. The structures of interest to a conceptual evolutionist presumably are the conceptual environments (problem-contexts), problems, results, and solutions usually expressed in scientific texts. These are much more complex structures than citations or occurrences of words in keyword lists, so the richness and meaningfulness of variation would appear to be of much more interest. They may, for example, be interpreted as the text structures embodying scientific theories. The trade-off for the of intrinsic interest of such data is most importantly a loss of manageability, affecting both the complexity and reliability of procedures for the analysis of such units, and also the effort which must be devoted to the analysis of each case (Griesemer 1983), and secondarily a substantial decrease in sample size. Consider, for example, the graphical metaphor of a map either as a 'data structure' or as an extension of co-word analysis.

The graphical image of a map with links representing relationships and nodes representing concepts (like cities linked by roads on a roadmap) suggests ways to study and measure variation. For instance, the density of links connecting a node to others could be interpreted as a measure of the centrality of a concept in a system. Different connection densities for the same node in related systems could be a way to compare conceptual systems, especially when those concepts are central to the framing of problems. But it is not obvious how even so canonical a representational scheme as directed graph theory could be applied here since there is no known way to reliably code such features of conceptual systems specified in natural language in a knowledge base (see Griesemer 1983 for further discussion). In other words, one cannot turn natural language into node-and-arrow diagrams without an informed interpretation which really does all the hard, inferential work normally involved in reading a text from a particular point of view.

It might be imagined that the difficulties of operationalizing the map metaphor in this way stems from its pictorial basis. But consider again the notion of co-word analysis, which uses the co-occurrence of pairs of words in keyword indexes to build indicators of conceptual changes in scientific literatures (via changes in the structure of key word usage over time). Suppose we were to generalize co-word analysis to occurrences of n-tuples of words, where the sets of words are just arbitrary text strings chosen from scientific texts. Then, in principle, we could build indices of any sized text structure, including whole texts.

As the model structures get more complex (from citations to co-words to conceptual maps), the analytical machinery and theoretical justification needed to measure and compare them becomes more complex. Secondly, the available sample size goes down, both because there are fewer units of this larger size, and also because, with limited resources, the increased time required to analyze a unit means that fewer can be considered. The elaboration of the models in terms of their structure and measurement generates complexity, uncertainty and unreliability in inferences about conceptual differences. The problem of representing conceptual structures in scientific texts thus presents a series of trade-offs between the interest, manageability, coding reliability, sample size, and measureability of simple vs. complex data.

The upshot of this discussion is that in order to study conceptual variation, we need very simple model data structures which are: significant, potentially variable in some relevant population, have a significant number of instances, and are reliably coded. Neither current practices in text analysis, nor even attempts to characterize ideal conceptual structures present much of a guide to what we can reasonably hope to gain from empirical case studies. But there is a class of text structures, diagrams, which are fairly common in modern scientific literature and meet a number of the above criteria.

Diagrams are often expressed in 'visual language' (Alpers 1983; Rudwick 1985; Tufte 1983; Varnes 1974) which is much more manageable than written text for the purposes considered here. In the sections to follow, we discuss some virtues of diagrams as units for studying conceptual variation. We will then go on to consider the case of 'Weismann diagrams', diagrams expressing relationships between the germinal and somatic cells of the body. August Weismann (1834–1914) is widely credited with discovering the significance of the polar bodies in meiosis, proposing the doctrine of the continuity of the germ-plasm, establishing the non-inheritance of acquired characteristics, and proposing with Wilhelm Roux a now discredited 'mosaic' view of development. Like Lamarck before him, Weismann's views have been widely misinterpreted in the scientific literature. Lamarck's legacy seems to have been reduced to the view that acquired characters are inherited (see Hull 1984). Weismann's legacy has been reduced to the opposing view that acquired characters are not inherited, plus his doctrine of the continuity of the germ-line. Both are important tenets of 'neo-Darwinism', the term coined by Romanes to characterize Weismann's view of evolution.

We will trace the changing conception of Weismannism through an analysis of diagrams as conceptual maps which fix conceptual environments for a variety of problems about heredity and development. We wish to show not only that diagrams can be useful indicators of conceptual change in science and to illustrate their potential role for evolutionary analyses of scientific change, but to make the case that they are instances of a much more general conception of structured data which should be of interest to historians, philosophers, and sociologists of science interested in conceptual change.

IV. Are the conceptual units of analysis 'traits' or 'organisms'?

Above we spoke alternately of the units we wish to study as organisms or as characters. The choice of mode of description of these units has significant further implications. There are arguments favoring either decision, and it seems likely that in different contexts they may have to be viewed as one or the other or perhaps even both, from different perspectives simultaneously applied. This complexity is a product of at least three factors:

1. The first is the probable existence in cultural evolution of a variety of different level units of transmission and selection. A similar problem arises in biology, for similar reasons. Thus, if both individual and group selection are affecting the evolution of a trait of an organism, that organism is simultaneously a whole phenotype (or interactor, in Hull's terms) with respect to organism level selection and a part or complex trait of the group phenotype with respect to group selection.

2. With cultural evolution there is a greater difficulty in individuating these units, and less agreement on what the relevant units are. There are no well individuated universally applicable individual units which have the perceptual and causal unity, and relatively clear boundaries to provide the natural starting point for an evolutionary theory which organisms have provided for the biological theory. We argue below that diagrams have these features, but even if this is so, they are not common to all theories or theoretical contexts, and so lack the requisite generality as units. A consequence of this is that while diagrams may provide a particularly good example of the right kind of unit, they are one among many – while they might serve to demonstrate the utility and effectiveness in one case of a theory of cultural evolution, they do not provide a generally applicable unit which provides the structure for a general theory. (Nonetheless, we think that our treatment here provides many heuristics which will prove useful in the analysis of other kinds of cases.)

3. Finally, the different modes of cultural inheritance provides some additional complexities which make it more difficult to individuate cultural genotypes and phenotypes than in biological cases, a problem which in consequence, makes it also harder to identify whether a given complex is to be regarded as a replicator, an interactor, or as a trait complex – a part of an interactor.

The alternative pictures can be presented simply, but complications will follow:

1. On the view of conceptual evolution as the evolution of the beliefs of scientists, it is natural to view conceptual structures as conceptual characters of the scientists' phenotypes. To be sure, then the mode of acquisition and

transmission of these characters then commits one to the view that cultural inheritance is Lamarckian, (roughly, that it involves the inheritance of acquired characters, but see Hull 1982, 1984, 1988, for more extended discussions on this point) rather than Weismannian, but this is a small price which almost every cultural evolutionist is willing to pay. (We ignore here those who believe that cultural variation is genetically determined to any significant degree, and thus that cultural inheritance is Weismannian at the biological level.) This is Hull's view, or at least the view which he most often advocates.

2. An alternative view is possible however: Given the adaptive complexity of the conceptual structures which must be considered (certainly for theories or models, but also arguably for diagrams), it seems natural to regard them as kinds of conceptual organisms, with a wealth of co-adapted features designed to aid their acceptance, promulgation, and application. It fits better with the way many philosophers of science talk about theories as abstract entities, but theories require theorizers, and abstract entities, entifiers. On this view then, to take account of the fact that theories (unlike paradigmatic organisms) do not propagate or replicate themselves, these conceptual organisms have to be regarded as kinds of conceptual viruses or plasmids which can be transmitted between and infect scientists who then become agents of their further transmission and replication. The problem with the first view on this account is that it leads us to look to the scientist as the beneficiary of the adaptations of the conceptual structure, and indeed, Hull has ably exploited this feature. But it appears that there are adaptations of conceptual structures which can also be viewed as means to their own replication, and even more importantly, that there are gene and phenotype (or replicator and interactor) analogues within conceptual structures.

Biological genes both replicate themselves, and also produce an adaptive structure (the phenotype) which is acted upon by selective forces in the environment, and whose action mediates the differential reproduction of the genes which they carry. The self-replicating and other replicating activities of genes are called, respectively, their autocatalytic and their heterocatalytic functions. It is tempting, as many cultural evolutionists have (see, e.g., Dawkins 1976) to say that anything which is replicated is a gene analogue, but this will not do, since biological phenotypes are replicated as well as genotypes. The key in distinguishing the two is to note that the genotype plays a generative role in making the phenotype, but (with some complications not to be discussed here!) not conversely, and that the genes (together with a minimal support part of the phenotype) are what is transmitted to the next generation.

To see how this model can apply to scientific change, we need only to suppose

1. that scientific theories have, in their application and development, a generative structure which is roughly hierarchical,
2. that they are modified adaptively to improve their 'fit' with the environment, and
3. that they are transmitted or replicated by passing on a relatively small generative structure, from which the descendant theories are generated.

This third element is the analogue, for conceptual structures, of the genotype. The genotype then is recognized (for cultural evolution), primarily in terms of its heterocatalytic function. Of these assumptions, the first is implicit, not only in the traditional formalist account of scientific theories, but also in the post-positivistic accounts which are advanced to replace it (see, e.g., Kuhn 1962, 1970; Lakatos 1970; and Laudan 1977). The second assumption is common to all 'evolutionary' accounts of scientific change, and to many accounts which are not explicitly evolutionary. The third is based on an observation which perhaps explains the formalist's mistake of identifying a theory with its axiomatic basis. It is the simple observation that we often begin to teach theories by presenting the most basic laws or assumptions of that theory. The formalist in effect stops there, trusting that, when given the axioms, the rest of the theory and its applications will follow. But also important (as pointed out by Kuhn and largely missed by more formalistic accounts) is the fact that the meaning of these axioms and their use is conjointly taught through application to paradigmatic problems or examplars, which on this account would also have to be included among the generators or gene-analogues of the theory. Axioms are neither self-interpreting nor self-applying.

The third assumption gives us one way of recognizing the 'gene-analogues' in scientific evolution, though in cases where the axioms are not easily delimited or there are significant generators which are not included in the axiom sets (such as problem solving heuristics or experimental paradigms). *It is better to characterize the gene analogues as any elements which are significantly generatively entrenched – that is, any elements which play an important role in generating the adaptive or phenotypic structure of the theory* (see Wimsatt 1981a, 1986b, 1986c, and 1988 for further discussion). The phenotype analogue here is the generative apparatus of axioms or principles, the heuristics for formulating problems and applying them, and the theoretical and experimental paradigms of their application, (comprising the genotype analogue), together with the generated theoretical structure of principles, generalizations, results and applications (which comprise the rest of the adaptive structure of phenotype of the theory). It is ultimately through its applications that a theory realizes its fitness, and through a powerful and broadly applicable structure with many applications that it achieves high fitness by solving many problems and thus winning many adherents and displacing its competitors. This entire structure is its phenotype. (Here, as in biology, the genotype or generative structure is properly considered as an aspect or part of phenotype.)

One remarkable consequence of this way of framing the genotype-phenotype distinction in this context, is that a good part of conceptual transmission can be regarded as Weismannian in character: theory-phenotypes do not directly make more theory-phenotypes in the scientists which they infect. Rather what is passed on is a relatively compact generative structure which, in the right conceptual and social environment, generates the theory-phenotype of the next generation. Wilson's diagram of Weismannism, to be discussed further below in section VIII, captures this mode of transmission for conceptual structures as well as for biological structures (Wilson 1896, figure 3). The fact that conceptual transmission can take place serially at separate successive times in a scientist's conceptual ontogeny complicates *the description* of this process, since successive layers of increasing sophistication in the understanding and application of a theory may be transmitted during the education of a scientist, but it seems possible in each of these cases to argue that what is passed is a generative structure which further ramifies (for the later transmissions, in the context of the theory already elaborated to receive these additional generators) to complicate and elaborate the phenotypic structure of the theory-application complex which constitutes the conceptual phenotype. Because of the generative role of the gene-analogues, and the fact that only the generators, not the whole of the phenotypic structure is transmitted, conceptual inheritance is also Weismannian. Because multiple generators are transmitted, to fit into the appropriate place in an elaborating conceptual-phenotypic structure, we have a kind of hierarchical multi-level Weismannism which has no parallel in the case of biological inheritance (see Figure 1).

In the context of what will follow below, we argue that Weismann diagrams have been transmitted, both with and without modifications, and that they, and diagrams in general, play a significant generative role in the transmission, understanding, and elaboration of theory.

This important parallel with the biological case, the possibility of making a genotype-phenotype distinction, is missing from most evolutionary accounts of scientific change. This distinction is important for at least two reasons:

1. It is implicit in and essential for the further development of analogues to population genetics for scientific and cultural change, such as those of Boyd and Richerson (1985). This distinction was not present in Darwin's theory. Making it not only enormously increased the power of the biological theory, but removed important conceptual confusions. In the present context, it gives an important new way of identifying conceptual 'genes' – through their generative role.

2. Making it brings a number of developmental phenomena within the scope of the theory, phenomena which have often been ignored in the absence of a theoretical perspective for their analysis. Most striking is an explanation of

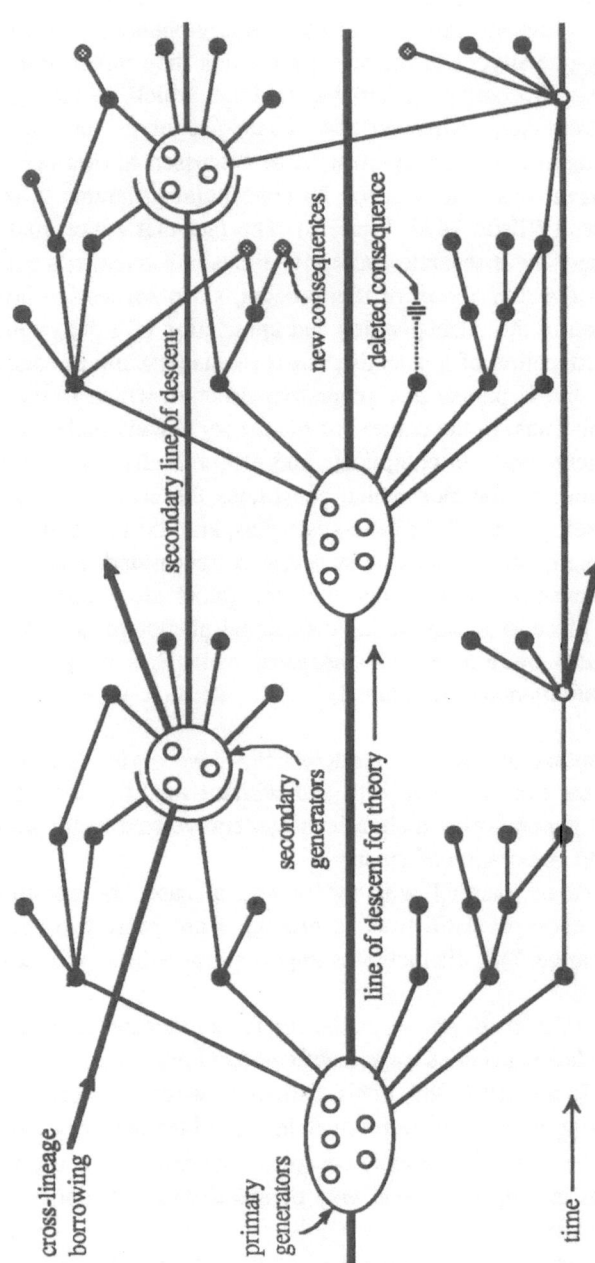

Fig. 1. Inheritance and evolution of scientific theories. Inheritance in scientific theories is via a kind of hierarchical Weismannism. The primary generators (axioms, heuristics, exemplary problems, represented by 5 unfilled circles) are transmitted to a scientist and provide the initial basis for the elaboration of a theory (represented by other (black) nodes) which is generated from them. Some of these consequences provide a natural context or receptor site (the semicircle) for the incorporation of other generators (the smaller circle with 3 unfilled circles) which may be transmitted later from other sources, but subsequently transmitted and taught as part of the theory. (They may, as here, remain separate from the primary generators of the theory, either because they are thought of as part of a separate subject matter, or because they cannot be taught until after other things are taught first.) Some consequences (the unfilled circle at the bottom) may be sufficiently noteworthy that they may themselves be transmitted independently (with or without modification) either to later generations of that theory, or exported to other lineages. Thus, think of Darwin's principles, as applied to organisms as the primary generators, Mendelian genetics as the imported secondary generative complex at the top, which becomes population genetics, and the idea that selection maximizes fitness at the generative node at the bottom (revised in the second generation, with input from population genetics) to become maximization of inclusive fitness. This diagram is grossly oversimplified, and cannot, at this level of magnification capture the fine structure of historical detail. Three kinds of inheritance are represented here: 1. derivational linkages within the theory, by the thinnest lines, 2. the inheritance of the primary generators by the thickest lines, and 3. transmission of secondary generators, both within and across theoretical lineages by medium lines. All kinds of inheritance can be multi-parental, since a consequence may be derived from the conjunction of several others and the generative complexes may be derived from a variety of sources. Also represented in the diagram is the production of new consequences (denoted by grey nodes) in the second generation, and loss of problematic consequences. Not represented in the diagram is the role of the conceptual environment (both intra- and inter-theoretical) in selection, mutational changes in the meanings of key elements, and perhaps most importantly, major changes in the structure of the theory. Thus, e.g., it is arguable that population genetics has in many contexts replaced or become co-equal with Darwin's principles as the primary generators of the modern theory. This complexity should be compared with the simplicity of Wilson's diagram of Weismannism (below), but the point of both is the same: what is transmitted is generators, not the whole phenotypic structure of the theory.

---

the differential effects and rates of evolution of characters expressed at different stages of development which has direct application to the prediction and explanation of rates of scientific change of different sorts. (See Wimsatt 1986b, 1986c, Rasmussen 1987, and Schank and Wimsatt 1988 for further elaboration and a variety of applications of this claim and the 'developmental lock' model upon which it is based. The last paper cited applies this model specifically to scientific change.)

In spite of this extended explication and defense of the 'organismal' approach, we do not at this stage consider the two views of diagrams, as organisms or as complex characters, as alternatives. Which way they should be viewed depends upon the level and the perspective of the analysis. For some problems (e.g. Hull's (1988) application of the concepts of conceptual inclusive fitness and the demic structure of science to the explanation of the behavior of scientists), it seems most natural to regard their theories as complex conceptual characters of scientists, and diagrams as complex sub-traits within these complexes. Or for other purposes, it might be appropriate to view theories or models as conceptual organisms which exhibit differential replication, in which case, the diagrams associated with them are complex traits of these theory-organisms. At our level of analysis, and particularly because the diagrams are exported and modified in different theoretical and dydactic contexts – often to contexts where their original parent-theory is under critical examination or attack, and they are employed in the service of a competitor, it seems more natural to treat the diagrams as conceptual organisms, and the theories which they serve as part of the conceptual environment. Finally, whether the organismal level is taken to be at any of these levels (a scientist, his theory, or the diagrams he takes to illustrate and to argue a part of that theory) the diagram will have traits, and can be treated as a trait-complex. Indeed, taxonomists often treat species as if they were trait-complexes, and talk about the evolution of trait complexes as if they were organismal lineages.

V. Diagrams as good 'model organisms' for studying cultural evolution

It has often been observed that in biology, the solution to a particular problem is facilitated, and in some cases only made possible through the right choice of an organism for study or experiment. In such cases the organism in question becomes a 'model organism' or paradigm for the study of that problem. If we are to argue that 'conceptual maps' meet the conditions for units of selection in cultural evolution, we must identify properties which make them useful entities to study through time for tracking evolutionary change. One problem which has infected earlier attempts to do so, however, is the need to develop a canonical mode of representation for them (Griesemer 1983). Such a canonical mode of representation would ideally:

1. be readily extracted from the text,
2. have a significant degree of inter-coder reliability,
3. have easily and unambiguously scored properties or characters whose stability and change provide the data for an evolutionary analysis, and
4. would also be such as to allow the assignment of a metric to its properties so as to allow the application of theories similar in form and intent to those of quantitative genetics.

There are at least two kinds of cultural objects which meet the first 3 desiderate directly without need for further analysis, and which have characteristics which are scored with relative ease. They therefore provide ideal 'model organisms' for the study of cultural evolution. These are equations and diagrams, both of which can be thought of as special kinds of conceptual maps, and which have the advantage that they are already explicitly demarcated in (and thus do not need to be extracted from) the text. We do not plan to discuss equations as objects subject to cultural evolution here, though many of the claims made for diagrams below apply to them as well.

A. Properties of diagrams as good units

Diagrams have the following useful properties which, when compared with other ideational entities, make them particularly perspicuous objects of study for cultural evolution:

1. *Locality*. Unlike ideas, concepts, or even propositions, diagrams are paradigmatically well-bounded and local in character. Anyone can tell the boundaries of a picture, and where it is delineated from text. By contrast, representations of important ideas or concepts usually occur in a variety of places or contexts in a text, and are used for a variety of ends and in a variety of applications. These contexts are often crucial to the understanding of an idea, and an apparently monolithic concept may have multidimensional and often conflicting meanings. In one particularly well studied example, Masterman (1970) documented 22 uses or senses of 'paradigm' in Kuhn's *Structure of scientific revolutions*, which she clustered into two main families, and argued that confusion between which of these meanings was intended in any particular case had caused considerable confusion in the discussion of this central idea.

This multiplicity of uses and meanings, and the inevitable role of judgement in abstracting and individuating them, makes it difficult to get unambiguous and widely accepted characterizations of them, and theoretical disputes often turn on arguments as to how key concepts are to be defined and interpreted. Even when authors take care to present explicit definitions of key concepts, unless other authors cite them explicitly, it is difficult to tell whether subsequent developments in the concepts take their work as a starting point or use other perhaps different (and too often implicit, rather than explicit) partial characterizations or definitions in the literature. Moreover, the same idea, concept, or proposition may be expressed by the same or by different authors in different words, and the same word may for the same or for different authors denote, on different occasions of use, different ideas or different variants of the same idea. This fact further complicates the task of identifying, individuating, and tracing changes in concepts. Doing this, as historians of science are well aware, often requires analysis of a large portion of an author's work, and continuing and often contentious judgements as to whether variant expressions at different times represent different partial slices through the same, largely unchanged concept, or whether they signal changes in the author's views. (Furthermore, most authors probably overestimate and do Whiggish rewriting of their train of thought to demonstrate the constancy, or at least the continuity, of their ideas, further complicating the theoretical task of analysis.) The locality of diagrams is by contrast refreshingly simple and clear, and suggests several further properties:

2. *The clarity of the 'organism/environment' boundary.* The preceding complexity and non-locality of meanings gives the conceptual evolutionist a much more difficult task than the biologist in distinguishing between the 'organism' under study and its environment. Biological organisms give at least the appearance of being well-bounded (sometimes misleadingly so), and it becomes correspondingly easier to determine what are characters of the organism and what are properties of the environment. Characters of the organism are things which must be traced and evaluated for evolutionary changes, and characters of the environment are aspects of the niche of the organism which are crucial for evaluating their fitness and determining the selection pressures governing their evolution. The well-bounded-ness of a diagram (or an equation) gives this clarity to the conceptual evolutionist. The icon is itself clearly well bounded, though the boundaries of the system necessary for interpretation of the diagram usually extend a little further into the text. The relevant unit must be taken to include both any figure labels, and any explication of them in the text if we want to know as well the significance of the diagram. Both of these however tend to be relatively well bounded, and it is rare that one must look any further than the surrounding page or two of text for the necessary explication. Even where references occur elsewhere in the text, the very modularity of the figure facilitates their unambiguous identification: figures and equations, unlike ideas, are usually numbered, and characteristically referred to by number.

3. *Easy scoring, evaluation, and comparison of characters.* While one may need to reach into the text (however locally) to understand the interpretation of a diagram, the vast majority of the character traits of the diagram as organism can be read off from the diagram as icon itself. At most, a general knowledge of the intended significance of a diagram's properties are needed to score and interpret it. In the case of diagrams of Weismannism, one can easily count nodes or connections, or analyze the topology and arrangement of the components of the diagram. *Another way of putting this is to note that the vast majority of characters of a diagram are intrinsic properties of it – whereas, by contrast, most of the characters of an idea, even where intrinsic, must be evaluated by looking at the relational properties of the idea.* (This is in part one of the reasons for the popularity of functional theories of meaning – cf. e.g., Putnam 1975.) An analysis of the diagrams of Weismannism has so far revealed over 20 characters which are easily read off from the diagrams alone. Scoring of these characters is for the most part unambiguous and highly

reliable, and there are enough of them that it would not be unreasonable to use numerical taxonomy to analyze their descent relationships. We doubt whether this would be possible for any of the conceptual entities normally studied by conceptual evolutionists – and, to our knowledge, no-one has even tried.

4. *(Relatively) easy determination of ancestry.* The determination of ancestry for diagrams is much easier than for ideas, for several reasons. Before going into them, we need to distinguish different kinds of evidence for descent, from stronger to weaker:

– The strongest possible evidence is the reoccurrence of the diagram in identical form down to arrangement and type face, together with explicit citation of the ancestral source.

– The same evidence, without a citation is nearly as strong, since it makes it virtually certain that the diagram is a lineal descendant of the first diagram of that type, though if that diagram has been replicated elsewhere in the literature, it is unclear whether the diagram in question is an immediate descendant, or through one of the other copies. Sometimes diagrams are redrawn, and here the evidence, while still strong, (and mediated by high degrees of similarity) is somewhat weaker.

– Citation to a text which contains no identical diagram is very strong evidence, with the only qualification being, that sometimes only an author is given, and the identification of the text is left up to the reader. In some cases, the bibliography disambiguates the reference, but with either multiple possible source works by a given author, or multiple possible source diagrams in the cited text, ambiguity of ancestry is still a real problem – one that has limited our determinations of ancestry in several cases.

– Similarity in degree r $(0 < r < 1)$. With readily scored and individuated characters, this could be made quite quantitative, in the spirit of numerical taxonomy, with greater similarity taken as evidence for relatively recent evolutionary divergence. Here (as in numerical taxonomy) the question of the relative importance of different characters in determining similarity arises, and the answer is the same in both cases: theoretical considerations and background information (in this case, about the ideas represented) provides whatever basis we have for weighting some characters more strongly than others. Thus, for example, to even count as a diagram of Weismannism, a diagram must contain a representation of the continuity of the germ line, but it is less important, for these purposes, exactly how the soma or phenotype is represented.

There are two special features of diagrams which give these studies of ancestry a special advantage even over comparable cases in biological evolution. The first is the obvious one that dating of diagrams is almost always relatively unambiguous, and normally not itself the subject of theory-mediated determinations. Since ancestors cannot occur later than their offspring, this gives a strong constraint on determinations of ancestry, and one which is not as open to argument as it is in biology, where common ancestors are inferred from contemporaneous organisms, or from fossil remains whose relative age is sometimes in doubt. (This advantage of course holds for published text as well as for diagrams.) The second is that citations (indicating ancestry) are much more commonly given for the sources of diagrams than for ideas, and we have no comparable data for organisms over evolutionarily significant spans of time.

Citations are commonly given for two reasons:

1. Ideas are not copywriteable, but diagrams are. (In this way, they are similar to text.) Thus writers are compelled (under the threat of legal action, at least since the origins of copyright conventions) to give citations for diagrams drawn from other sources.
2. It seems likely that in most cases, authors have more invested in claiming originality for ideas than for claiming originality for diagrams. (It also seems likely that false claims for originality of diagrams would be more easily detected, because of the relatively ready and unambiguous scoring of characters. False claims for the originality of quotes or text are also easy to uncover).

For both of these reasons, the lines of descent for diagrams are less likely to be hidden than those for ideas, since there is seldom a reason for, and too great a risk in, doing so.

5. *(Relative) context-independence.* We have already commented on the relative ease in interpreting diagrams, when compared with ideas. This flows in two ways from their relatively small and well-delimited dependence on context:

1. The vast majority of character traits of diagrams are intrinsic properties, which can be read off from the icon itself, without reference to context at all, or at most to the generic context of diagrams of that type. (Thus the conception of a diagram as a cellular descent tree is sufficient to read off a large number of the properties of most diagrams of Weismannism. While this is contextual knowledge, it is not knowledge that is specific to the context to any one of the diagrams.)
2. The relevant context for interpreting specific diagrams is generally much smaller and more compact than the relevant context for interpreting ideas. The only situations where this would be expected to break down is where the interpretation of a feature of a diagram is itself strongly dependent on

the interpretation of an idea or ideas about which there is some argument. Thus what is said to be continuous, and the criteria for continuity may be in need of further explication in understanding the evolution of Weismann's doctrine of the continuity of the germinal material but that a diagram depicts continuity of something germinal is not.

While the interpretation of the diagrams may be quite context-independent, the fitness of the diagrams to illustrate the points in question, or their correctness when placed in a new context, is not. (This would parallel the observation that traits of organisms may be readily observed and scored in a variety of contexts, but their adaptive significance in these different contexts may vary considerably.) This variability of fitness or correctness in different contexts can induce strong selection pressures, or reasons for evolutionary changes in the diagrams. This last fact points to an important role for context in the evaluation of such diagrams, for it is the context, particularly the problem context which leads to the introduction of the diagram, which determines and should be used in evaluating its fitness. It provides relatively direct explanation for changes made in the diagram to accomplish new purposes of the context. Thus, after the rediscovery of Mendelism, a modified diagram of Weismannism was introduced to illustrate the principles of Mendelian segregation operating in gametogenesis (Thomson 1908, p. 344, fig. 35). The modifications in this diagram are easily understood in the light of the different aims of this discussion. *The contexts of the diagrams thus provide the relevant features of their ecology to identify the selection pressures affecting their evolutionary change.* Desiderata of clarity and salience in argument lead authors to mark these selective factors much more clearly than one would find in nature, where the identification of the salient selective factors is often the most difficult and hotly contested part of an adaptationist analysis (see Gould and Lewontin 1979).

6. *Easy portability*. So far, we have focused on how the relative context-independence of diagrams and equations makes them ideal objects for study. But there is a deeper reason why they are worth our attention which also springs from this context-independence. These same properties which confer context-independence and make diagrams easy things to study also make them easy and useful things to borrow. If a diagram is more readily extracted from a text than an expression of an idea in order to study it, it is also more readily extracted from a text to borrow it. Similar but not identical considerations render equations as readily borrowed elements. Equations are easily localizeable, and though not as readily understandable as diagrams, are almost invariably items of high importance since they both are central reflections of the assumptions and structure of a theory or model, but are also crucial tools for its application to new problems. If an element is less context-dependent, it is easier to understand, which can provide a differential motivation for taking it. To draw the organic analogy (misleading here for the active voice,

since diagrams do not move themselves), *it has a greater potential for migrating out of its context.* (One is hopefully more likely to borrow things which one understands, or thinks one understands, than things which seem more obscure or difficult to tease out.) If it less context-dependent, there is also a greater chance that it will prove to be useful in other contexts – i.e., *it has a greater chance of being able to migrate into and successfully colonize new contexts.* In other words, its heritability of fitness is higher. *These two consequences of context-independence naturally make diagrams good conceptual organisms – they should both serve to increase their reproductive rates, and also to increase their probability of evolution through adaptive radiation into new contexts.*

A parallel process occurs for ideas which are imported through quotation, (which stylistically isolates the quote and provides warning that the quoted material is not in its natural context) but quotation is frowned upon as a way of expressing one's own ideas (as opposed to accurately expressing the ideas of others – particularly if they are about to be attacked.) It is interesting that no similar onus appears to be attached to the borrowing of diagrams, except perhaps by the publishers of profusely illustrated elementary textbooks, who no doubt are strongly influenced by images of large royalty costs, and perhaps also by desires for a unity of style. For these reasons, such books are probably the biggest single source of redrawn illustrations, with journals having strong style requirements a likely close second. (See, e.g., the redrawn version of Wilson's (1896, figure 5) as figure 1, page 243 of Gilbert's (1985) textbook on developmental biology.)

B. Properties as heuristics – heuristics as generalized adaptations

Diagrams can be effective means of communication as is attested to by the old saw that 'A picture is worth a thousand words'. Furthermore, they are also effective tools for thought, and as tools for visualization can provide handles for a mode of conceptualization which language lacks. Until relatively recently, the role of visual means of communication has been virtually ignored by philosophers of science. (But see Hanson 1970). Philosophers of science have traditionally been concerned primarily with questions of logical structure, justification, logical deduction and proof, and truth-preserving algorithms. Along with this has gone a conception of theories as linguistic structures, and an emphasis on defineability and deriveability relations holding among elements of these structures. Since no-one has proposed a way in which these relations could be explicated for pictorial representations, they have simply been ignored.

Several new currents in psychology and philosophy of science have facilitated a change in perspective. During the heyday of linguistic philosophy, that which could not be expressed linguistically was held to be meaningless, and it was supposed that verbal or written language was also the form of thought. But since Roger Shepard's classic work in 1971, it has become increasingly clear that a great deal of thinking in at least some of us is visually, rather than linguistically mediated (Shepard and Metzler 1971). This still didn't solve the problem of how to analyze meaning relations, and defineability and derivability relations for visual representations, but another trend has made these relations seem to be less than all-important. In the last decade there has been a rapid growth of interest in tools of discovery and problem-solving techniques which are not readily amenable to analysis from this perspective (see Nickles 1980). These heuristic tools are quite unlike the former ones in that heuristic procedures, unlike truth-preserving algorithms, do not guarantee correct results, even when they are correctly applied. In spite of this, they are used because they generally or frequently produce correct answers with far less effort or computational demands, and are thus cost-effective solutions (Wimsatt 1980). This is particularly (but not exclusively) the case for dealing with ill-structured problems, for which there may be no algorithmic solutions.

It is also important that these heuristics are conceived of as procedural rather than propositional in form, with an emphasis on the activity of using them and on the transformations which they produce. (The fact that they do produce transformations means that in many cases they can be studied propositionally, but, most importantly, there is no canonical need to do so. They are most fundamentally conceived of as tools for achieving given ends, and one can, for example, recognize that a diagram may be an effective means for communicating an idea, and be able to explain why this is so, without being able to analyze its propositional structure.)

Wimsatt (1980, 1981b, 1986a) has analyzed and provided a list of general features of these heuristic procedures, of which the following is a slight elaboration. (See also Lenat 1982 for a penetrating discussion of heuristic procedures). Although these properties are expressed in propositional mode, the talk of solutions, or correct solutions (which suggest that the end product is a proposition) can also be reconceptualized as actions or adaptive (or maladaptive) behaviors. The most important properties of heuristic procedures are as follows.

1. *Properties of heuristics.* (1) By comparison with truth-preserving algorithms or with other procedures for which they might be substituted, heuristics make *no guarantees* (or if substituted for another heuristic procedure, weaker guarantees) that they will produce a solution or the correct solution to a problem. A truth-preserving algorithm correctly applied to true premises *must* produce a correct conclusion. But one may correctly apply a heuristic procedure to correct input information without getting a correct output.

(2) By comparison with the procedures for which they may be substituted, heuristics are very '*cost-effective*' in terms of demands on memory, computation, or other limited resources. (This of course is why they are used.)

(3) The errors produced by using a heuristic are not random, but *systematically biased*. Talk of systematic bias should be taken to imply two things: a) The heuristic will tend to break down in certain classes of cases and not in others, but

not at random. Indeed, with an understanding of how the heuristic works, it should be possible to *predict* the conditions under which it will fail. b) Where it is meaningful to speak of a *direction of error*, heuristics will tend to cause errors in a certain direction, which is again a function of the heuristic and of the kinds of problems to which it is applied. These systematic biases can be useful in two ways:

- Their analysis provides a 'calibration' of the heuristic; an evaluation of the conditions under which it can be safely used.

- If different heuristics leave characteristic 'footprints' (heuristic-specific biases), the detection of systematic biases can provide clues as to the heuristic reasoning processes which produced them. This procedure was pioneered by Tversky and Kahneman (1974) in their classic study of biases in probabilistic reasoning, and has been discussed and applied further in Wimsatt (1980, 1986a).

(4) The application of a heuristic to a problem yields a *transformation* of the problem into a non-equivalent but intuitively related problem. This means that answers to the transformed problem may not be answers to the original problem, even though, if the new problem formulation leads to an adaptive solution, various cognitive biases operative in learning and science may lead us to ignore this. A problem formulation which yields a solution where none was before possible may be taken as the correct formulation of the problem. Even if different, we may say that 'this is what they were looking for all along'.

(5) Heuristics are useful *for* something – they are *purpose relative*. Tools which are very useful for one purpose may be very bad for another. This often gives a useful way of identifying or predicting their biases: one would expect a tool to be relatively unbiased for the applications it was designed for, and perhaps quite biased for others. One might also expect that increases in performance in one area will be accompanied by decreases elsewhere.

We wish to argue that diagrams are heuristic tools for the effective communication of visual information, and as such, exhibit all of these properties. (We will provide specific examples of each of them below, when we discuss the use and evolution of diagrams of Weismannism.) It is interesting also to note that the six advantages for diagrams listed above may be viewed as heuristics for the identification of objects for study of cultural evolution. In addition, some of them can be thought of as heuristics to facilitate communication and understanding of the subject matter (e.g., locality, ready identification of characters in the diagrams, and ready identification of selection pressures in their conceptual environments – i.e., the reasons for choosing and modifying the diagrams). Other features of the diagrams, e.g. relative context-independence and portability of diagrams may be thought of as heuristics to aid in their propagation. We do not suppose that these heuristics will be consciously applied in the design of the diagrams (in this they are more like biological adaptations). However, they may be functional side effects of other choices or strategies which are consciously applied.

The five properties of heuristics listed above not only provide useful handles for the analysis of the uses (and misuses) of diagrams, but they also provide a direct link between the analysis of cognitive tools and the analysis of adaptations, and thus forge an immediate connection between the study of heuristics and a general model of evolution of the sort advocated by Campbell, Hull and others. This link is direct in two ways, because both biological adaptations and cognitive adaptations have all of the above properties, and can thus be regarded as special kinds of heuristics. Campbell (1974) describes a special class of biological, psychological, and cultural adaptations which he calls 'vicarious selectors', which are crucial to his account of evolutionary epistemology. Wimsatt (1980, 1981a, 1981b, 1985) has elsewhere argued that vicarious selectors have all of the properties of heuristics, but they will not be further discussed here. We will provide however a summary argument that biological adaptations have these properties. The following claims are numbered so as to correspond to the preceding list of properties of heuristics.

2. *Properties of adaptations.* (1') It is widely accepted that the proper performance and use of an adaptation, even in its normal environment, does not guarantee the survival of an organism or its production of viable offspring.

(2') Adaptations are however cost-effective ways of contributing to that end, which it is assumed (on the 'adaptationist program') is the reason for their selective incorporation and maintenance.

(3') Any adaptation can be made to malfunction under the appropriate circumstances, and the conditions under which an adaptation will fail to confer its advantage are systematic. In fact the use of experimental conditions to cause malfunctions is one of the most powerful tools for discovering how a system functions, providing not only clues as to how the system is organized and works, but also an analysis of what conditions are required for it to function properly.

(4') This condition is easiest to demonstrate for sensorimotor functions, but a recognition of how it applies in these cases suggests that it is indeed generalizeable. Consider the problem of how to detect seasonal changes in species whose morphology or behavior must change to allow survival or proper functioning in the changed environment. As Levins (1968) observes, this is characteristically done by sensing, tracking, or responding to an indicator variable which is a reliable predictor of the oncoming change. It may be that temperature change or food availability may be the survival-relevant parameter which necessitates the change in an organism, but it is far easier to detect changes in day length. This change is only contingently correlated with the adaptively relevant variables, a correlation which may break down under unusual circumstances, either in nature or as deliberately produced in the laboratory, but it use as an indicator to generate the appropriate changes transforms and enormously simplifies the problem of 'deciding' when to make the appropriate changes.

(5') Adaptations are clearly adaptations to or adaptations for something – ultimately for maximizing fitness and its heritability, but more specifically for detailed tasks which are determined by the role of that adaptation in the functional organization of the organism. These are what we describe as its functions. Adaptations can acquire other tasks through evolutionary time (they are then called 'exaptations', Gould and Vrba 1982), but in doing so, they are commonly differently elaborated and pruned under the new selection pressures, indicating that an adaptation designed for one purpose is not generally good (and must be modified for) other purposes.

These properties of heuristics and of adaptations as heuristics find frequent parallel instantiations in the use and modifications of diagrams of Weismannism, as will be illustrated further below.

C. Properties as units of selection

We have already discussed 'Darwin's principles' (Lewontin 1970), which are widely agreed by biological and conceptual evolutionists to play a central role (for many as separately necessary and jointly sufficient conditions) in the characterization of entities which can act as units of selection which are undergoing evolutionary change. The units in question must show 1. variation 2. which is heritable, and this heritable variation must induce 3. heritable differences in the fitnesses of the variants.

The variety of diagrams of Weismannism which we will discuss (and the much larger variety that we have found whose discussion must be deferred to a later occasion) clearly meet these conditions. There is obvious and readily characterized variation among the diagrams. These diagrams have descendants. Sometimes these descendants are exact copies (descent without modification) and sometimes they are significantly modified in a variety of minor and major ways (descent with modification), demonstrating total or partial inheritance of their characters. Finally, different diagrams have different numbers of descendants, and it is usually quite clear from context why the diagrams are replicated exactly or changed selectively. Thus Moore (1972) and Gilbert (1985) replicate E.B. Wilson's diagram of Weismannism exactly (literally, as a photographic copy in the first case, and redrawn in the second) for reasons of historical accuracy. The occurrence of the diagram in Moore is in the context of a reprint collection which includes the relevant extract from Wilson's book. (Interestingly, the text, including the legend of the figure, was reset by the publisher so that all of the reprinted materials would occur in a consistent typeface, but the diagram was not redrawn, indicating possibly different standards for what counts as accurate replication for diagrams and text (a standard weakened still further in Gilbert's textbook, where the diagram too was redrawn). The occurrence of the diagram in Gilbert's (1985) textbook on developmental biology is more complex, for we need to know not only why the figure was redrawn, but also why Gilbert should have chosen Wilson's diagram as representative of Weismannism. This latter question almost certainly has to do with the signal role that Wilson's book had in the education of three generations of geneticists, cytologists, and developmental biologists. Alternatively, this diagram might simply have been the easiest one to find, given the number of exact replications of it, not only in the 2 later editions of Wilson's book, but also in many other books which took over the diagram.

Part II: Analysis of the Diagrams of Weismannism

VI. The ancestry of Weismann's diagram

In a number of essays throughout the 1880s August Weismann developed a theoretical picture of development and heredity which explained both the process by which genetically transmissible determinants cause the differentiation of the soma and the means by which they are passed to the germ-cells (collected in Weismann 1889). His views are much more complex than, and touch on a number of topics and issues which are not included in, the representations one finds of them today (see also Churchill 1968; Mayr 1985). One feature left out of modern discussions is that Weismann advocated a mosaic or 'dissection' view of differentiation, in which the qualitative division or dissection of germinal material (Nageli's idioplasm) led to the presence of one kind of determinant in each mature somatic cell.

To explain regeneration, Weismann supposed that some determinants of distal structures were carried in a dormant state (called 'accessory idioplasm' after Nageli's term) in the not-quite-so-distal cells, ready to be transmitted to newly dividing cells to replace the loss of the original material. Likewise, idioplasm which was to be transmitted to offspring would be passed unaltered through a sequence of somatic cells via cell division until it reached primordial germ cells. Offspring resemble their parents because there is a continuity of germplasm preserved through the process of development from fertilized zygote to mature gametes.

Weismann's doctrine of the continuity of the germ-plasm explained why acquired characteristics could not be inherited. Effects on the cells of the soma are not transmitted to the germplasm lying dormant in the nuclei of somatic cells on their way to differentiating into germ cells. Weismann's mosaic theory led him to distinguish between nuclear material responsible for differentiation of the soma through qualitative separation of determinants, and nuclear material which does not engage in that process and which is not divided. His view that germ-plasm is a material substance passed through a series of somatic cells distinguished his view of the continuity of the germ-*plasm* from a view common in the mid 1880s of the continuity of the germ-*cells,* e.g. Jäger and Nussbaum. The process of differentiation separated the germ-cells of one generation, the egg and sperm which fuse to form a zygote, from the germ-cells which emerge from a process of differentiation from mesodermal and/or endodermal precursors in the offspring.

In his famous book of 1892, *Das Keimplasma,* which appeared in English a few months later, Weismann presented the fullest treatment of his views, including a historical treatment of the field and discussions intended to show the differences between his views and those of contemporaries like Hugo de Vries (1889). A noteworthy feature of Weismann's book is his use of diagrams to illustrate his theories. These diagrams abstracted selected features of his views and presented them in pictorial form.

We contend that subsequent interpretations of Weismann's views depend critically on readings of Weismann's doctrine through diagrams patterned after his diagrams or their descendants, and that changes in the diagrams are indicative of important conceptual change in biological research. Not only do conceptions of Weismannism change, but the *use* of Weismann diagrams to frame problems changes as Weismann's problems are transformed by consideration of Mendelian transmission genetics and the separation of problems of heredity and development. The conceptual changes marked by the diagrams are thus indicative of changes in what we called conceptual maps above, and their analysis can help explain important problem-shifts associated with the changing conceptual picture.

We can find no diagrams in Weismann's works before 1892 which are suggestive of the sort of abstract theoretical treatment which is characteristic of Weismann diagrams, and so we will focus on the 1892 work. Before considering the paradigmatic Weismann diagram, figure 16 on p. 196 (1893 English translation), we will discuss several diagrams preceeding it in the text. A consideration of the sequence of figures 3, 13, 14 and 15 suggests a presentation of abstract, separable features of Weismann's view which are combined in figure 16 to illustrate his central doctrine of the continuity of the germ-plasm. On p. 102, Weismann presents figure 3, 'Diagram of the cell-generations in the fore-limb of a triton' (see Figure 2). This diagram is a cellular descent tree which shows a pattern of differentiation of cells in the limb of a particular species. Differentiation is indicated in the figure by labels indicating the state of histological differentiation (e.g., 'radiale Metacarpus + Digiti'). But more significantly, Weismann illustrates his theory of mosaic development by listing a set of hypothetical determinants supposed to be present in each cell and the successive separation of these determinants in subsequent cell generations.

The association of the state of histological differentiation with the extent of qualitative division of determinants is shown by the progressive separation of histological labels. The cell labelled 'radiale Metacarpus + Digiti' with determinants 8–20 is shown to have divided into two cells labelled (respectively) 'radiale Mittelh', with determinants 9–11 and 'radiale Digiti' with determinants

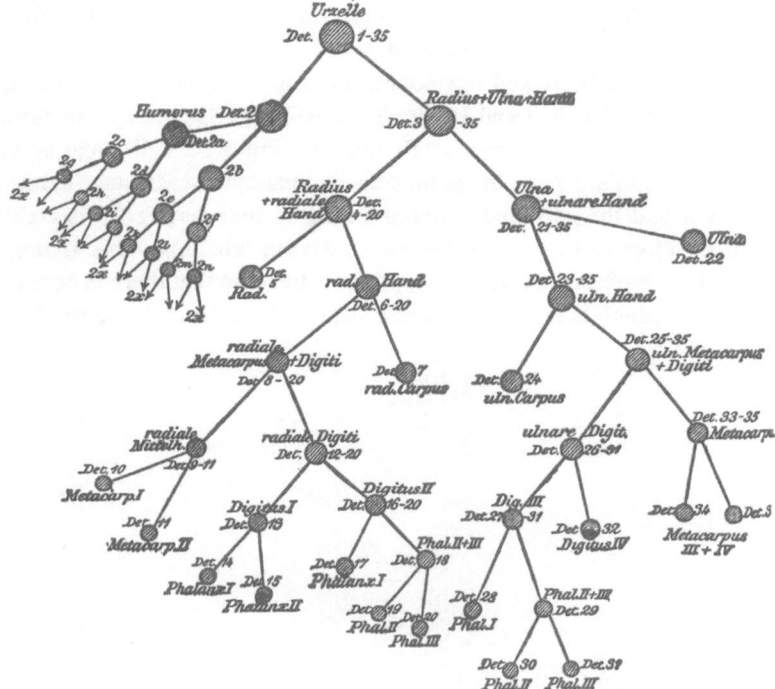

FIG. 3. — DIAGRAM OF THE CELL-GENERATIONS IN THE FORE-LIMB OF A TRITON.

Fig. 2. Diagram of the cell-generations in the fore-limb of a triton (Weismann 1893; Figure 3, p. 102).

12–20. Determinant 8 was not passed on because it is the unique determinant of the parent cell's histological state. Because cells developmentally 'upstream' retain a greater proportion of determinants (in our example the parent cell contained 8–20), regeneration can occur by sending the right determinants into new daughter cells, should the old ones be lost due to amputation, disease or other damage.

Figures 13 through 16 appear in the section 'The germ-tracks' of Weismann's chapter on formation of the germ-cells. Figure 13 is a standard developmental stage diagram for three early stages in the development of a particular species. The diagram is reproduced from a textbook by Lang, who fashioned it after a diagram of O. Hertwig (according to Weismann's figure caption). The time scale of the diagram is short, showing three successive cell divisions in which the number of germ cells goes from 2 to 4 to 8. The diagram clearly shows the differentiation of germ cells from endoderm and their subsequent division and separation in relation to cells of different germ layers. Weismann uses this diagram to illustrate the continuity of the germ-cells by their relative positions in the side-by-side sub-diagrams. More importantly, the widely discussed notion of the continuity of the germ-*cells* is demonstrated with a conventional stage diagram borrowed from

106

another biologist. When Weismann presents his own views (figure 16), he illustrates them with his own novel diagram.

Figure 14 is also a borrowed developmental stage diagram for another species (from a textbook by Korshelt and Heider, patterned after Grobben). The time scale is much longer than in the previous figure, showing a 32 cell stage embryo, a blastula, and a gastrula. Again, the germ cells are marked in each stage subdiagram, and the notion that the germinal material packaged into germ cells is continuous throughout development is shown by the consistent labelling. This figure, when considered along with the previous one, presents the case that there is continuity of the germ-cells through later developmental stages where the cells are too numerous to count.

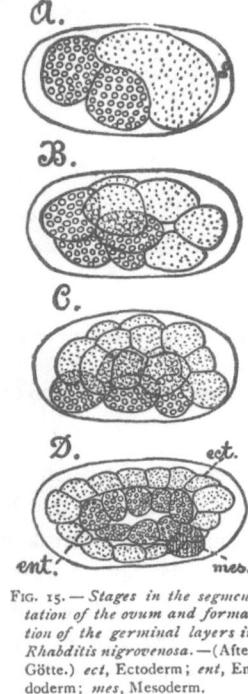

FIG. 15. — *Stages in the segmentation of the ovum and formation of the germinal layers in Rhabditis nigrovenosa.* —(After Götte.) *ect*, Ectoderm; *ent*, Endoderm; *mes*, Mesoderm.

Fig. 3. Stages in the segmentation of the ovum and formation of the germinal layers in *Rhabditis nigrovenosa* (Weismann 1893; Figure 15, p. 195).

Figure 15 is also a developmental stage diagram, but more abstract than the preceeding two figures, showing only the formation of the three germinal layers (see Figure 3). Again, the figure is borrowed (this time from Götte), and depicts embryonic stages from a particular species, *Rhabditis nigrovenosa*. Since germ cells are not shown, unlike the preceeding two figures, Weismann is probably trying to highlight different information than in the others. Unlike the previous two, figure 15 labels all three germinal layers (endoderm, mesoderm and ectoderm). While figure 13 indicated that the germ-cells in the species represented there arose from endoderm, no indication of the origin of germ-cells is given in figure 15. Since the following figure 16 is a diagrammatic re-representation of stages in the same species, we suggest that figure 15 functions to orient the reader to the major developmental features of the important figure 16.

Figure 16 shows '… the genealogical tree of the cells and the germ-track' diagrammatically (Weismann 1893, p. 195). This figure differs from the others in a number of interesting ways (see Figure 4). First, it is more explicitly genealogical, showing lines connecting cells represented by circles. Each generation of cells (through generation 9 when primitive germ-cells appear in the organism) is labelled with a number. Moreover, the differentiation of all three germinal layers shown in the previous figure are shown in their genealogical, as opposed to histological, relationship to one another and to the cells of the germ-track. Additionally, the description of the diagram in the figure caption gives the meanings of symbols in the diagram whereas the captions to the previous figures indicate structures whose diagrammatic representation is taken for granted. Weismann clearly knew he was introducing something novel and of great theoretical significance in figure 16. He takes pain to note what each pictorial feature – open, solid, dotted circles, arabic numbers, thick and thin lines – means and what the limitations of the diagram are, e.g., that cells are only delimited as to germinal type and are only shown up to generation 12.

One of the most important features of this diagram, and the one which is usually lost in subsequent borrowings from Weismann's views via diagrams, is his idea that the relevant link between development and heredity stems from the fact that germ-*plasm* is continuous but germ-*cells* are not. Germ-cells are not continuous in Weismann's diagram because germ-cells are products of differentiation of the soma, like any other cells of the body. He writes,

> So much is certain, and does not depend on any hypothesis. Opinions may differ as to whether the cells situated in the germ-track are to be described as real somatic cells. I have called them so, because in many cases the germ-tracks extend far beyond the period of embryogeny into the fully-developed functional tissues, and because it can be proved that *even cells which are histologically differentiated may produce germ-cells under certain circumstances.*
>
> … it is certain that real somatic cells are situated along the germ-tracks; in all cases the cells of the germ-track *are not germ-cells from the first*, and they always take part in the construction of the body.
> (Weismann 1893, p. 197, emphasis in original.)

Germ-cells fuse to form zygotes which produce the histologically differentiated cells of the germ-track which, in turn, are the carriers of the material substance of heredity, the complete complement of determinants inherited from the parents

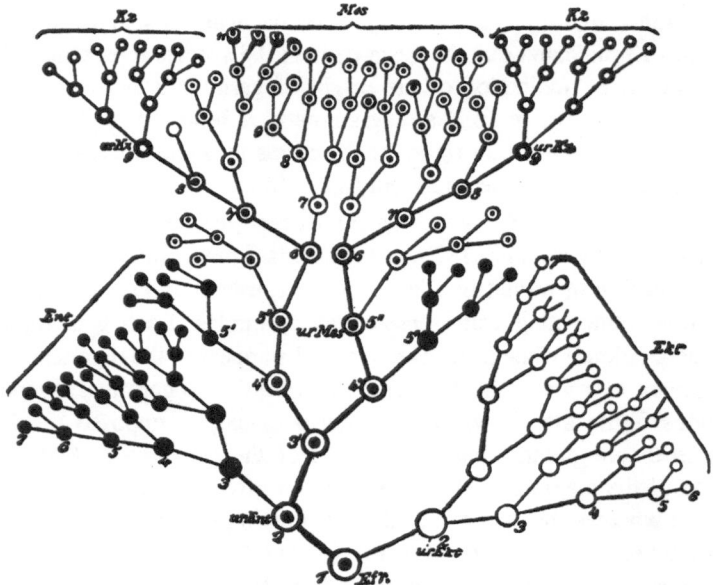

FIG. 16. — *Diagram of the germ-track of Rhabditis nigrovenosa.* — The various
generations of cells are indicated by Arabic numbers, the cells of the germ-track
are connected by thick lines, and the chief kinds of cells are distinguished by various markings: — the cells of the germ-track by black nuclei, those of the meso-
blast (*Mes*) by a dot in each, those of the ectoderm (*Ekt*) are white, those of the
endoderm (*Ent*) black; in the primitive germ-cells (*ur Kz*) the nuclei are white.
The cells are only indicated up to the twelfth generation.

Fig. 4. Diagram of the germ-track of *Rhabditis nigrovenosa* (Weismann 1893; Figure 16,
p. 196).

ready to be passed on to the offspring. Just as the characteristic of being a cellular
descent tree is an important property for identifying Weismann diagrams, the
depiction of the germ-cells differentiating from somatic cells is an important
property which distinguishes Weismann's diagrams from those of most his
successors. In subsequent years the interest in the germ track lost its developmental
focus and emphasized instead its hereditary role – the continuity of germinal
material or, in modern terms, of genetic information. It is likely that this change in
the diagrams occurs because it no longer mattered for this latter interest what the
somatic form of this information was at different stages of development.

It is also significant that Weismann's diagram illustrates the continuity of the
germ-plasm during a single generation within a single organism. The origin of the
zygote in a process of sexual reproduction is probably not depicted because that
would represent transgenerational information not germane to Weismann's attempt
to *explain hereditary continuity as it is mediated by development*. The continuity of
the germ-plasm is represented with respect to a particular species because the

pathway of the germ-track in development is species-specific. We suggest, in short, that the significant features of Weismann's abstract diagram stem from his interests in material problems of the process of development. We will see that the re-representations of Weismann's doctrine with still more abstract diagrams change the conceptual context of discussion of his doctrine by transforming the problems formulated within that changing context. However, before we do that, we will briefly consider a precursor to Weismann's own diagrams in the work of Geddes and Thomson.

VII. Geddes and Thomson: is the first Weismann diagram an ancestor?

In their book, *The Evolution of Sex*, Geddes and Thomson 'take up an altered and unconventional view upon the general questions of biology' (Geddes and Thomson 1889, p. v). They acknowledge their purpose is both to present a novel view to specialists and to popularize a field. Their intention is thus to offer an early textbook in evolutionary biology as a field of natural knowledge. In this context, they take up Weismann's views as presented in his 1889 collection of essays titled *Essays upon heredity and kindred biological problems*. In a section discussing the history of views on the nature of the ultimate sex elements, Geddes and Thomson summarize ideas on the separation in development of body cells and reproductive cells. They trace the idea to Balbiani and consider the views of Owen, Haeckel, Brooks, Jäger, Galton, and Nussbaum as well as Weismann.

In presenting Weismann's view of the continuity of the germ-plasm, Geddes and Thomson give what is probably the first diagrammatic representation of Weismann's doctrine (see Figure 5). The figure first appears on page 94 in the context described above, and is reprinted in a different orientation with the figure caption rearranged but otherwise unchanged on page 261, in a section on organic immortality from the chapter on the physiology of sex and reproduction. The figure has a number of interesting features, some of which appear in Weismann's own diagram three years later and some which do not. It is clear from Weismann's 1893 text that he had read Geddes and Thomson's account of his views because he disputes their interpretation that Jäger and Nussbaum had the same idea independently.

The Geddes and Thomson diagram is, like Weismann's figure 16, a cellular descent tree, although in the former descent is indicated by direct cell contact along a continuous time dimension rather than by connecting lines. Geddes and Thomson's diagram is midway in abstraction between Weismann's developmental stage diagrams (figures 13–15) which clearly depict the histological relations among germinal layers and his genealogical tree showing the germ-track within the abstract topology of cellular descent (figure 16). Among developmental properties, Geddes and Thomson indicate germ cell differentiation; ova are large stippled circles in the diagram and sperm are small ovals with a dot and a 'tail'. After

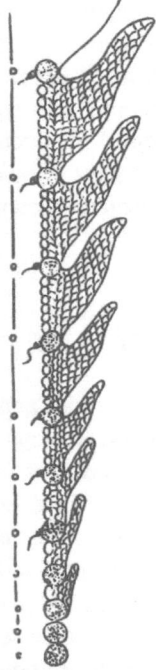

The relation between re-
productive cells and the
body. The continuous
chain of dotted cells at
first represents a suc-
cession of Protozoa;
further on, it represents
the ova from which the
" bodies " (undotted)
are produced. At each
generation, a sperma-
tozoon fertilising the
liberated ovum is also
indicated.

Fig. 5. The relation beween reproductive cells and the body (Geddes and Thomson 1889; p. 94).

fertilization, a sequence of open circles indicates the germ-track cells extended through time and from which a new ovum appears. Somatic cells of the abstractly pictured body branch to the right in the diagram and overlap the next fertilization event temporally, perhaps indicating that the soma is mortal but extends into the next generation a short way.

The differentiation of germ-cells from the soma is a key feature of Weismann's doctrine. The depiction of sexual reproduction (gamete fusion) in a multi-genera-tional diagram represents two important features which do not appear in Weis-mann's diagram. Since the context of Geddes and Thomson's verbal discussion is the larger picture of evolutionary theory, they have apparently used Weismann's earlier conceptual map, also presented verbally, as a problem-context for their problems about the role of sex in evolution.

Geddes and Thomson's diagram also indicates evolution. At the bottom of the diagram, the succession of single-celled protozoa depicts the direct continuity of reproductive cells in which there can be no distinction between soma and germ-line, because the body *is* the germ in a single-celled organism. But the separation of reproductive and somatic cells in *evolution* is shown by the emergence of multicellular organisms with a soma and reproductive tract. Note that evolutionary and ontogenetic temporal scales are indicated in the same diagram. In the latter part of the diagram, the somatic 'bodies' change over time by getting larger and changing shape, presumably indicating morphological change over evolutionary time.

The figure caption is also revealing. None of the figures in Geddes and Thomson's book are numbered. Figures are not referred to in the text, but serve to illustrate points made nearby. The title of the figure in the caption reads, 'The relation between reproductive cells and the body', and is thus ambiguous in picking out developmental, hereditary or evolutionary relationships, since in its hybrid, abstract conventions the diagram picks out all three. In this respect it is reminiscent of some of Haeckel's tree diagrams, though Haeckel had a more direct theoretical reason for conflating ontogenetic and phylogenetic time scales (Robinson 1979, pp. 65–67, discusses plate 81 (reproduced on her p. 66) from Haeckel's (1876) *Die Perigenesis der Plastidule*).

Geddes and Thomson's diagram is clearly a precursor of Weismann's. It illustrates some of the most important features of Weismannism: continuity of the germ-plasm without continuity of germ-cells, and by implication, the non-inheritance of acquired characteristics. The diagram does not illustrate Weismann's mosaic theory of development, though this is described in Geddes and Thomson's text two pages earlier. But Geddes and Thomson's diagram also illustrates processes (sexual reproduction and evolution) not addressed in Weismann's diagram, which leads us to raise the interesting and difficult question of the parentage of Weismann's own diagram: is Geddes and Thomson's diagram an ancestor of Weismann's or not? In their preface, Geddes and Thomson discuss Weismann's book of essays on heredity which appeared the same year, 1889, as their book. But Weismann's views trace back at least to his paper of 1883 ('On heredity'). It seems plausible that Geddes and Thomson had read the earlier paper and were citing the 1889 book rather than the original as a convenience.

Thus, on the one hand, if we accept the claim that Weismann diagrams function as conceptual maps by supposing that Weismann's earlier verbal discussion and later diagram are parts of a single historical sequence, then it is plausible to consider Geddes and Thomson's diagram a descendant of Weismann's conceptual map presented between 1883 and 1889. This leaves the interesting and likely unresolvable question whether Weismann's 1892 diagram is a descendant of Geddes and Thomson's diagram, with simplifications removing the sex and evolution elements. On the other hand, if Weismann developed his diagrams independently of Geddes and Thomson, perhaps both their diagram and Weismann's should be considered descendants of Weismann's original verbal conceptual map. Finally, it is conceivable though not likely that Geddes and Thomson's diagram was developed independently of Weismann's earlier views and that they applied it in discussion of the latter. Thus, it is unresolved whether both diagrams trace directly to Weismann's earlier work or whether Geddes and Thomson's diagram is the descendant of Weismann's textual map and is the ancestor of Weismann's diagram.

VIII. E.B. Wilson's diagram of Weismannism

In the two decades following its first publication in 1896, there was probably no book more influential to the relevant community than E.B. Wilson's *The cell in development and inheritance*. Even after the publication in 1915 of the Morgan school's textbook, *The mechanism of Mendelian heredity*, on what later came to be known as transmission genetics – a book which served to move students of heredity away from developmental concerns – the influence of Wilson's book continued unabated, though its audience gradually changed. Wilson's book went through two later editions, in 1900, and in 1925, and numerous printings in between. Indeed, it is still sold today, now as a classic reprint. Through the period while the fused study of heredity and development was differentiating into the separate disciplines of genetics and embryology, Wilson was read by practitioners of both. After the development of transmission genetics, Wilson was no longer read in genetics courses, but continued to be read in cytology and histology courses, as well as in many embryology courses. (My father used it as a student in courses in embryology and in histology in 1940, and continued to assign it for selected readings and as a reference in his histology course until he died in 1985 – WW.) It served first as the authoritative text, and later as the classic review (incomplete perhaps, but not incorrect) for three generations of biologists in a variety of sub-disciplines.

Wilson's representation of Weismannism (figure 5, on page 13 of the 1st and 2nd editions) occurs in the introduction, imbedded in unchanged text through the first two editions (see Figure 6). In the third edition of 1925, the diagram, labelling, and surrounding text are all changed, in various ways indicating the change in status of Weismann's ideas. This third diagram shows an additional cell generation, and extended lines indicating further descent in the soma, and the label (contra Weismann) marks the diagram as 'the Nussbaum-Weismann theory of heredity' – a label applied also in the textual discussion of the view. In this edition also,

Weismann's views have been demoted from manifesto (see quotes below) to a historically important given, and followed by a further review of the development of cytology along the path from the discovery of mitosis, through the development of Mendelism and beyond. The diagram of the first two editions had many descendants – that of the third edition, none that we have found.

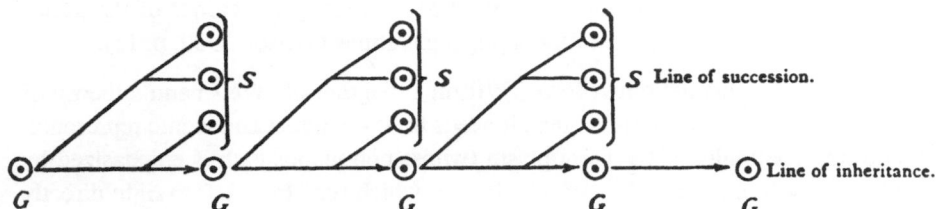

Fig. 5.— Diagram illustrating Weismann's theory of inheritance.

G. The germ-cell, which by division gives rise to the body or soma (S) and to new germ-cells (G) which separate from the soma and repeat the process in each successive generation.

Figure 6. Diagram illustrating Weismann's theory of inheritance (Wilson 1896; Figure 5, p. 13).

From the beginning, Wilson recognizes the controversial nature of Weismann's views, many of which were later rejected (e.g., Weismann's views on the architecture of the genome, and the Roux-Weismann hypothesis of mosaic development). These are neither diagrammed nor talked about in the introduction. The diagram Wilson gives us is an enormously simplified presentation of selective aspects of Weismann's views, and is a paradigmatic conceptual map (in the fullest sense of Griesemer 1983) which explicitly sets the context for the text which follows. As Wilson says in the 1896 and 1900 editions (but not in the 1925 edition), in the text immediately following discussion of the diagram:

... aside from the truth or error of his special theories, it has been Weismann's great service to place the keystone between the work of the evolutionists and that of the cytologists, and thus to bring the cell-theory and the evolution-theory into organic connection. It is from the point of view thus suggested that the present volume has been written (Wilson 1900, pp. 13–14).

Which aspects of Weismann's views did Wilson choose to emphasize, and to illustrate in his diagram? Wilson's target is the theory of the inheritance of acquired characters, as discussed by Lamarck, Darwin, Brooks, and Weismann – the Weismann of the 1883 inaugural lecture 'On heredity' (*Über Vererbung*) – is his oft quoted and paraphrased point man:

If we turn to the facts, we find, Weismann affirms, that not one of the asserted cases of transmission of acquired characters will stand the test of rigid scientific scrutiny. It is a reversal of the true point of view to regard inheritance as taking place from the body of the parent to that of the child. The child inherits from the parent germ-cell, not from the parent-body, and the germ-cell owes its characteristics, not to the body which bears it, but to a pre-existing germ cell of the same kind. Thus the body is, as it were, an off-shoot from the germ-cell (Fig. 5). As far as inheritance is concerned, the body is merely the carrier of the germ-cells, which are held in trust for coming generations (Wilson 1900, p. 13).

In keeping with this delimited focus, Wilson's diagram of 'Weismann's theory of inheritance' is no more complex than it needs to be – a direct and iconic representation of the continuity of the germ plasm (whose continuous path is emphasized by straightening it out into a 'line of inheritance' which runs from left to right directly across the page) and of the 'line of succession' (a visual misnomer, since it is neither a line, nor a causal succession). Wilson in effect visually accuses those who believe in the inheritance of acquired characters of mistaking correlation for causation, for in this diagram neither material particles (such as Darwin's gemmules) nor causal influences (forbidden by the Roux-Weismann hypothesis of mosaic development) pass from somatic cells to affect the content of the germ line.

What is the ancestry of this diagram? It is Wilson's own creation, but it seems likely that Weismann's figure 16 from *The germ-plasm* is a major source, as judged by consideration of the following ten characteristics: 1. It is the only one of two tree diagrams of cellular descent (the other being figure 3 from the same essay) which 2. shows the germ-track. The only other possible ancestor of which we are aware, on which is in fact iconically and conceptually closer, is the 1889 figure of Geddes and Thomson, from pages 94 and 261 of *The evolution of sex*, which Wilson has obviously read and cites elsewhere in his book. Unlike Weismann's diagram, but like Wilson's, 3. their diagram is transgenerational, 4. the germ line is represented as linear 5. with the soma as 'offshoots', and 6. the germ line appears to be represented as a continuous line of germ cells. (This last is ambiguous in the diagram of 1896 and 1900, but clear in the diagram of 1925.) Unlike Wilson's diagram, 7. they include sex (at least after it evolves) and 8. their 'transgenerationality' is deliberately ambiguous between a micro-evolutionary time scale of successive generations (like Wilson's) and a macro-evolutionary time scale on which morphological evolution becomes detectable. 9. This suggests (like Weismann's figure 16, but unlike Wilson's) a significant species-specific phenotypic pattern of complexity. 10. A final complexity is that the Geddes and Thomson diagram shows cellular descent in the germ line, but not (at least not clearly) in the soma – moving it in this respect further from the Wilson diagram than the latter is from Weismann's figure 16.

Surely in this last case, one might reason, all of these writers believed in the cell-theory (that all cells come from cells) and in the monozygotic origin of soma and germ line (the latter is represented in all three figures), so perhaps we should give Geddes and Thomson the benefit of the doubt on this score, and attribute to them cellular descent in the soma. But this charity is misplaced: it is not what these authors believe, after all, which is at issue, but what they choose to represent in their diagrams. *We are tracking icons, not ideas – except later, below, where we shall consider whether ideas may change as a result of cross-context borrowing of the icons used to represent earlier ideas. In the latter case, the inheritance of ideas is through an icon-lineage, so a consideration of what writers believed before and after is legitimate. The charity proposed above would not be legitimate here however, for it would be to treat the icons as epiphenomenal to the ideas of the authors – much as some modern writers have (mistakenly) taken Weismann to be asserting that the soma is epiphenomenal to the germ line.* All things considered, it is thus unclear whether the ancestry of Wilson's diagram is biparental or asexual (and if the latter, which of the earlier diagrams is the parent.) Given the similarities among the possible parents and Wilson's offspring, perhaps the only things that can be asserted are:

1. that there seem to have been no *other* possible parents, and
2. that, as in modern biology, spontaneous generation, no parentage at all, seems unlikely.

IX. Simplification in E.B. Wilson's diagram of Weismannism

Monophyly is an almost universal assumption among taxonomists (occasional exceptions are recognized, but they are clearly marked as such), but it should be far less common for conceptual evolution. (Few people, or diagrams have a single source of inspiration – see Boyd and Richerson (1985) for a discussion of modes of inheritance in cultural evolution.) Nonetheless, we suggest, that Wilson's diagram of the 1896 and 1900 editions provides as close to a monophyletic ancestor of a large fraction of the subsequent diagrams of Weismannism as we are likely to find anywhere in conceptual evolution. It is, at least nearly, a bottleneck in the conceptual phylogeny of such diagrams. We have already seen some of the reasons why this should be so in the pervasive and extended influence of Wilson's text throughout at least the English speaking world, and the focal importance of this diagram and the text which explains it to the architecture of Wilson's text.

Another important pair of reasons is to be found in the centrality of the two themes (the non-inheritance of acquired characters, and the continuity of the germ-plasm), which Wilson chose to extract from Weismann's cluster of ideas and hypotheses, and consequent simplicity of the doctrines (and the diagrams) which were the result. This modular simplification and packaging enormously facilitated the presentation and acceptance of what was to come to be called Weismannism pure and simple. In the next sections we will first outline the simplifications of Wilson's diagram. After the next 30 years, in which various parts of Weismann's doctrine came under discussion and attack, it was, with relatively few exceptions, Wilson's streamlined, simple, and attractive doctrine which was passed to subsequent generations as Weismannism or neo-Darwinism.

The following simplifications of Weismannism are made in Wilson's diagram, and became instrumental in producing other misunderstandings when the diagram and simplified ideas were exported to other contexts:

1. Wilson's diagram would be accurate only for asexual reproduction, since the next generation organism is derived solely from the germ line of a single organism. (Weismann made fundamental contributions to our understanding of meiosis and would never have diagrammed his theory in this way except perhaps for this limited purpose. See the diagram of Weismannism in Geddes and Thomson (1889) which includes sex, in the form of a sperm introduced at each generation.) A better representation is found in Walter (1922), figure 3, where both sexes (and their phenotypes) are represented symmetrically, and no germ line is represented as a straight line (see Figure 7). (In other respects, the simplicity of Walter's figure suggests the influence of Wilson's diagram, even though cells and cellular descent are not represented explicitly.) A diagram by Thomson (1908, figure 35), represents segregation, and Conklin (1915, figure 41), represents the formation of gametes (but without making reduction division explicit) but neither of these are trans-generational, so their effects on the continuity of the germinal material are not represented. *Net effect: the continuity of the germ plasm is exaggerated by ignoring reduction division and sexual recombination.*

2. Wilson's diagram schematizes the phenotype/soma as 3 cells which are identical from generation to generation. *Net effect: the pattern of the phenotype, which is clearly apparent in Weismann's original diagram,* (see also his figures 3 and 13–15) *is ignored.* Weismann has different diagrams for different species, and probably would have recognized differences even for different individuals within a species.) This is even more extreme in Maynard Smith's diagram (1965, 1975, figure 5) which however serves a somewhat different purpose. *Here the stability of phenotypic pattern,* through inheritance, *is ignored,* leading to the mistaken conclusion, expressed by Williams (1966) and Dawkins (1976) that *the phenotype is not inherited and cannot be a unit of selection.*

3. Since the environment is not represented in any of these diagrams (the only exceptions being Maynard Smith's figures 2 and 3, pp. 39 and 40), neither are its effects, thus suggesting not only that the pattern of the phenotype arises geneologically, through cellular descent in ontogeny, but that that is the only cause of the phenotypic pattern. *This suggests in various ways a strong nativist position,* one which (given the Roux-Weismann hypothesis of mosaic development) Weismann

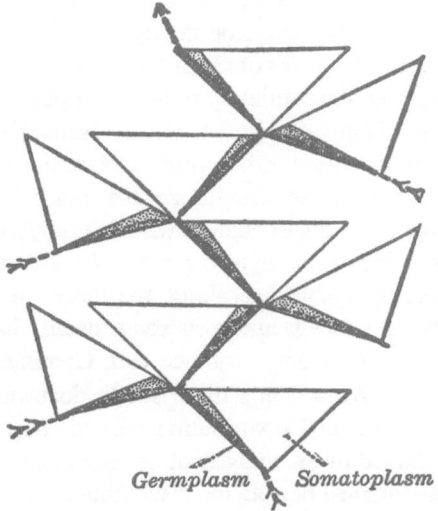

FIG. 3.—Scheme to illustrate *the continuity of the germplasm.* Each triangle represents an individual made up of *germplasm* (dotted) and *somatoplasm* (undotted). The beginning of the life cycle of each individual is represented at the apex of the triangle where germplasm and somatoplasm are both present. As the individual develops each of these component parts increases. In sexual reproduction the germplasms of two individuals unite into a common stream to which the somatoplasm makes no contribution. The continuity of the germplasm is shown by the heavy broken line into which run collateral contributions from successive sexual reproductions.

Fig. 7. Scheme to illustrate the continuity of the germ-plasm (Walter 1922; Figure 3, p. 14).

might have favored, but one which is in any case incorrect. (Interestingly, in Konrad Lorenz's first expressions of the innate-acquired distinction, he drew on this hypothesis, and regarded innate traits as products of mosaic development.)

4. Since this is a diagram of heredity, not of evolution, *the operation of selection* (through differential reproductive rates and differential viabilities of *phenotypes*) *is not represented.* (Compare Maynard Smith's 1972 diagrams of heredity, figure 2, p. 39, and of evolution, figure 3, p. 40.) The former includes the effect of environment on the development of the phenotype, and the latter includes selection, though, revealingly, selection is treated as if it acts directly on genotypes.) *Net effect: selection is treated as if it acts on genes rather than on phenotypes.*

5. The effects of mutation are ignored, further *overemphasizing the stability of the germ plasm*. (Mutation is not depicted in any of the diagrams we have found, and is at best implied by the diagrams of Geddes and Thomson and of Garstang.)

6. The distinction between the continuity of the germ-plasm and the continuity of the germ-cells is ignored, Wilson's diagram misrepresents the fundamental fact of development that germ cells are differentiated somatic cells. By simplifying Weismann's complex diagram to emphasize the non-inheritance of acquired characters, Wilson's diagram fails to capture what is arguably Weismann's central point. Weismann took pains to distinguish germ-cell continuity from germ-plasm continuity because, developmentally speaking, the former would only hold in the special case (which Weismann discusses in some detail) in which the first cell division separates a germ cell from a somatic cell. Continuity of germ cells had already been claimed and Weismann's more subtle doctrine was often confused with it. Weismann's fundamental contribution was to show that hereditary continuity can only be explained by an analysis of development. That is, the continuity of the germ-plasm is maintained despite the discontinuity of the germ-cells through the course of development of the soma. Weismann's figure 16 distinguishes these doctrines by clearly marking the appearance of germ-cells in *Rhabditis nigrovenosa* only in cell-generation 9, while the germ-plasm's continuity is indicated by highlighting the somatic cells of the primoridal germ-track. Without the continuity of the soma, there could be not continuity of germ-plasm because germ-cells are products of development like any other differentiated somatic cells.

These simplification and the conceptual biases they introduced were unproblematic for the functions served by Wilson's diagram in his text. Their effects have not been so benign in recent contexts however. We will now turn to a discussion of some of the most interesting examples of the enormous adaptive radiation of species of Weismann diagram in the years that followed.

X. The subsequent evolution of diagrams of Weismannism

The importance of Wilson's diagram is reflected directly in the diagrams we have found and in the plausible phylogenetic relations we have been able to infer. Of the 41 diagrams we have found and analyzed to date, 7 of them (the various diagrams of Weismann and Geddes and Thomson) antedate Wilson's diagram, so this leaves 33 possible descendants. Of these, 5 are clearly in other 'phyla' so different as to suggest independent invention, 5 trace an alternative lineage (through a diagram by Conklin) back to Weismann's figure 16, and 2 are copies of the Geddes and Thomson diagram. One case (a diagram from Wells, Huxley, and Wells 1930) represents an iconic organismal metaphor for a Weismann diagram, and is very difficult to classify: though it suggests most strongly either the Geddes and Thomson or the Wilson diagram, it could also be classified as an independent invention. Virtually all of the rest are traceable to Wilson's diagram, as direct copies (5 cases), or slight modifications (3 cases), or with larger modifications but with features of the diagrams and sometimes their context that strongly suggest a lineage tracing back to Wilson (13 cases).

Of the 5 diagrams classified in other phyla, all but the first look quite different from 'main line' Weismann diagrams:

1. A diagram from Jordan and Kellogg (1907) looks superficially like a Weismann diagram (indeed, quite like that of Conklin 1915), but is derived from Boveri, and designed to illustrate homologies between the ontogenies of male and female gametes. It looks like a Weismann diagram, and might well be said to include Weismannism as a constraint on its mode of representation (though even this is not uncontestable), but it is clearly not functioning as a diagram which either illustrates or applies Weismannism, and one could draw this diagram without presupposing anything more than the cell theory and sexual dimorphism in the differentiation of gametes. (One might for example (though the authors did not) consistently include this diagram while believing in the inheritance of acquired characters.) This classification reflects Hull's injunction that we consider descent rather than similarity (in cases where the two conflict) in the construction of lineages, since the diagram is remarkably like many later Weismann diagrams, e.g., those of Conklin (1915), or Wilson (1925, figure 135), though its function and cited ancestry are quite different. The classification could be changed if it turned out that Boveri's diagram was traceable back to that of Weismann, Wilson or another early Weismann diagram. (This question is presently unanswerable without a great deal of further work, since through a numbering error, references to this chapter are deleted in the bibliographic appendix. Sometimes there are gaps in the record of conceptual fossils as well!)

2–4. The remaining diagrams are very different in character from other diagrams of Weismanism. Three are from James Mark Baldwin (1902), in which he seeks to illustrate the differences among 'Neo-Darwinism or Weismannism', 'Lamarckism or Neo-Lamarckism', and his own theory of 'Orthoplasy', a culturally mediated form of bio-cultural evolution. In these three, he does not distinguish germ line and soma, and indeed does not use cellular representations at all. Instead, the theories are represented from a quantitative (biometrical or 'quantiative genetic') perspective on the evolution of the phenotype, with phenotypic characters partitioned into genetic and environmental contributions. These diagrams appear to represent a wholly independent invention (with possible ancestry in diagrams of Galton, Pearson, or the biometrical school, though we have not looked), and as far as we can tell, left no replicates or modified descendants.

5. The last, figure 14 from Walter, 1922, (which is a rich stratum, as it includes 4 of the other diagrams) is a high level representation of 'the theoretical results, in the offspring, of parental acquisitions'. In this rich and highly confusing diagram, Weismannism appears (unnamed as such, but described only as 'the non-in-

heritance of acquired characters) as the second of 5 theoretical alternatives which are compared and contrasted in the diagram. In each of the alternatives, germ and soma are represented with 2 pairs of linked inclusive circles in parental and offspring generations with a variety of confusing lines and arrows representing descent and causal influence. Walter's text was not particularly influential, and this diagram, though very interesting and packed with information, was apparently not at all influential, as it left no visible descendants or influences of any kind. In its complexity, it violates a condition of easy exportation, since it is almost impossible to understand. One is tempted to say that it is maladapted even in its original context.

Of the remaining 28, two are exact replicates (with citation) of the original Geddes and Thomson diagram: Thomson (1908, figure 43) copies his own diagram (with a slightly changed legend which makes the interpretation of the diagram slightly more abstract and general), and Herbert (1910) copies it (from Geddes and Thomson). This leaves 26.

One branch is extremely interesting, since it is a lineage which appears to trace back directly to Weismann's figure 16. This is Conklin's (1915) figure 41, which rearranges Weismann's somatic lineages somewhat, and extends the germinal lineages further to show the differentiation of gametes of both sexes in another species – that of a hermaphroditic worm (see Figure 8). This diagram in turn has two further descendants (by explicit citation), in Walter's (1922) adaptations of it to illustrate the differences between Weismann's and De Vries's theories of the determination of somatic characters (figures 79 and 80). Here the general features of what we have been calling a Weismann diagram are further elaborated to illustrate another component of Weismann's theory (also illustrated in Weismann's figure 3), that different determinants are passed to different somatic descendants, and to contrast it with the alternative (and now widely accepted) theory of De Vries, which was the ancestor of our modern theories of differentiation. Both diagrams are cellular descent trees, with factors represented as present in both the nucleus and in the cytoplasm of the cells. In the diagram illustrating De Vries' theory, all nuclear factors are passed on to all cells, and it is the cytoplasmic factors which are divided among somatic cells, causing differentiation. In this, even Weismann's critics are represented within the constraints of a Weismann diagram!

Conklin's diagram, or its ancestor, Weismann's figure 16, also has two other possible descendants. Walter's (1922), figure 58, is similar to Conklin's, but adds the fusion of the gametes to produce the zygote of the next generation. Walter does not credit Conklin for this figure, and without access to the 1913 edition of Walter's work, independent invention, influence from Weismann's figure 16, or even the prior parentage of Walter's diagram cannot be ruled out. Wilson's 1925 edition adds a new diagram (figure 135, p. 311) which resembles Conklin's very strongly, though sperm and egg come from two different organisms, rather than

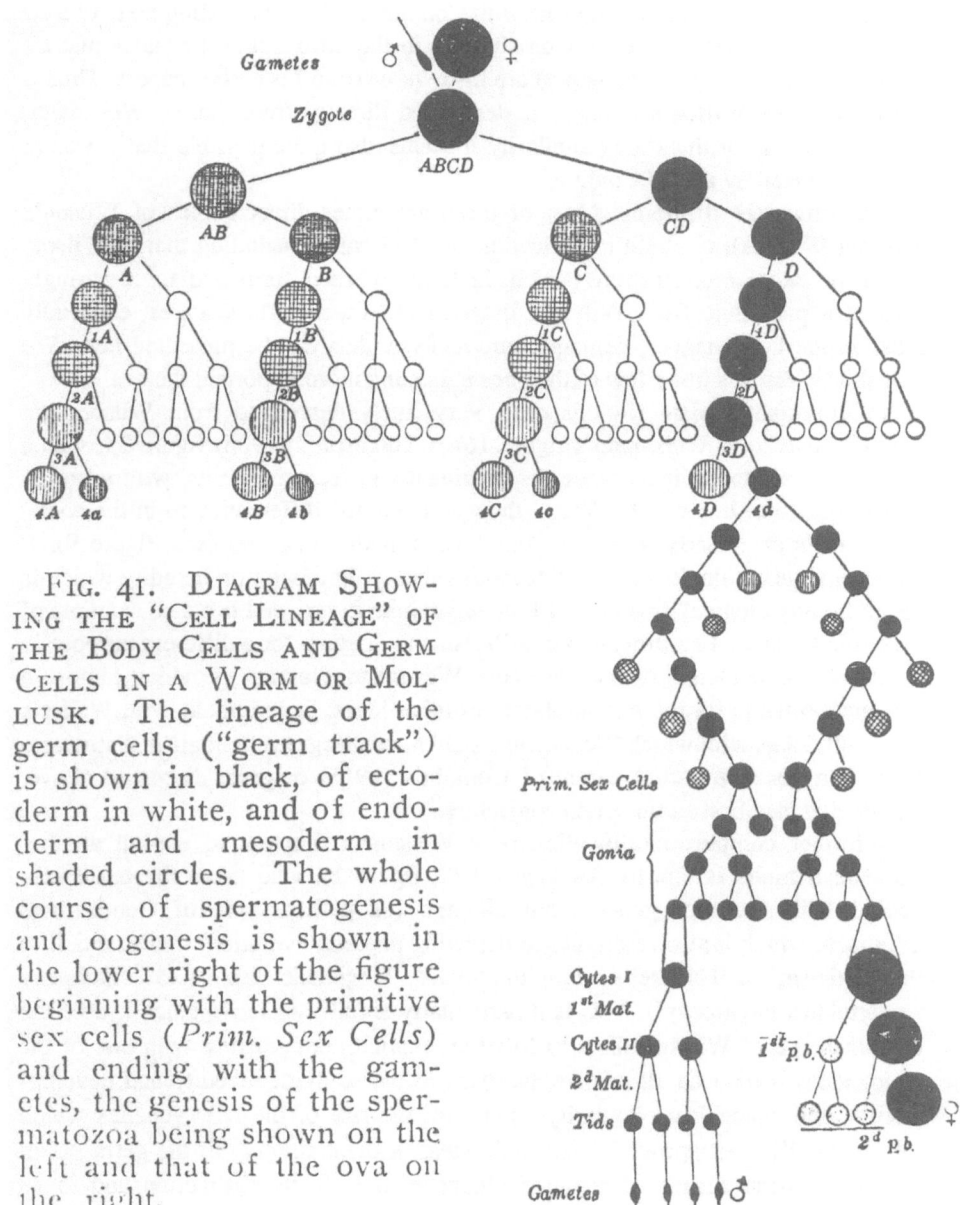

FIG. 41. DIAGRAM SHOWING THE "CELL LINEAGE" OF THE BODY CELLS AND GERM CELLS IN A WORM OR MOLLUSK. The lineage of the germ cells ("germ track") is shown in black, of ectoderm in white, and of endoderm and mesoderm in shaded circles. The whole course of spermatogenesis and oogenesis is shown in the lower right of the figure beginning with the primitive sex cells (*Prim. Sex Cells*) and ending with the gametes, the genesis of the spermatozoa being shown on the left and that of the ova on the right.

Fig. 8. Diagram showing the 'cell-lineage' of the body cells and germ-cells in a worm or mollusk. Reprinted from *Heredity and Environment in the development of men* by Edwin Grant Conklin (2nd ed., 1920; Figure 41, p. 126) with the permission of Princeton University Press, Princeton. Copyright 1915, 1943 (©) renewed by Princeton University Press.

from a single hermaphroditic one, as with Conklin. In the surrounding text, Wilson cites Weismann, 1892, and not Conklin, and in the introduction he states that all figures not credited (this one is not) are his own or from his earlier papers. Thus it is possible that Wilson's diagram is descended directly from that of Weismann, though in virtue of the strong similarity it seems also quite possible that it was at least influenced by that of Conklin.

This leaves 21 diagrams. Most of these are either direct copies of Wilson's diagram (5 cases), or slight modifications of it (3 cases, including that of Wilson, 1925), or larger modifications which, in terms of their form and force strongly suggest a parentage from Wilson's diagram (13 cases). These cases, especially those in the last category, demand more analysis than can be presented here. We will shortly discuss just a few of the ones indicating more important themes.

An important variant (in this case, very likely descended from Wilson, but possibly also from Weismann's figure 16) is Thomson's (1908) figure 35, which represents Mendelian inheritance for dominant and recessive traits, with segregation in the germ line and De Vries' theory of somatic differentiation in the soma, whose cells are clearly represented as having both characters (see Figure 9). It represents the cytological basis of Mendel's theory, in a book on heredity which is full of non-cytological diagrams of Mendelian inheritance, and is a nice example of the exploitation and modification of a Weismann diagram for a different purpose in a different conceptual niche. Here the Weismann diagram provides almost a canonical scheme for the representation of other ideas, including, as with Walter's figure 80, ideas with which Weismann would have disagreed. Its date and structure make it a possible partial parent of Conklin's (1915) diagram discussed above, though that is at best an uncertain conjecture.

A further complex modification of a Weismann diagram to exploit another conceptual niche is due to Garstang (1922, figure 1), who takes the basic idea behind Wilson's trans-generational diagram (or perhaps that of Geddes and Thomson, which it also resembles in depicting phyletic evolution) and adapts it to the criticism of Haeckel's recapitulationist 'biogenetic law' (that 'ontogeny recapitulates phylogeny'). This is a particularly significant move since it uses the central theme of Weismannism to forge an explicit contradiction with one of the major ideas behind an alternative approach to the study of heredity and development – one which lost popularity soon after the rise of the new genetics which fundamentally presupposed Weismann's ideas of the continuity of the germ plasm and the non-inheritance of acquired characteristics. In its abstraction and in its major features, it is closer to Wilson's diagram than to any other. The diagram is arranged to show a zygotic lineage and a phylogenetic lineage, with causal arrows clearly indicating that phylogenetic succession is the causal (and in effect, epiphenomenal) result of a succession of ontogenies (see Figure 10). In it, as with Wilson's diagram, the hereditary lineage bypasses the adult phenotypes, and as a

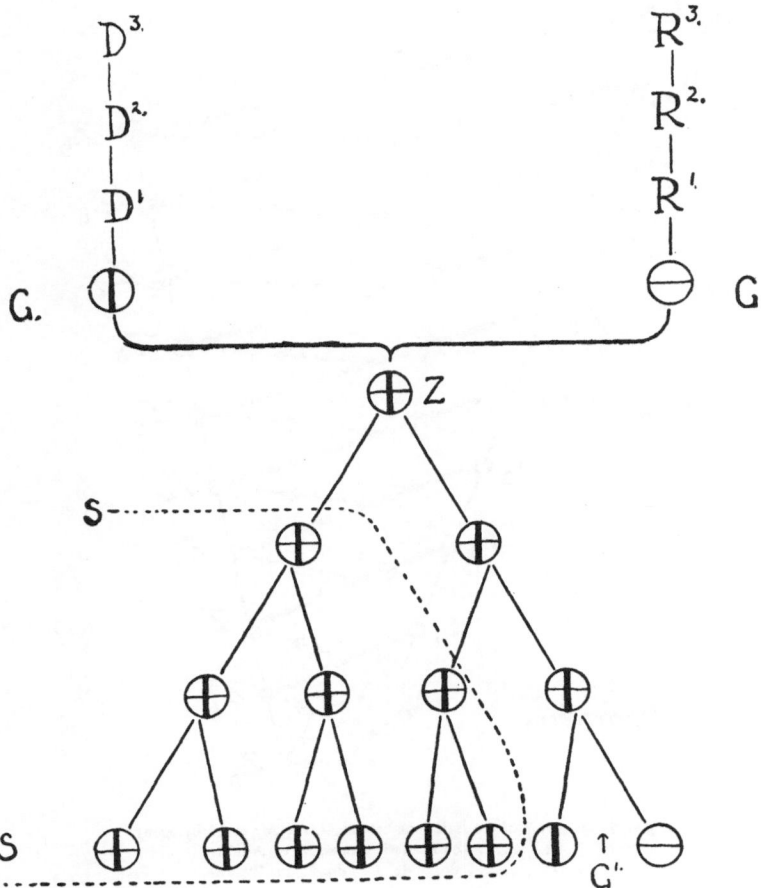

FIG. 35.—Diagram illustrating segregation of germ-cells.

D', dominant parent, its ancestry—D², D³; R', recessive parent, its ancestry—R², R³; G and G, germ-cells; Z, the zygote or fertilised egg-cell; enclosed in the dotted line S . . . . . .S, the somatic cells of the developing body; G' two germ-cells, one with a dominant character and one with a recessive character; dominance is indicated by the strong vertical stroke; recessiveness (latent in the body S . . . . . S) is indicated by the light horizontal stroke.

Fig. 9. Diagram illustrating segregation of germ-cells (Thomson 1908; Figure 35, p. 344).

result, it is readily seen that ontogeny is in no sense the product of phylogeny. This diagram is copied exactly in the second and third editions of De Beer's book, *Embryos and ancestors* (1945 and 1958), but does not appear in the first (1930) edition.

124

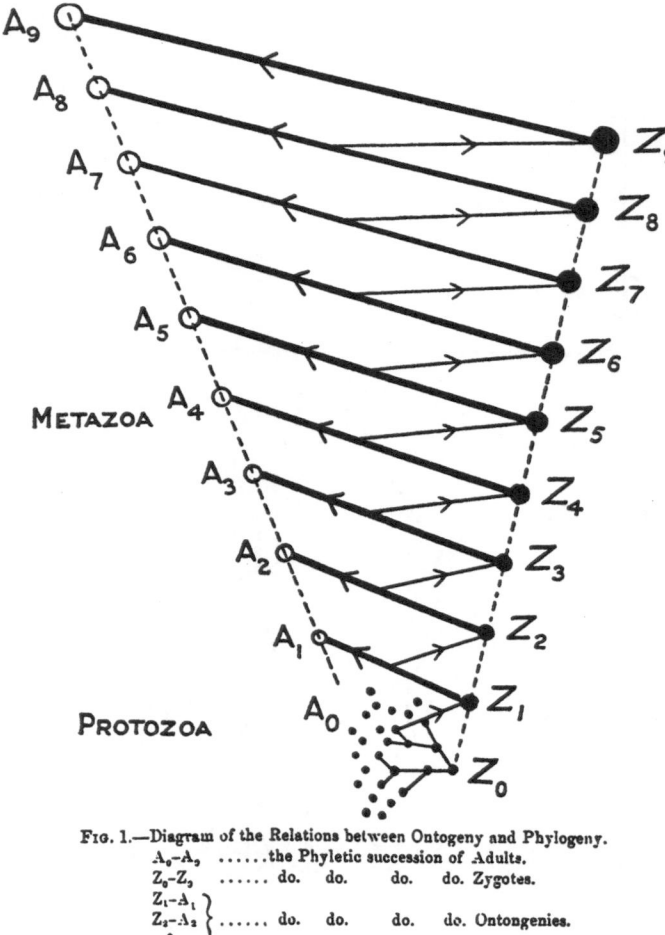

FIG. 1.—Diagram of the Relations between Ontogeny and Phylogeny.
A_0-A_9 ......the Phyletic succession of Adults.
Z_0-Z_9 ...... do. do. do. do. Zygotes.
Z_1-A_1 ⎫
Z_2-A_2 ⎬ ...... do. do. do. do. Ontongenies.
&c. ⎭

Fig. 10. Diagram of the relations between ontogeny and phylogeny. Reprinted from W. Garstang (1922, *Journal of the Linnean Society of London: Zoology* 35: 81–101; Figure 1, p. 83) with the permission of Academic Press Incorporated, Ltd., London.

Also particularly interesting in Garstang's figure is the explicit diagrammatic representation of the somatic character and morphological discontinuity of the germ line, which is represented as branching off from the soma at midstream in each generation of ontogeny to fund the next generation's zygote. As a morphologist, Garstang could be expected to be particularly sensitive to this point – more so than the evolutionists and students of heredity who were more interested in emphasizing the continuity of the hereditary material. As noted above, this is left out of Wilson's diagram and subsequently out of virtually all other diagrams of Weismannism.

For an amusing exception to this generalization which serves to further reinforce the impact of this misrepresentation of Weismann's views, see McLaren (1981). Her diagram of Weismannism (in figure 2) is like that of Wilson. Her figure 3, which presents her alternative theory, combines a *correct* representation of Weismann's view on this score with a denial of the mosaic theory of inheritance of the Roux-Weismann hypothesis. This mutation of Weismann's view as represented in Wilson's diagram has apparently become sufficiently well established that another view which represents Weismann's view correctly (in this respect) is presented as a competitor!

XI. Modern representations of Weismannism

By contrast with most of the diagrams of Weismannism up through about 1930, all of the modern diagrams of Weismannism seem starkly simple – even simpler (for comparable parts) than Wilson's diagram, and all preserving the two fundamental ideas of his representation – the continuity of the germ line and the non-inheritance of acquired characters. Most of them also represent a further idea suggested iconically in his representation but not explicitly in his text or in those of Weismann's contemporaries – that of the discontinuity of the soma. They represent specialized further simplifications of Wilson's representation of the core ideas of Weismannism in a context where cellular descent and developmental concerns are no longer at issue. In this context, we have Maynard Smith's (1965, 1975, figure 8) representation of parallels between Weismannism and the central dogma of molecular biology, which is thereby giving the latter a Whiggish but honorable ancestry (see Figure 11). Just as Weismann's germ line perpetuates itself and also makes the phenotype, so also DNA reproduces itself and also makes proteins. In neither case is there flow of information or causal influence from phenotype to genotype or from protein to DNA. In both of his two parallel diagrams, germinal continuity is emphasized, and the whole of the phenotype has been reduced to an unstructured dead-end. McLaren (1981) and Arthur (1987) give similar diagrammatic representations of Weismann's views, and all three represent iconically the major features and arrangement with further abstractions and simplifications of Wilson's diagram of Weismann's views.

Two other diagrams by Maynard Smith represent a theory of heredity (1972, figure 2), and a theory of evolution (1972, figure 3), and are more complex than his earlier representation because they add roles for the environment in phenotypic expression, and in differential selection (see Figure 12). They nonetheless retain the essential simplifications of Wilson's diagram in that they ignore sex, and show no structure or patterning in the phenotype, which is represented as an unstructured circle branching off from the hereditary line. Maynard Smith's latter diagram adds a crucial further modern abstraction (and misrepresentation!) in the presentation of selection as acting on the genetic material rather than on the phenotype. This is a

126

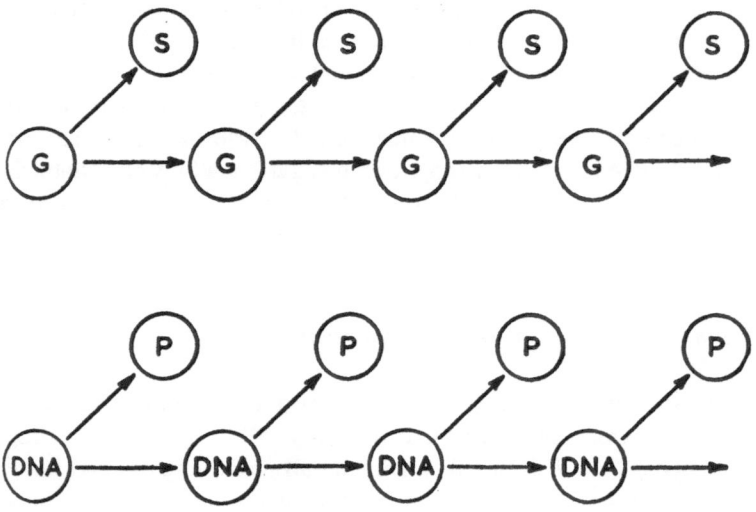

Figure 8. Weismann and the central dogma

Fig. 11. Weismann and the central dogma. Reprinted from *The theory of evolution* by J. Meynard Smith (2nd ed., 1965; Figure 8, p. 67) with the permission of Penguin Books Ltd., London, copyright (©) John Meynard Smith 1958, 1966, 1975.

common, and indeed perhaps the predominant modern heresy, one advocated widely among modern genetic reductionists (who have included George Williams (1966) and Richard Dawkins (1976) as well as Maynard Smith). It has been criticized by many authors – see, e.g., Brandon (1982), reprinted in Brandon and Burian (1984). Williams and Dawkins both pay hommage to Weismann for the origin of their perspective, but neither text includes a diagram of Weismannism, though in a recent paper, Williams (1986, figure 2, p. 182) represents the phenotypes as leaves ('temporary manifestations of the activity of genetic information') on the tree of genetic descent. All of these show an illegitimate extension of Weismann's ideas to support a genetic reductionism in evolutionary theory.

These last simplifications provide strongly seductive suggestions indeed for the view that selection acts on the genes and that the phenotype is a temporary spin-off of genetic activity which does not have the stability or continuity to act as a unit of selection. As noted above, these simplifications, emanating from Wilson's original diagram which was designed for another purpose, serve to overemphasize the continuity of the genotype or genetic material (by ignoring sex, mutation and recombination), and to mask the continuity of the phenotype (by schematizing its structure so severely that its heritable patterns are not represented), and its role in evolution as the object of selection. Indeed, not only does Maynard Smith represent selection as acting on the genes, rather than on the phenotype, but the structure of

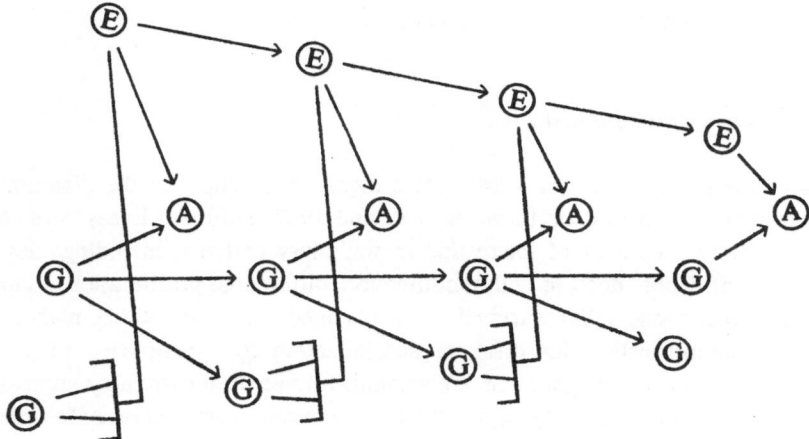

Figure 3. Diagram of a theory of evolution

Fig. 12. Diagram of a theory of evolution. Reprinted from *John Meynard Smith on evolution* by J. Meynard Smith (1972; Figure 3, p. 40) with the permission of Edinburgh University Press, Edinburgh.

his diagram (figure 3) would make it difficult (without substantial increases in complexity which would reduce the desired parallels with his other diagrams) to represent selection as affecting hereditary transmission through action on the phenotype.

In this lineage of diagrams, from Wilson's depiction of the continuity of the germ-plasm and the non-inheritance of acquired characters, we see diagrams which, in a changing conceptual context acquire a new significance as these original points move from foreground issues which must be illustrated and argued for to background assumptions which are taken for granted. A diagram originally representing the continuity of the germ-plasm comes to convey as its primary message the mistaken idea of the discontinuity and mortality of the phenotype, and provide iconic support for the views of Williams and Dawkins. (Even if one accepts that the soma, or 'phenotoken', is discontinuous and mortal, the phenotype – the pattern of the soma – is not thereby shown not to be inherited. Indeed, evolution requires heritability of fitness, which in a stable environment in turn requires heritability of the phenotype. No evolutionist can consistently deny its substantial stability and heritability; see Wimsatt 1981a.) We have here a case in which a diagram which is marvelously adapted to its original niche, becomes the dominant variety, and is taken as the canonical representation of Weismann's views. These views are taken up in a new context by Williams and Dawkins in which the diagram has new and different implications than it did originally –

incorrect implications which lend illegitimate support to a modern reductionistic misinterpretation of the force of Weismann's doctrine.

XII. Summary and conclusions

We have discussed the characteristics of a significant sample of the diagrams of Weismannism found in texts between 1889 and 1987. Different lineages of these diagrams show a number of interesting evolutionary patterns, including: descent without modification, descent with modification, differential proliferation, adaptive radiation, extinction, relict survival in a changed and specialized niche, and successive simplification for efficient specialization to a simplified niche. As conceptual organisms, diagrams of Weismannism have been extremely successful. Our discussion, however, only begins the task of evolutionary analysis. In order to present a full evolutionary account of these diagrams, we would need to:

1. develop a precise, quantitative analysis of the character states of the diagrams (work currently in progress) and,
2. incorporate the character analysis in selective explanations and a 'phylogeny reconstruction' showing the relationships among various lineages of representations of doctrines concerning theories of heredity, development and evolutions.

Our work on diagrams of Weismannism suggests that at the turn of the twentieth century, as problems of heredity and problems of development began to diverge, diagrams of Weismannism began to converge on diagrams depicting the structure of Mendelian genetic principles. The relation between diagrams of Weismannism and Mendelism would not only provide an interesting context in which to explore the divergence of problems of heredity and development, but may also serve as an appropriate focus for the search for a 'common ancestor' of both sets of representations and problems – probably in 19th century tree diagrams of phylogenetic descent. In any case, we believe that our focus on diagrams in this chapter serves to illuminate the complex problem of understanding the separation of problems of heredity and development.

We have argued that diagrams have a number of advantages over other conceptual structures as 'model organisms' for the study of conceptual evolution. These stem in various ways from three crucial properties which allow diagrams to function as heuristics:

1. the clarity of the boundary between diagram and environment, which renders them readily localizeable, simplifies the identification of selection pressures in their textual environments, and renders them easily exportable

to new contexts;

2. the relative ease of scoring their characters, arising from readily understood iconic conventions which serves to generate a multiplicity of easily identifiable features; and

3. a variety of factors which makes the construction of reliable geneologies a relatively straightforward task.

These properties would make diagrams worthy objects of study even if they were epiphenomenal to the processes of conceptual evolution – they could then be useful 'marker genes' for tracking the evolution of 'functional loci'. But the evidence presented here suggests that diagrams are much more than that; the particular array of properties which diagrams share with other heuristics permit diagrams to function as conceptual maps. The visual and conceptual diagram/text boundary also serves to demarcate a problem-context for reasoning about germinal continuity – in short, the diagrammatic features function as a set of boundary conditions for what the reader is meant to take *as* germinal continuity. It is plausible to say that usually a great deal of thought goes into the design, adaptation, and selective borrowing of illustrations. They are usually well adapted to their cognitive tasks. Often, as with the crafting of Weismann's figure 16 and the selective simplifications involved in the synthesis of Wilson's figure 3, they have played a crucial role in aiding and directing the conceptualization of a theory or doctrine. Finally, we have seen that in new contexts, these diagrams were ingeniously adapted to serve new functions, and even that, in the context of modern reductionistic views of evolutionary theory, the 'genetic inertia' of a singularly popular variety of diagram tended to predispose subsequent thought in an unfortunate direction. In this last case, if we are right, a lineage of diagrams has not only accurately reflected the conceptual lineage of an important theory, but may even have influenced its evolution in a certain direction.

Finally, the properties adduced in section 5 which make diagrams good 'model organisms' for the study of conceptual evolution provide more specific kinds of criteria or desiderata which can be used as heuristics in the search for other fruitful objects to study in tracing and documenting the fact of conceptual evolution. Are there other objects of study which meet these conditions? We think so, and on the basis of our analysis we think that there are some new questions which deserve close attention in this search:

1. Just as in biology, we expect there to be a variety of levels of entities which undergo transmission, are differentially selected, and evolve as a result. Some of these are bound to be more readily analyzed than others. Thus, rather than looking at whole theories, or even the smaller models adduced in their support and elaboration, the identification of selection pressures on diagrams was simplified by their local presence in the text – as often explicitly specified problem-contexts. Are there other units of analysis of parts of theories which have this feature? As suggested

above, we think that equations may provide one such example. David Hull (personal conversation) has suggested that explicit definitions may provide another.

2. The issue of how to conceptualize the units of conceptual evolution remains an important and unfinished task. Part of this is accomplished in Boyd and Richerson's masterful delineation (in chapter 1 of their 1985) of modes of cultural transmission. Another part of the task, concerning how to delineate, both theoretically and operationally, conceptual genotypes and phenotypes, is begun in section 5, but this task is far from finished. If theories are not transmitted whole, but successively, during a scientist's conceptual ontogeny, in smaller units (such as the diagrams of Weismannism), is there anything more general to be said about these units, and how they act as generative structures to make the theory-application complex which is the conceptual phenotype?

3. What aspects of a theory are likely to be more readily communicated, and how does this differential proliferation affect its evolutionary transformation? Perhaps most importantly, are there systematic strategies to be followed in the presentation and packaging of a theory which aid in its reliable propagation? Are there optimal strategies to be followed in learning a theory? Most complicated scientific theories are learned through successive recapitulations at increasing levels of sophistication over a number of years. Is there anything to be said about the order in which principles and topics are or should be introduced? Can the parts of a theory be made more modular (either in conception or in presentation) to aid in its propagation?

4. Philosophers tend to think of the evaluation of theories in terms of the traditional scientific norms – predictive and explanatory power, simplicity, fruitfulness, etc. But the obvious importance of ready exportation, a product of locality, context-independence, and ease of understanding, raises additional desiderata not considered in the standard list. How often is a better theory less successful than one of its competitors simply because it fails on one of these latter considerations? Provine (1986) documents how Wright's mathematical theory failed to spread rapidly because few evolutionary biologists had the mathematical ability to understand it, or to understand Wright's self-taught mathematics, often idiosyncratic notation, and sometimes torturous derivations. The substantial influence it had (through Dobzhansky and a few others) was often propagated without transmission or understanding of its mathematical apparatus, and this led many evolutionary biologists to believe incorrectly that the evolutionary synthesis was accomplished without need or benefit of the mathematical theory, or led them to often bowdlerized and incorrect renditions of Wright's views. How often do inadequate popularizations (or other re-representations) of theories lead to their misunderstanding and impede their acceptance, and how often, conversely, might the work of a talented popularizer and advocate push the outcome of a scientific debate in directions which it would not otherwise have followed? How does the audience of a theory or a part of it aid or impede its promulgation? How do theories

win allies (or spawn enemies) in related disciplines? (See Gerson 1987 for an important start on this problem.)

Many of these are new questions, though perhaps more so to philosophers than to historians, but they also point to an important role for psychologists, educational psychologists, social psychologists, and the new sociologists and anthropologists of science in the future forging of a broader evolutionary synthesis of the tree of life, embodiments of mind, the evolution of science, and the elaborations of culture. The resulting tree is likely to make Darwin's tangled bank seem simpler than an arctic lichen. We must hope, when the task is finished, that the familial resemblance will still be detectable.

Request

We would appreciate communication from readers who have discovered additional Weismann diagrams. References and if possible, photocopies of diagrams and the relevant few pages of surrounding text, should be sent to Bill Wimsatt.

Acknowledgements

The order of the authors is alphabetical. This paper represents a natural fusion of prior work by Griesemer on the nature and function of conceptual maps and a growing collection and analysis by Wimsatt of diagrams of Weismannism. We have both been conceptual evolutionists for as long as either of us can remember. The primary authorship of various sections by one or the other of us should not obscure the substantial symbiotic interpenetration of our ideas over the last decade and in the development of this paper. In addition to David Hull, significant other undocumented influences have been provided (not always under the banner of conceptual evolution) by Bill Bechtel, Robert Boyd, Donald Campbell, Fred Churchill, Elihu Gerson, Stephen Gould, Ross Kiester, Richard Lewontin, Richard Levins, Bob McCauley, Bob Olby, Bob Richards, Bob Richardson, Peter Richerson, Frank Rosenblatt, Marty Sereno, Herbert Simon, Leigh Star, and Mike Wade. (Cultural inheritance is nothing if not complicated!) We wish to thank the Department of Philosophy, University of California Davis, for providing travel and word-processing funds, and the History and Philosophy of Science program of the National Science Foundation (Grant No. SES-8709856) for support without which this project could not have been completed. We now represent perhaps two generations in a growing conceptual and social lineage in which David Hull has been from the beginning, a significant progenitor, a friendly and supportive symbiotic interactor, a prolific replicator, and above all, a worthy type specimen.

David, we thank and salute you for two decades of friendship, support, and intellectual stimulation!

Bibliography of Diagrams of Weismannism

Arthur W. (1987). *Theories of life: Darwin, Mendel, and beyond.* Middlesex: Penguin. Figure 9, p. 63: a. Organisms as lines through time. b. The Weismannian view. Figure 19, p. 135: A morphogenetic tree showing the distinction between germ-line and soma. Figure 20, p. 137: A nested morphogenetic tree system of insect development.

Conklin E. (1920). *Heredity and environment*, 2nd ed. Princeton: Princeton University Press. (1st ed. 1915.) Figure 41, p. 126: Diagram showing the 'cell lineage' of the body cells and germ cells in a worm or mollusk.

Conn H. (1906). *The method of evolution.* New York/London: G.P. Putnam. Figure 10, p. 167: Diagram illustrating the principle of heredity.

De Beer G. (1958). *Embryos and ancestors*, 3rd ed. London: Oxford University Press. Figure 1, p. 10 (copy from W. Garstang 1922) (also in 2nd ed., 1945, but not in 1st ed., 1930).

Dendy A. (1928). *Outlines of evolutionary biology.* New York/London: Appleton. (1st ed. 1912.) Figure 83, p. 200: Diagram to illustrate the contrast between Darwin's Theory of pangenesis and Weismann's theory of the continuity of the germ-plasm.

Garstang W. (1922). The theory of recapitulation: a critical re-statement of the biogenetic law. *Journal of the Linnean Society of London: Zoology* 35: 81–101. Figure 1, p. 83: The relations between ontogeny and phylogeny.

Geddes P., Thomson J. (1889). *The evolution of sex.* London: Walter Scott. Unnumbered figure, p. 94: The relation between reproductive cells and the 'body' (vertical). Unnumbered figure, p. 261: The relation between reproductive cells and the 'body' (horizontal).

Gilbert S. (1985). *Principles of embryology.* Sunderland: Sinauer. Figure 1, p. 243 (redrawn from Wilson, 1896).

Goldschmidt R.B. (1929). *Die Lehre von der Vererbung.* Berlin: Springer. Abbildung 18, p. 67: Darstellung der Unsterblichkeit der Keimzellen.

Herbert S. (1910). *The first principles of heredity.* London: A&C Black. Figure 36, p. 61 (copy of Weismann 1892, fig. 16). Figure 37, p. 62 (copy from Geddes and Thomson 1889, p. 94).

Jordan D., Kellogg V. (1907). *Evolution and animal life.* New York: D. Appleton and Co. Figure 151, p. 266: at left, diagram illustrating the development of the spermatazoon; at right, diagram illustrating the development of the egg. (After Boveri.)

Kerr J.G. (1926). *Evolution.* London: MacMillan. Figure 28, p. 109: Diagram to illustrate the continuous strand of gonad associated with an ancestral chain of individuals.

Lock R.H. (1906). *Recent progresss in the study of variation, heredity and evolution.* New York: Dutton. Figure 1, p. 68: Diagram illustrating Weismann's theory of inheritance. (Copied from Wilson, 1896.)

Lull, R.S. (1917). *Organic Evolution.* New York: MacMillan. Figure 17, p. 144: Diagram to illustrate the continuity of the germ-plasm. (Copied from Walter 1913, fig. 3.)

Maynard Smith J. (1958). *The theory of evolution*, 1st ed. Middlesex: Penguin. Figure 5, p. 63: Chains of causation for three kinds of inheritance in which the egg cytoplasm is important: A. Delayed gene action B. Transmission of environmentally induced changes C. Cytoplasmic inheritance. (Not in 2nd or 3rd editions.)

Maynard Smith J. (1965). *The theory of evolution.* 2nd. ed. Middlesex: Pengun. (3rd ed. 1975). Figure 8, p. 67: Weismann and the central dogma. (Not in 1st ed., 1958.)

Maynard Smith J. (1972). *John Maynard Smith on evolution.* Edinburgh: Edinburgh University Press. Figure 2, p. 39: Diagram of the theory of heredity. Figure 3, p. 40: Diagram of a theory of evolution.

McLaren A. (1981). *Germ cells and soma.* New Haven: Yale University Press. Figure 1, p. 2: Two contrasting views of the relation between germ cells and soma. Figure 2, p. 4:

An alternative view of the relations between germ cells and soma.

Moore J. (1972). *Readings in heredity and development*. London: Oxford University Press. Figure 5, p. 79 (reprint from Wilson 1896).

Thomson J. (1908). *Heredity*. London: John Murray. Figure 9, p. 43: Diagram illustrating the idea of germinal continuity. (copied from Wilson, 1896.) Figure 35, p. 344: Diagram illustrating segregation of germ cells. Figure 43, p. 434: The relation between reproductive cells and the 'body'. (From Geddes and Thomson 1889.)

Walter H. (1922). *Genetics*. New York: MacMillan. (1st ed. 1913.) Figure 3, p. 14: Scheme to illustrate the continuity of the germ-plasm. Figure 14, p. 91: The theoretical results in the offspring of parental acquisitions. Figure 58, p. 224: Diagram to show typical maturation and fertilization. Figure 79, p. 256: Differentiation in somatogenesis according to Weismann. (After Conklin.) Figure 80, p. 258: Differentiation in somatogenesis according to De Vries. (After Cronklin.)

Weismann A. (1892). *Das Keimplasma, Eine Theorie der Vererbung*. Jena: Gustav Fischer. English translation (1893) by Parker W., Ronnfeldt H. *The germ-plasm, A theory of heredity*. New York: Charles Scribner's Sons. Figure 3, p. 102: Diagram of the cell-generations in the forelimb of a Triton. Figure 13, p. 193: Three early stages in the development of Sagitta. Figure 14, p. 194: Three early stages in the development of the summer eggs of Moina. Figure 15, p. 195: Stages in the segmentation of the ovum and formation of the germinal layers in Rhabditis nigrovenosa. Figure 16, p. 196: Diagram of the germ-track of Rhabditis nigrovenosa.

Wells H., Huxley J., Wells G. (1929). *The science of life*. New York: The Literary Guild. Figure 164, p. 458: The continuity of the generations.

Williams G. (1986). Comments by George C. Williams on Sober's 'The nature of selection'. *Biology & Philosophy* 1: 114–122. Figure 2, p. 117: ... leaves on the phylogenetic trees ... represent recurring physical effects of the continuity of information.

Wilson E.B. (1896). *The cell in development and inheritance*. London: Macmillan. (2nd ed. 1900.) Figure 5, p. 13: Diagram illustrating Weismann's theory of inheritance.

Wilson E.B. (1925). *The cell in development and heredity*. 3rd ed. New York: MacMillan. Figure 5, p. 13: Diagram illustrating the Nussbaum-Weismann theory of heredity. Figure 135, p. 311: General diagram of the germ-line in animals.

References

Alpers S. (1983). *The art of describing, Dutch art in the seventeenth century*. Chicago: University of Chicago Press.

Arthur W. (1987). *Theories of life: Darwin, Mendel, and beyond*. Middlesex: Penguin.

Baldwin J. (1902). *Development and evolution, including psychophysical evolution, evolution by orthoplasy, and the theory of genetic modes*. New York: Macmillan.

Boyd R., Richerson P. (1985). *Culture and the evolutionary process*. Chicago: University of Chicago Press.

Bradie M. (1986). Assessing evolutionary epistemology. *Biology & Philosophy* 1: 401–459.

Brandon R. (1982). The levels of selection. In Asquith P., Nickles T. (eds) *PSA-1982* Vol. 1, pp. 315–323. East Lansing: Philosophy of Science Association.

Brandon R., Burian R. (eds) (1984). *Genes, organisms, populations: controversies over the units of selection*. Cambridge: MIT Press.

Callon M., Courtial J., Turner, W. Bauin S. (1983). From translations to problematic networks: an introduction to co-word analysis. *Social Science Information* 22: 191–235.

Callon M., Law J., Rip A. (eds) (1986). *Mapping the dynamics of science and technology, Sociology of science in the real world*. London: Macmillan.

Campbell D. (1965). Variation and selective retention in socio-cultural evolution. In Barringer H., Blanksten G., Mack R. (eds) *Social change in developing areas: a reinterpretation of evolutionary theory*; 19–49. Cambridge: Schenkman.

134

Campbell D. (1974). Evolutionary epistemology. In Schilpp P.A. (ed.) *The philosophy of Karl Popper*; 413–463. La Salle: Open Court Press.

Churchill F. (1968). August Weismann and a break from tradition. *Journal of the History of Biology* 1: 91–112.

Conklin E. (1915). *Heredity and environment, in the development of men*. Princeton: Princeton University Press.

Conklin E. (1920). *Heredity and environment* 2nd ed. Princeton: Princeton University Press.

Conn H. (1906). *The method of evolution*. New York/London: G.P. Putnam.

Darwin C. (1871). *The descent of man and selection in relation to sex*. London: John Murray.

Dawkins R. (1976). *The selfish gene*. New York: Oxford University Press.

Dendy A. (1928). *Outlines of evolutionary biology*. New York/London: Appleton. (1st ed. 1912.)

De Beer G. (1930). *Embryology and evolution*. Oxford: Clarendon Press.

De Beer G. (1945). *Embryos and ancestors*, 2nd ed. Oxford: Clarendon Press.

De Beer G. (1958). *Embryos and ancestors* 3rd ed. London: Oxford University Press.

De Vries H. (1889). *Intracellular pangenesis* (C. Gager's translation, 1910). Chicago: Open Court.

Edge D. (1979). Quantitative measures of communication in science: a critical review. *History of Science* 17: 102-134.

Eigen M., Gardiner W., Schuster P., Winkler-Ostwatitsch R. (1981). The origin of genetic information. *Scientific American* 244: 88–118.

Falconer D. (1981). *Introduction to quantitative genetics*, 2nd ed. London: Longman.

Fujimura J. (1987). Constructing 'do-able' problems in cancer research: articulating alignment. *Social Studies of Science* 17: 257–293.

Garstang W. (1922). The theory of recapitulation: a critical re-statement of the biogenetic law. *Journal of the Linnean Society of London: Zoology* 35: 81–101.

Geddes P., Thomson J. (1889). *The evolution of sex*. London: Walter Scott.

Gerson E. (1987). Audiences and allies: the transformation of American zoology, 1880–1930. Paper read at the Summer Conference on history, philosophy and social studies of biology, Blacksburg, Va, June 1987.

Gerson E. (1988). *Computing and methods of social science research* (book manuscript in progress).

Gerson E., Star S. (1986). Analyzing due process in the workplace. *ACM Transactions on Office Information Systems* 4: 257–270.

Gilbert S. (1985). *Principles of embryology*. Sunderland: Sinauer.

Goldschmidt R.B. (1929). *Die Lehre von der Vererbung*. Berlin: Springer.

Goodman N. (1976). *Languages of art, An approach to a theory of symbols*. Indianapolis: Hackett.

Gould S., Lewontin R. (1979). The spandrels of San Marco and the Panglossian paradigm: a critique of the adaptationist programme. *Proceedings of the Royal Society of London* B 205: 581–598.

Gould S., Vrba E. (1982). Exaptation – A missing term in the sicence of form. *Paleobiology* 8: 4–15.

Griesemer J. (1983). Communication and scientific change: an analysis of conceptual maps in the macroevolution controversy. Unpublished Ph.D. dissertation, University of Chicago.

Griesemer J. (1984). Presentations and the status of theories. In Asquith P., Kitcher P. (eds) *PSA 1984* Vol. 1, pp. 102–114. East Lansing: Philosophy of Science Association.

Hacking I. (1982). Language, truth and reason. In Hollis M., Lukes S. (eds) *Rationality and relativism*; 48–66. Cambridge: MIT Press.

Hanson N. (1970). A picture theory of theory meaning. In Radner M., Winokur S. (eds) *Minnesota studies in the philosophy of science, Vol. IV, Analyses of theories and methods of physics and psychology*; 131–141. Minneapolis: University of Minnesota Press.

Herbert S. (1910). *The first principles of heredity*. London: A & C Black.

Hull D.L. (1975). Central subjects and historical narratives. *History and Theory* 14: 253–274.

Hull D.L. (1978). Altruism in science: a sociobiological model of cooperative behavior among scientists. *Animal Behaviour* 26: 685–697.

Hull D.L. (1980). Individuality and selection. *Annual Review of Ecology and Systematics* 11: 311–332.

Hull D.L. (1982). The naked meme. in Plotkin H. (ed.) *Learning, development and culture*; 273–327. London: John Wiley.

Hull D.L. (1983). Exemplars and scientific change. In Asquith P., Nicles T. (eds) *PSA 1982*, Vol. 2, pp. 479–503. East Lansing: Philosophy of Science Association.

Hull D.L. (1984). *Lamarck among the Anglos, Introduction to Lamarck's zoological philosophy*; xi-lxvi Chicago: The University of Chicago Press.

Hull D.L. (1985). Darwinism as an historical entity: a historiographic proposal. In Kohn D. (ed.), *The Darwinian heritage*; 773–812. Princeton: Princeton University Press.

Hull D.L. (1988). A mechanism and its metaphysics: an evolutionary account of the social and conceptual development of science. *Biology & Philosophy*, 3: 123–155.

Jordan D., Kellogg V. (1907). *Evolution and animal life*. New York: D. Appleton.

Kerr J.G. (1926). *Evolution*. London: MacMillan.

Kuhn T. (1962). *The structure of scientific revolutions*. Chicago: University of Chicago Press.

Kuhn T. (1970). *The structure of scientific revolutions*, 2nd ed. Chicago: University of Chicago Press.

Lakatos L. (1970). Falsification and the methodology of scientific research programmes. In Lakatos I., Musgrave A. (eds) *Criticism and the growth of knowledge*; 91–196. Cambridge: Cambridge University Press.

Latour B. (1987). *Science in action*. Cambridge: Harvard University Press.

Laudan L. (1977). *Progress and its problems*. Berkeley: University of California Press.

Lenat D. (1982). The nature of heuristics. *Artificial Intelligence* 19: 189–249.

Levins R. (1968). *Evolution in changing environments*. Princeton: Princeton University Press.

Lewontin R. (1970). The units of selection. *Annual Review of Ecology and Systematics* 1: 1–17.

Lloyd E. (1988). *The structure and confirmation of evolutionary theory*. Greenwood Press.

Lock R. (1906). *Variation, heredity and evolution*. New York: Dutton.

Lull R.S. (1917). *Organic evolution*. New York: MacMillan.

MacRoberts M., MacRoberts B. (1986). Quantitative measures of communication in science: a study of the formal level. *Social Studies of Science* 16: 151–172.

Masterman (1970). The nature of a paradigm. In Lakatos I., Musgrave A. (eds) *Criticism and the growth of knowledge*; 59–89. London: Cambridge University Press.

Maynard Smith J. (1958). *The theory of evolution*, 1st ed. Middlesex: Penguin.

Maynard Smith J. (1965). *The theory of evolution*, 2nd ed. Middlesex: Penguin. (3rd ed. 1975.)

Maynard Smith J. (1972). *John Maynard Smith on evolution*. Edinburgh: Edinburgh University Press.

Mayr E. (1983). Comments on David Hull's paper on exemplars and type specimens. In Asquith P., Nicles T. (eds) *PSA 1982*, Vol. 2, pp. 504–511. East Lansing: Philosophy of Science Association.

Mayr E. (1985). Weismann and evolution. *Journal of the History of Biology* 18: 295–329.

McLaren A. (1981). *Germ cells and soma*. New Haven: Yale University Press.

Mills S., Beatty J. (1979). The propensity interpretation of fitness. *Philosophy of Science* 46: 263–286.

Moore J. (1972). *Readings in heredity and development*. London: Oxford University Press.

Morgan T., Sturtevant A., Muller H., Bridges C. (1915). *The mechanism of Mendelian heredity*. New York: H. Holt.

136

Nickles T. (ed.) (1980). *Scientific discovery: case studies*. Dordrecht: D. Reidel.

Provine W. (1986). *Sewall Wright and evolutionary biology*. Chicago: University of Chicago Press.

Putnam H. (1975). The meaning of 'meaning'. In Gunderson K. (ed.) *Minnesota studies in philosophy of science, vol. 7, Language, mind and knowledge*. Minneapolis: University of Minnesota Press. Reprinted in Putnam H. *Mind, language and reality, philosophical papers*, Vol. 2, pp. 215–271. Cambridge: Cambridge University Press.

Rasmussen N. (1987). A new model of developmental constraints as applied to the *Drosophila* system. *Journal of Theoretical Biology* 127: 271–301.

Richards R. (1977). Discussion: The natural selection model of conceptual evolution. *Philosophy of Science* 44: 494–501.

Richards R. (1981). Natural selection and other models in the historiography of science. In Brewer M., Collins B. (eds) *Scientific inquiry and the social sciences*; 37–76. San Francisco: Jossey-Bass.

Robinson G. (1979). *A prelude to genetics. Theories of a material substance of heredity: Darwin to Weismann*. Lawrence, Kans. Coronado Press.

Rudwick M. (1985). *The great Devonian controversy*. Chicago: University of Chicago Press.

Schank J., Wimsatt W. (1988). Generative entrenchment and evolution. In Fine A., Machamer P. (eds) *PSA 1986*, Vol. 2: 33–60. East Lansing: Philosophy of Science Association.

Shepard R., Metzler J. (1971). Mental rotation of three dimensional objects. *Science* 171: 701–703.

Small H. (1977). A co-citation model of a scientific specialty: a longitudinal study of collagen research. *Social Studies of Science* 7: 139–166.

Small H., Griffith B. (1974). The structure of scientific literatures I: identifying and graphing specialties. *Science Studies* 4: 17–40.

Sullivan D., White D.H., Barboni E.J. (1977). The state of a science: indicators in the specialty of weak interactions. *Social Studies of Science* 7: 167–200.

Thomson J. (1908). *Heredity*. London: John Murray.

Toulmin S. (1972). *Human understanding, the collective use and evolution of concepts*. Princeton: Princeton University Press.

Tufte E. (1983). *The visual display of quantitative information*. Cheshire, Conn.: Graphics Press.

Tversky A., Kahneman D. (1974). Judgment under uncertainty: heuristics and biases. *Science* 185: 1124–1131.

Varnes D. (1974). The logic of geological maps. *U.S. Geological Survey, Professional Paper* 837.

Walter H. (1922). *Genetics*. New York: MacMillan. (1st ed. 1913.)

Weismann A. (1889). *Essays upon heredity and kindred biological problems* Poulton E., Schonland S., Shipley A. (eds). Oxford: Clarendon Press. Reprinted and with an introduction by Mazzeo J. (1977). Oceanside: Dabor Science Publications.

Weismann A. (1892). *Das Keimplasma, Eine Theorie der Vererbung*. Jena: Gustav Fischer. English translation (1893) by Parker W., Ronnfeldt H. *The germ-plasm, A theory of heredity*. New York: Charles Scribner's Sons.

Wells H., Huxley J., Wells G. (1929). *The science of life*. New York: The Literary Guild.

Williams G. (1966). *Adaptation and natural selection*. Princeton: Princeton University Press.

Williams G. (1986). Comments by George C. Williams on Sober's 'The nature of selection'. *Biology & Philosophy* 1: 114–122.

Wilson E.B. (1896). *The cell in development and inheritance*. London: Macmillan. (2nd ed. 1900.)

Wilson E.B. (1925). *The cell in development and heredity*, 3rd ed. New York: MacMillan.

Wimsatt W. (1980). Reductionistic research strategies and their biases in the units of selection controversy. In Nickles T. (ed.) *Scientific discovery: case studies*; 213–259. Dordrecht: D. Reidel.

Wimsatt W. (1981a). Units of selection and the structure of the multi-level genome. In Giere R., Asquith P. (eds) *PSA 1980*, Vol. 2, pp. 122–183. East Lansing: Philosophy of Science Association.

Wimsatt W. (1981b). Robustness, reliability, and overdetermination. In Brewer M., Collins B. (eds) *Scientific inquiry and the social sciences*; 124–163. San Francisco: Jossey-Bass.

Wimsatt W. (1986a). Heuristics and the study of human behavior. In Fiske D., Shweder R. (eds) *Metatheory in social science; pluralisms and subjectivities*; 293–314. Chicago: University of Chicago Press.

Wimsatt W. (1986b). Developmental constraints, generative entrenchment, and the innate-acquired distinction. In Bechtel W. (ed.) *Integrating scientific disciplines*; 185–208. Dordrecht: Martinus Nijhoff.

Wimsatt W. (1986c). Generative entrenchment, scientific change, and the analytic-synthetic distinction. Unpublished manuscript.

Wimsatt W. (1987). False models as means to truer theories. In Nitecki M., Hoffman A. (eds) *Neutral models in biology*; 23–55. New York: Oxford University Press.

Replicators and Interactors in Cultural Evolution

C.M. HEYES and H.C. PLOTKIN

Trinity Hall, University of Cambridge and University College London, U.K.

Several substantial analyses of cultural change have been published in recent years (Boyd and Richerson 1985; Cavalli-Sforza and Feldman 1981; Pulliam and Dunford 1980). At minimum they all share the assumption that there are important parallels between the ways in which cultures change and biological systems evolve. Indeed, both Boyd and Richerson and Cavalli-Sforza and Feldman go so far as to use formal mathematical analyses derived from population genetics as a tool for modelling cultural change. In his essay 'The Naked Meme' (Hull 1982), which serves as a vehicle for presenting many of David Hull's major contributions to biology – species as historical entities, conceptual systems as historical entities, the replicator-interactor-lineage conception, and sociocultural evolution – Hull subjected this assumption of similarity between biological evolution and cultural change to some scrutiny. In the following pages we want to carry on where Hull left off. We too want to examine the assumption that there are parallels between cultural change and biological evolution and, in the process, to probe certain aspects of Hull's own analysis.

The ways in which cultures change, and the conditions under which culture evolved in the first instance, have become matters of considerable interest over the last decade. It is not our intention to present a balanced review of this area. Our purposes can be served by a highly restricted view of the literature. Not just Hull though. We believe Boyd and Richerson's (1985) *Culture and the evolutionary process* to be the most significant work that has been published since Hull's essay appeared. One cannot write on cultural evolution and not take Boyd and Richerson into account. These then, Hull (1982) and Boyd and Richerson (1985), are the essential texts for this chapter.

The assumption that is shared by all the work cited above is, in fact, stronger than that of parallels between the way in which cultures change and biological systems evolve. It is that cultural change occurs through evolutionary processes that are the same as those responsible for biological evolution, and that both may therefore be understood within the same conceptual framework. That is, certain entities – let us follow Dawkins (1976) and call them memes – are present in variant forms; selection acts upon these memes either directly or indirectly via

M. Ruse (editor), What the Philosophy of Biology is. pp. 139–162.
© 1989 *Kluwer Academic Publishers, Dordrecht*

higher-order entities that comprise aggregations of memes or 'vehicles' that contain the memes; and a copying and transmission system exists by which memes are moved about in space and maintained over time. The combination of selection and transmission processes results in differential propagation of memes. The net effect is a change in the frequency of memes in a cultural pool over time. Furthermore, memes are not immutable. For whatever reason they may change, and their changed forms are subject to further selection which results in further changes in the constitution of the cultural pool. The result is descent with modification; in short, cultural evolution. Hence cultural change is a manifestation of cultural evolution.

In the 'The naked meme' Hull accepted this general premise. This allowed him to embark on an examination of the analogies and disanalogies between cultural and biological evolution. Hull was especially concerned with the transmission system by which memes move between individuals. He assumed that some form of social learning was responsible for such transmission. That, then, is our focus as well. We too want to examine the purported analogies that hold between biological and cultural evolution, and especially their respective transmission systems.

Dawkins (1982a) presented the notion of replicator selection as defining any kind of evolution, genetical or cultural, earthly or other-worldly, and Hull (1980) expanded Dawkins' scheme from an emphasis on replicators to a three-term scheme: replicators, interactors and lineages. Thus, Dawkins and Hull, in common with Lewontin (1970) and Campbell (1974), define evolution in terms of processes that are not tied to specific mechanisms or entities, such as genes, genetic transmission or phenotypes. The advantage of such abstract formulations is that evolution operating at any locus or level should conform to them. With their guidance the search for evolutionary loci need not be conducted in a piecemeal fashion, through the drawing of isolated analogies between specific mechanisms. Rather, it might occur through a process of template matching whereby an abstract, mechanism-free formulation of evolution is tested for goodness-of-fit against the mechanisms associated with a particular domain suspected of being an evolutionary locus. In the case of genetical and cultural evolution, for example, genes and memes are analogues because both are replicators in the Dawkins-Hull scheme of evolution. If cultural evolution can be better understood in this way, by pin-pointing analogues through the use of an intermediate conception that is mechanism-free, then the exercise might be usefully extended to other potential loci of evolutionary activity, such as development (Thompson 1986).

In this chapter we will discuss cultural transmission using the replicators-interactors-lineages framework, and primarily from the perspective of experimental psychology. In the context of the Dawkins-Hull scheme, questions about cultural transmission become questions about cultural replication; and to an experimental psychologist, questions about cultural replication are questions about how information is acquired from other people, how it is stored or retained, and how it is expressed in behaviour. We have some doubts about both the way in which cultural evolutionists have interpreted the psychological literature to date, and the extent to which that literature supports the current conception of cultural evolution. We hope to communicate these doubts by discussing meme acquisition, meme retention and meme expression, in turn. Meme acquisition will be discussed at the greatest length since it is the topic to which cultural evolutionists have given most attention.

Before going further a cautionary note must be entered. We are centering our discussion on Hull on the one side and Boyd and Richerson on the other because between them they cover so many of the important issues. However, these authors have arrived at the claim that culture evolves from very different positions and with quite different aims, and, at minimum we can expect their perceptions of the weakest point in the evolutionary analogy to diverge as a consequence of these differences. It is likely that for Hull, a philosopher of science, the notion of cultural evolution is a potential means of saving the claim that science is to some extent referential and progressive, despite the theory-ladeness of observations, paradigm shifts, 'band-wagons' and perpetual squabbling among scientists. In the face of such manifest instability, the danger is that the evolutionary analogy will break down for want of a process of sufficiently faithful replication. While scientific anti-realism in general, or the strong social constructivist program in particular, may be Hull's *bête noir*, Boyd and Richerson are labouring before the spectre of radical sociobiology. Rather than a scientific theory, Boyd and Richerson's typical cultural artifact might be a recipe or a stone arrowhead. They are particularly interested in those products of human social interaction that just might reduce inclusive fitness, and a prominent feature of many such survival-relevant beliefs, practices and structures is their invariance across generations. Preoccupied by what they perceive to be this 'cultural inertia', Boyd and Richerson go to some lengths to demonstrate that the evolutionary analogy need not be flawed by lack of variants due to low mutation rates or blending inheritance.

While it may turn out that there are different kinds of cultural evolution, each with its own entities and dynamics, in this essay we will adopt the simplifying and perhaps simple-minded assumption that all forms of cultural evolution are one; that Hull and Boyd and Richerson are merely probing different parts of the same elephant.

Meme acquisition: social learning and imitation

Both Hull and Boyd and Richerson consider learning of a particular kind to be crucial in cultural evolution. For this reason we will discuss at some length in this section certain issues relevant to the classification of learning into types. Our discussion here will focus on Boyd and Richerson's assumptions about learning, and the uncertainty that has arisen in this field from the conjunction of behaviourist and cognitivist analyses of learning. We will argue that contemporary cognitive psychology undermines Boyd and Richerson's assumptions about the properties of imitation, one of the processes that they take to be responsible for meme acquisition, and that Hull's analysis, while less explicit than Boyd and Richerson's, is more compatible with the psychological literature.

Boyd and Richerson assume that there is a subset of learning distinguishable from other learning, not just on the basis of the role that it plays in cultural evolution, but also in terms of some psychological variables. They usually label this subset 'social learning', but they make it clear that they are not thereby referring to all the learning processes that psychologists are willing to call 'social learning'. Instead, they wish to distinguish learning in general from:

> Social learning in the narrow sense. Animals may acquire behavior directly from conspecifics by imitating their behavior or because conspecifics (often parents) teach naive offspring by reinforcing appropriate behavior (p. 35).

Thus Boyd and Richerson, in common with other authors (e.g. Lumsden and Wilson 1981) isolate imitation and teaching, processes that they assume to effect cultural transmission or replication, from other forms of learning which, under the headings of 'guided variation' and 'direct bias' are assumed to be the interactor processes contributing to cultural selection or differential replication. This distinction is very important in the context of Boyd and Richerson's models of cultural evolution. It is only by assuming that 'social learning [imitation and teaching] and individual learning are *alternative* ways of acquiring a particular behavioral variant' (p. 97) that they can talk about the relative importance of individual and social learning, and, as Boyd and Richerson point out:

> The models in this book assume that culture has the properties of an inheritance system ... culture can have these properties only if individual learning is not too important in determining behavior (p. 98).

Boyd and Richerson claim that imitation and teaching are 'direct' and 'cheap' methods of acquiring information compared with other forms of learning, and that these characteristics qualify them as cultural transmission processes. In suggesting that imitation and teaching are 'direct' Boyd and Richerson are assuming that they involve the acquisition of information without that information being evaluated for its utility or accuracy of reference on the way. The information may be so evaluated *after* it has been acquired (guided variation), and the credibility of a potential model or teacher may be assessed *prior* to acquisition but, according to Boyd and Richerson, the acquisition processes themselves, imitation and teaching, are direct and consequently, under many conditions, constitute cheap methods of adaptation compared with other forms of learning (trial-and-error and rational calculation).

These assumptions about imitation and teaching, that they are distinctive psychological processes which are direct and cheap, are at odds with the contemporary view in cognitive psychology. Towards the end of this section we will outline that view, but the principal question that we want to address is: why do Boyd and Richerson make these assumptions? We suggest two reasons, and consider them worth discussing at some length because they may inform us about the value of using an abstract characterisation of evolution in investigating cultural evolution, and warn of some specific pitfalls of interdisciplinary research in this area. First, we think that Boyd and Richerson may be assuming that cultural transmission is direct because replication in biological evolution happens to involve direct transmission, and they are seeking cultural analogues of genetic mechanisms, rather than an alternative mode of evolution. Second, Boyd and Richerson may take certain forms of social learning to be psychologically distinctive, direct and cheap because the literature on social learning is a dangerous place for the uninitiated. It apparently attests to the existence of a type of cognitive activity which involves direct information transmission between conspecifics, but a degree of insider information is needed to probe the weakness of that claim. Each of these possibilities will be discussed in turn.

Replication and transmission

The 'active germ-line replicators' (Dawkins 1982b) in genetic evolution replicate via two processes; mitosis and meiosis. These processes tend to interest evolutionists most when they are involved in the production of new 'interactors' (Hull 1988a), usually organisms. There is a comforting concreteness about genetic replication, involving as it does a continuity of physical structure through the medium of an especially inert macromolecule, DNA. As a consequence, it seems perfectly legitimate to describe the production of new interactors via mitosis and meiosis as examples of genetic 'transmission'. The *effect* is one of transmission (physical structures, chunks of DNA, are in one place at the beginning and another at the end), and the sequence of events (movement through space of a largely changeless physical entity) seems so likely to achieve this effect that we might be tempted to call it a transmission *process*. A further connotation of the word 'transmission' would be tapped if, as is common, one were to overlook the complex interactions between genes and environment that constitute development. If one thinks of genes as providing a 'blueprint', and of interactors as 'vehicles', then the 'receivers' of genetic transmissions are as passive as radio ariels.

While parts of this description might fit, it misses the point. 'The physical details of replication are irrelevant' (Hull 1982, p. 276), what is important is that a copy of some *information* has been produced through the agency of the original, or a pre-existing copy. Whether this occurs through movement of a physical entity in space (the central connotation of 'transmission' as the name of a process) matters only to the extent that it has a bearing on the fidelity of the copy produced.

144

We cannot help wondering whether Boyd and Richerson have insisted that the social learning involved in cultural replication is 'direct' because the mechanisms involved in genetic replication are direct, and they are seeking precise analogues between biological and cultural mechanisms rather than abstract, formal resemblances of process. In fact, unlike genetic replication, where the nature of the physical structures involved ensures a certain continuity in space and time, with social learning of whatever kind such physical continuity is absent. In its place are a series of transformations from covert mental states and their concommitant neural network states, through their expression in overt behaviour, their passage in environmental media – sound waves, light waves, direct mechanical deformation of the body surfaces of the receiver – to their recoding into internalized states of the receiver. It seems extraordinary that such profoundly *indirect* processes might ever have a direct outcome, i.e. that they might achieve faithful replication of information. If imitation and teaching do, in fact, effect information transmission, it is clear that they do not do so by virtue of being transmission processes; that is, processes that regularly effect information transmission because they were exclusively designed to, or selected for their capacity to, do so.

This idea, that while social learning may occasionally result in information transmission it is not a transmission process, is explored further in the next section. There we hope to track down the origin of the belief that imitation is a process of information transmission, and to make plausible the view that all social learning is not only indirect at the level of physical mechanisms, but also in terms of the extent to which it involves filtering or cross-validation of incoming information. To these ends, we will be taking a look at the history of research on social learning, particularly that culminating in the views sampled by Boyd and Richerson. Research on imitation will be singled out for consideration both because it has been the focus of Boyd and Richerson's attention, and because it is apparently the more direct of the two putative processes that effect meme acquisition. If imitation is indirect, we would argue, then teaching, whether or not it involves language, must be yet more indirect.

The 'onerous concept of imitation'

The words 'transmission' and 'imitation' are similarly ambiguous; both can name an effect, or a process that regularly terminates in that effect. Thus, 'imitation' can name a thing which is a replica of another thing, or a sequence of events likely to, or intended to, produce such a replica. The ambiguity of the term 'imitation' has caused havoc within the literature on social learning.

Imitation was a subject of great interest to Ancient Greek dramatists and rhetoricians, Darwin was intrigued by the phenomenon, and it was discussed intelligently by Baldwin and some Continental philosophers at the end of the last century, but the first disciplined, modern classification of 'imitative phenomena' was produced by Lloyd Morgan in 1900. Following Darwin, who referred to the *principle* of imitation', Lloyd Morgan treated imitation as an effect which might be achieved through various processes. He confined the effect to one of behavioural, rather than morphological, identity across individuals. This was an advance on previous usage which subsumed similarity of markings or colouration; what we would now call 'mimicry'.

Lloyd Morgan distinguished three types of imitation according to the process thought to bring about the observed topographical identity of behaviour. In 'instinctive imitation', evolution by natural selection was thought to result in an animal being structured such that it exhibited a certain behaviour whenever it encountered a conspecific exhibiting that behaviour. In 'intelligent imitation' a behaviour introduced into an individual's repertoire by instinctive imitation is developed or maintained through the action of intelligence, i.e. non-imitative, or what were subsequently called trial-and-error, learning processes. Finally, 'intentional imitation' was the production of a behavioural copy as a result of an observer desiring to achieve a certain goal, and reasoning that if it were to do the same thing as another individual who is in possession of that goal, then it too would gain satisfaction with regard to the goal. That is, imitation of the last sort was thought to be distinctive in being 'directed to a special end more or less clearly perceived as such' (Morgan 1900, p. 193).

It was universally assumed, on the basis of introspective reports, that adult humans exhibit intentional imitation. Informal observations, collated by Romanes (1884) in the service of Darwin, suggested that some non-human species also exhibit this 'highest' form of imitation, but Thorndike (1898) was determined to address the issue in a rigorous manner. In the absence of introspective reports from the cats, dogs and chickens that were his subjects, Thorndike assumed that he could answer the question by a process of elimination. Wielding Occam's razor he proposed that if behavioural copies could be shown to be the result of either instinct or of trial-and-error learning by the observer, then they could not be constitutive of intentional imitation.

Following Thorndike's reasoning, and with the support of Lloyd Morgan's Canon, subsequent generations of experimentalists have contested the existence of intentional imitation in both humans and other animals, assuming that, if it does exist, it is something fundamentally different from trial-and-error learning. In this light, the issue of the existence of intentional imitation assumed some importance during the 1940s and 1950s when attempts were made to unify all behavioural study under some single, grand theory. Miller and Dollard (1941) argued that if it were possible to acquire behaviour simply by observing it being performed (Thorndike's operational definition of intentional imitation), then the Hullian theory within which they worked would not be a comprehensive account of behaviour. Resistant to this conclusion, they conducted a lengthy series of studies on rats and humans which, they claimed, showed that neither are capable of intentional imitation since the 'matched-dependent' behaviour of each could be explained by external contingencies of reinforcement, and hence by Hull's universal laws of reward.

In 1956, Thorpe saw fit to re-work Lloyd Morgan's classification. In fact, he made only minor alterations, but he introduced a new terminology which converted 'imitation' from an effect to a process, a noun to a verb, and thereby entrenched certain distinctions made more tentatively by Lloyd Morgan. Thorpe did this by relabelling the three categories of imitation: 'instinctive imitation' became 'social facilitation'; 'intelligent imitation' became 'local enhancement'; and only 'intentional imitation' retained the title of 'imitation'. It was now said to be 'true' or 'genuine' imitation. The distinctions entrenched by Thorpe were those between instinct, trial-and-error learning and intention as processes influencing behavioural change. He required nothing less than immediate 'copying of a novel or otherwise improbable act or utterance of which there is no instinctive tendency' as evidence for imitation. Thorpe seems to have regarded imitation as the name of a process, involving intention, which is sufficient to produce this effect.

Galef (e.g. 1976, 1988) Boyd and Richerson's principal source of information about social learning, is the contemporary representative of the Morgan-Thorndike-Thorpe line. His was the first influential general discussion of social learning since Thorpe, and there (Galef 1976), as well as in his experimental papers, he cites these authors extensively. Like his predecessors, Galef is primarily interested in animal behaviour rather than human cognition, and he continues to uphold the assumption that imitation and trial-and-error learning are qualitatively different, treating imitation as 'an onerous concept to be employed only when no other explanation of an observed social influence on behavior is possible' (Galef 1988). However, Galef works in the environment of a post-cognitive revolution psychology and, although he has not shifted the focus of his own attention away from overt behaviour, the influence of his surroundings is apparent in subtle departures that he makes from traditional accounts of imitation. The most conspicuous of these departures originates in Galef's sensitivity to the distinction between processes and effects, or, as he portrays it, between 'explanatory' and 'descriptive' terms. While he wishes to reserve the term 'imitation' for the same set of observable, behavioural phenomena that Thorndike and Thorpe called 'intentional' or 'genuine' imitation, he does not, like them, suppose that the phenomena can be explained by that label:

> ... imitation requires that the sight of an act be sufficient instigation to the act. It suggests purposeful, goal-directed copying of the behavior of one animal by another. Demonstration of true imitation would require a far more cognitive approach to the study of animal behavior than has generally been pursued by laboratory investigators (Galef 1988).

In Galef's view, behavioural copying of a kind appropriately described as imitation marks the boundary between social influences on behaviour which are and are not explicable in terms acceptable to a behaviourist. For Lloyd Morgan, Thorndike and Thorpe, to say that imitation is intentional is to give it an explanation. For Galef, to label something as imitation is to hand it over for explanation to another paradigm: cognitivism.

The only general, cognitive theory of imitation currently available is that formulated by Bandura (1977). Essentially, it suggests that the following sequence of events mediates the copying of one person's behaviour by another: the learner selectively attends to the behaviour of a model (notices some things and not others), encodes what is perceived into images or some sort of internal dialogue, rehearses and further organises this information, and then, if motivated to do so by rewards or the expectation of rewards, converts the information into behaviour by co-ordinating existing motor programmes and checking these against encoded information through self observation. As Bandura stresses, under a description like this, imitation occurs by way of cognitive faculties little different from those required for any form of information acquisition, social or otherwise. Of some relevance to the issue of cultural evolution is that each sub-process involves selection of, integration with, and interpretation by, the learner's pre-existing knowledge structures. The latter have been formed by unlearned endowments and the learner's direct interactions with its environment, as well as previous social interactions; and they influence which aspects of the model's behaviour are observed, how they are encoded, whether this information is retained, and whether and how it is translated into action. What it is important to understand is that for Bandura, as for Piaget and cognitive psychologists in general, all learning, including social learning, is an 'active, constructive process that proceeds somewhat independently of environmental constraint' (Rosenthal and Zimmerman 1978, p. 30).

If one takes Bandura's characterisation of imitation seriously, then as a psychological process there is nothing very distinctive, direct or cheap about it. It is possible that Boyd and Richerson may have mistaken a fuzzy distinction among psychological paradigms for a sharp boundary between psychological processes. They seem to have understood Galef to have been saying that imitation is a distinctive psychological process when he was, in fact, pointing out that imitation is distinctive in being a learning phenomenon which is beyond the explanatory power of behaviourism. Within a cognitive framework, imitation is seen to involve trial-and-error processes which, although covert, resemble those that are manifest in behaviour when other forms of learning are in progress. Furthermore, Boyd and Richerson have interpreted Galef's work as providing support for the notion that imitation involves direct transmission of *information* when, in fact, he regards imitation as the direct transmission of *behaviour*. For people like Thorpe and Galef who are preoccupied with behaviour, imitation is considered direct simply because nothing that they recognise as the stuff of a psychological explanation intervenes between the model's behaviour and the learner's behaviour. However, if one is willing to count events inside of individual's heads as part of an explanation, and this defines cognitivism, then social learning emerges as a twisted and tumultuous process in which memes are sifted and ground, and, perhaps, filtered and blended. Finally, a cognitive approach to imitation encourages consideration of the metabolic costs of establishing, maintaining and running the information processing apparatus that supports this process, and thereby casts doubt on the assumption that it is relatively cheap.

The consequences of a cognitive reappraisal of imitation for the structure of Boyd and Richerson's theory might be profound. If imitation is an indirect process that involves the doctoring of information from a model with other information supplied and verified by the individual's own interaction with its inanimate environment, then individual learning and social learning are not 'alternative ways of acquiring a particular behavioral variant' (Boyd and Richerson 1985, p. 97) and social learning cannot be fulfilling the purely replicative or transmissive role assigned to it in Boyd and Richerson's models of cultural evolution. More specifically, a cognitive account of imitation suggests that if inclusive fitness reducing characteristics are generated and sustained through cultural evolution it is not by virtue of the operation of a certain kind of transmission process, but rather as a consequence of the *content* of certain beliefs or information structures. Some beliefs are inherently untestable by the individual (e.g. certain metaphysical claims), others appear in groups where their fellows discourage testing (e.g. beliefs that certain things are dangerous accompanied by the belief that dangerous things should be avoided), and both of these types, along with many others, might be sustained across generations, despite inclusive fitness disadvantage, when their selective environment is composed primarily of social norms. Such norms are essentially beliefs about the social world, and they constitute both cultural variants and a major part of the environment to which culture effects adaptation.

A cognitive account of social learning also threatens Boyd and Richerson's assumption that social learning is as effective as, but cheaper than, individual learning in certain environments; and that the capacity for social learning evolved as a consequence of this economy, rather than in response to uniquely demanding patterns of change in the social environment. Boyd and Richerson emphasise the involuntary nature of social learning, the fact that under some circumstances one can hardly avoid learning from another person's example. However, in view of the cognitive complexity of the process of social learning, one should distrust such introspective evidence if it leads one to assume that just because social learning is subjectively effortless, it must also be cheap in the currencies of reproductive fitness.

Individual learning and social learning

Even if there is not a distinctive, cheap and direct social learning process, social learning may be distinguishable from individual learning, and as cultural replication social learning may still have properties in common with genetic replication. At the very least social and individual learning are distinguishable, not as processes, but in terms of what is learned (learning about the social versus the asocial world), and the involvement of conspecific behaviour in the learning process. However, we would take a minimal definition of social learning further. Social learning, we suggest, is a subset of individual learning in which the learned information embodied by one individual is causally related to the information acquired by another individual. So while the process of social learning may be little different from that of other learning forms, the contents, and hence the consequences are different. What one animal or person learns is in part a function of what some other animal or person has learned.

This seems to be compatible with Hull's view of what it is about the individual-social learning distinction that matters:

> In one instance [individual learning], an organism reacts to regularities in its environment (e.g. water putting out fire). In the other instance [social learning] it reacts to regularities in another organism's behavior which are systematically related to other regularities (e.g. a lioness teaches her cubs how to hunt). It makes little difference whether this difference is marked by the distinction between individual and social learning or between first and second order individual learning, but the distinction is crucial. Without it either all regularities become memes or else memetic regularities become regularities like any other (Hull 1982, p. 301).

Our characterisation of social learning is consistent with Hull's in stressing that it must be learned information that affects the observer's information acquisition. Although little is known about the relevant neurobiology, one can suppose that the structures of the entities embodying learned information exhibit not just 'lawful regularities', but also regularities reflecting 'the results of a long series of past interactions' (Hull 1982). This is the sense in which DNA molecules carry information, and it is what makes genetic and memetic systems historical entities.

It is not clear, though, whether our definition of social learning is compatible with Hull's in another respect; one which has a direct bearing on the fidelity of cultural replication. We specify that, in order to be of interest to the cultural evolutionist, a particular instance of learning must involve the information set of an observer being influenced by the learned information set of a model. We do not require that the influence be such as to result in the two information sets being the same or even similar in their content. However, Hull states that

> In the most important sort of social learning, memes encoded in one entity are transmitted to another entity ... symbolic structure is transmitted from one physical vehicle to another (Hull 1982, pp. 301-302).

What does Hull mean by transmission of symbolic structure? If he is saying that the only sort of social learning that could support cultural evolution is that which leads to the model and the observer having information with the same assertive content, then he is contradicting his strong claim that 'the identity that counts in evolution is identity through descent'. But if he simply means that the information in the observer's head must *come from* the learned information in the model's head, then our characterisations of social learning are not in conflict. Hull clarifies this issue a little when he tells us that

> Some similarity must exist in versions [of memetic systems] from generation to generation, but it is secondary to descent (Hull 1982, p. 292).

Hull seems to be saying that only a moderate degree of fidelity in cultural replication is necessary in order for the evolutionary analogy to fit the cultural domain. This is not too surprising since even Dawkins, who thought up the slogan 'Longevity, fecundity, fidelity' to characterise the qualities of a successful replicator, and who seems to prize fidelity as highly as any contemporary evolutionist, concedes that 'no copying process is infallible. It is no part of the definition of a replicator that its copies must all be perfect' (Dawkins 1982b, p. 85).

When does imperfection become inadequacy? How far can one bend the fidelity requirement and still have replication? At what point does it become more appropriate to talk of a process of production rather than replication? We cannot give answers to these questions, but we do have serious doubts about fidelity, fecundity and longevity in cultural replication. We will convey these by discussing, in addition to the stage of acquisition just considered, psychological literature relevant to the retention and expression of memes.

Meme retention: memory and blending

In the 1930s Bartlett developed a 'naturalistic' approach to the study of human memory. Unlike other psychologists at that time, who were giving their subjects lists of nonsense syllables to remember and plotting recall against time, Bartlett asked people to memorise passages of prose. He recorded not only what they recalled, but also the manner in which they recalled it. Instead of focusing on time as the crucial variable influencing memory, Bartlett also looked at the effects of content (technical, descriptive, fantastic, emotive), manner of presentation (written or spoken), and the conditions of recall (written or spoken, to one person or to a larger audience). In short, Bartlett's studies look like the very kind of research that a cultural evolutionist would prescribe for a better understanding of memetic replication and interactor-selection processes.

One of the procedures that Bartlett used may be of particular interest in the context of the present discussion. He called it the 'method of serial reproduction', and it involved one person reading some prose and then relating it to another person who, in turn, related it to a third person, and so on for ten or fifteen 'generations'. It was a formal version of the children's game 'Telephone' or 'Chinese whispers'. The difference is that, while children are rather hoping that the last one in the line will receive a hopelessly distorted message, Bartlett's subjects were instructed to reproduce the prose as faithfully as possible and there is no reason to suppose that they did not make every effort to do so.

In reporting his findings, Bartlett noted that

> ... the first notion to get rid of is that memory is primarily or literally reduplicative, or reproductive ... some widely held views have to be discarded, and none more completely than that which treats recall as the re-excitement in some way of fixed and changeless 'traces' (Bartlett 1932, pp. vi and 204).

The change brought about by serial reproduction which Bartlett found most striking was that of 'conventionalisation'. With retelling, the content of the prose passages became increasingly bland. Peculiarities of style and opinion were lost, and the whole tended to be recouched in terms familiar to the subject. Often this also involved 'rationalisation'; the assertive content of the passage was changed so that it became intelligible and acceptable to the reader. However, particularly emotive items, such as snakes or ghosts, would not only be retained, but their significance would be magnified. Similarly, some detail that amused or interested the subjects was often slavishly reproduced, even at the expense of the general coherence of the piece.

Radical changes in the course of serial reproduction did not only occur when the material was descriptive or emotive. In one of Bartlett's experiments, subjects with some training in logic were given the following passage:

The modification of species
One objection to the views of those who, like Mr. Gulick, believe isolation itself to be a cause of modification of species deserves attention, namely, the entire absence of change where, if this were a *vera causa*, we should expect to find it. In Ireland we have an excellent test case, for we know that it has been separated from Britain since the end of the glacial epoch, certainly many thousands of years. Yet hardly one of its mammals, reptiles or land molluscs, has undergone the slightest change, even though there is certainly a distinct difference of environment, both inorganic and organic. That changes have not occurred through natural selection is perhaps due to the less severe struggle for existence owing to the smaller number of competing species; but if isolation itself were an

efficient cause, acting continuously and cumulatively, it is incredible that a decided change should not have been produced in thousands of years. That no such has occurred in this and many other cases of isolation seems to prove that it is not itself a cause of modification.

Two particularly interesting changes occurred as this passage was serially reproduced. First, the proposition put forward by Mr. 'Garlick' (as he rapidly became), that isolation is a sufficient cause of modification of species, was reversed in the first two reproductions, and then reversed back until the end of the series. Second, all subtelty was drained from the argument by the third reproduction, when the nature of Irish flora and fauna was taken to make the species question an open and shut case.

If Dawkins is right in thinking that 'it is fundamental to the idea of a replicator that when a mistake or a 'mutation' does occur it is passed on to future copies.' (Dawkins 1982b, p. 85), then Bartlett's research offers some good news and some bad news about cultural replication. The good news is that certain kinds of memes do seem to be faithfully reproduced in memory. The bad news is that these tend to be details, funny or highly imageable items; propositions and mundane facts, things that we take to be important in cultural evolution, particularly science, are readily lost or radically transformed. (See Nisbett and Ross (1980) for contemporary arguments and evidence suggesting that 'vivid' information has a disproportionate impact on belief.)

When Bartlett retired, his approach to the study of memory was largely neglected by Anglo-American psychologists until about ten years ago. Perhaps the most interesting research to come out of the minor renaissance is Loftus' (1979) work on eyewitness testimony. Her findings have direct implications for the judicial system, but they are also of general interest because Loftus provides a more detailed and rigorously experimental analysis of many of the issues that Bartlett raised. Specifically, Loftus has shown that information is transformed while in 'storage', not just during acquisition and retrieval; and that the transformations may be fundamental, involving alteration, not just supplementation or replacement, of information.

Typically, subjects in Loftus' experiments watch a video of a crime (e.g. a bank robbery or a motoring offence), perform some filler task for twenty minutes, and are then asked a series of questions about the incident that they observed on tape. One of these questions, the target, makes casual reference to objects or events that were not present or did not occur in the film. For example, the target question might ask: 'Did another car pass the red Datsun while it was stopped at the stop sign?', when the film showed a yield sign but no stop sign. A few days or weeks later the subjects are asked further questions about the incident, and Loftus and her colleagues examine the answers that subjects give on this second questionnaire as a function of whether their target question contained a falsehood. At this stage, subjects usually base their answers on the misleading information provided by the experimenters. For example, they claim to have seen a stop sign on the video.

Variations on this procedure demonstrate that the provision of misleading information is not necessary to distort information in memory. The activities involved in attempting to remember, assigning labels to events or guessing what was observed, result in similar transformations. For example, if at an initial interrogation subjects report that a bank teller in a hold-up had access to an alarm bell, they will persist in claiming this, whether or not it was true, under subsequent questioning days or weeks later. In short, people tend to remember their own version of events rather well, but this version often bears scant resemblance to either the information that they were offered or the information that they initially received.

The studies in which subjects are fed misleading information by the experimenter show that information can be transformed while 'in store', and thus there is a legitimate stage two in the psychology of cultural replication. Information is not only lost or transformed during acquisition and in the course of retrieval or expression. However, the component of Loftus' work which may be of most interest to a cultural evolutionist concerns the nature of the information transformations in question. Loftus attempts to find out whether information gained through watching a film, and subsequently acquired misleading information, 'coexist', or whether the latter 'alters' the former. This seems to us to be formally equivalent to the question of whether cultural inheritance is particulate or whether it involves blending.

It is all but impossible to demonstrate unequivocally that memory does involve alteration, because it could always be argued that the original information remains in the system, and that the experimenter has just not found the right way to get the subject to use it. However, Loftus has used a number of techniques that would make this argument implausible. First, and most straightforward, she has offered people substantial financial rewards if they can answer questions correctly, i.e. disregard misleading information supplied by the experimenter after the witnessed event had occurred, or produced as a result of the subject's own cognitive processing. This did not significantly reduce the error rates of the subjects. Second, Loftus has examined the power of blatantly false information to prevent people from being misled by more subtle falsehoods. If the two are presented in the same questionnaire it has such power, but if the blatantly false information is presented two days later than the subtle falsehoods, the latter mislead the subject as usual. This is consistent with the notion that misleading suggestions are incorporated into the witness's memory, effectively transforming what the witness had previously stored. Finally, Loftus used a 'second guessing' method, the results of which were similarly consistent with the alteration hypothesis.

Whether these experiments are interpreted as evidence of meme-loss or meme-blending will depend, among other things, on what one judges to be the extent both of individual memes, the units of cultural replication, and of memotypes, the analogues of genotypes. If 'I saw a red car go through a stop sign', and 'Did the red car pause at the yield sign', might reasonably be regarded as single memes, then blending is apparent; but if these statements are held to more like memotypes, then Loftus' experiments suggest only information loss and replacement.

Given that memes are supposed to be replicators, and 'replicators are those entities that pass their structure largely intact through successive replications' (Hull 1988a), it seems that, by definition, we should accept the second interpretation of Loftus' research. However, persistent use of this strategy of reasoning may be unsatisfactory on two counts. First, it might prevent us from ever accepting evidence of memetic blending, an issue of such importance that rather stronger evidence for or against should be required. Second, it is not unlikely that such a strategy would result in the identification of memes with such small fragments of information that the resulting model of cultural evolution might cease to make contact with the phenomena that it was originally intended to explain. If memes turn out to be the kinds of things that edge-detectors detect, then they might not be very useful in explaining the rise and fall of phrenology, or tribal variations in the treatment of bitter manioc.

Fleeing a potential onslaught of questions concerning units of meaning and the reducibility of folk psychology to information theory/neuroscience, we will conclude this section by noting that, at minimum, the relevant psychological literature on memory suggests a want of longevity and fidelity on the part of cultural replicators. Of course, it may be that factors such as exosomatic storage in science act to stabilize cultural evolution. We wish only to point out the necessity for such extrinsic support, the likely insufficiency of the relevant psychological processes to achieve faithful cultural replication; and to note that it is the manner in which texts are used, not their mere existence, which determines the extent to which support is available.

Meme expression: language and thought

A point that Hull has gone to some lengths to make is that cultural evolution is not Lamarckian. He is willing to accept that the right kinds of correlations exist in cultural evolution (cultural offspring are better able to adapt to their cultural environments than their cultural parents by virtue of the adaptations that they acquired from those parents), but he claims that the mechanisms responsible for these correlations are all wrong. Cultural evolution is not *literally* Lamarckian because information acquired through learning is not passed on to the next generation via the genes. This is uncontroversial, but Hull also denies that cultural evolution is *metaphorically* Lamarckian, and his argument is roughly as follows.

In order for cultural evolution to be metaphorically Lamarckian, it must be possible to distinguish the cultural genotype from the cultural phenotype, with changes in the genotype being wrought by changes in the phenotype and then being passed on to the next cultural generation. This is not possible because the nature, or indeed the existence of cultural phenotypes, is in doubt. Conceptual systems have some of the properties of both biological phenotypes and biological genotypes; they are the things that evolve, and they are composed of replicators, the things that permit evolution. In Hull's view memes are unlike genes in that they have an autocatalytic but not a heterocatalytic function. That is, they replicate themselves, but they do not normally generate other, non-memetic bodies. Consequently, Hull believes that memes have 'manifestations' but not 'effects', and he regards only some of these manifestations as having the phenotypic property of being involved in interaction rather than replication:

> For a consistent metaphor .... two sorts of manifestations must be distinguished: those that retain the structure of initiating ideas and those that result from this structure but do not retain it. Examples of the first sort are the written and spoken word. They are as much memes as those transcribed in brains. Not until memes become translated into non-memetic action do we reach the phenotypic level (Hull 1982, p. 310).

In a later essay, Hull embodies the same distinction in the terminology of replicators and interactors:

> Natural languages serve many functions. One of them is communication. Another is description. And part of what are communicated are these descriptions. Communication is the analog to replication, while testing of descriptive statements is the analog to interaction (Hull 1988a, p. 41).

Thus, Hull's construal of the cultural genotype-phenotype distinction, and consequently his denial that cultural evolution is metaphorically Lamarckian, seems to rest on the assumption that the entities that function in cultural replication 'retain the structure of initiating ideas', i.e. the structure of the replicators themselves. (We recognise that genotypes are usually not considered to be replicators, but discussions of Lamarckian inheritance are usually couched in terms of phenotypic changes having their effects on the genotype rather than genes, and we have kept to that convention.)

Is the structure of initiating ideas retained in their communication, or is there, in the expression of ideas (memes) additional transformation and uncertainty added to that already injected by the processes of acquisition and retention?

There are two alternative interpretations of Hull's assumption regarding the retention of structure, both of which are rather dubious. First, Hull might be claiming that all behaviour is isomorphic with the information upon which it is based. If so, then any reference to information is superfluous, and Hull's position is formally equivalent to that of a radical behaviourist. Since this position is unacceptable to most contemporary psychologists, nearly all parties to the cultural evolution debate, and probably also to Hull, we will pass it over. It is much more likely that Hull is claiming that speech and script, but not other behaviour, is structurally similar to the information of which it is a product. If so, then Hull may have, uncharacteristically, given way to the twentieth century philosophical prejudice which assumes that thought is fundamentally propositional.

Fodor (e.g. 1975) is the present philosophical champion of the view that there is a language of thought. His writings on the subject were important in helping cognitivism, which can be virtually defined in terms of its root assumption that thought is a process of quasi-linguistic abstract symbol manipulation, to take over psychology. Now, after 20 years of domination with computer analogies, this conception of psychological functioning is under considerable threat from the direction of neuroscience and artificial intelligence. The 'parallel distributed processing' (PDP) or 'New connectionist' approach (see Rumelhart and McClelland 1986 and Churchland 1986, for overviews) seeks to show that complex mental processes such as concept formation can be accomplished by very simple networks of nodes, not unlike neural structures.

Before considering more tangible evidence that language is not structurally similar to thought, it is worth noting that PDP seems to have a commitment to psychological processes that would favour a blending model of cultural evolution. Rather than the 'billiard-ball' kinds of images that traditional information processing and behaviourist accounts of psychological functioning bring to mind, those of PDP are akin to absorption-diffusion, or compost heap-like, structures. Smolensky makes this point:

> In the symbolic paradigm [cognitivism/information theory], symbols are concatenated. Or maybe they're tree-like structures that embed in one another. In the subsymbolic paradigm [connectionism/PDP] its quite different. The semantically interpretable entities are activation patterns, and they *superimpose* upon each other, the way that wavelike structures always do in physical systems (Smolensky 1986, emphasis in original).

The connectionist approach is of interest because it offers a serious, contemporary challenge to the view that thought is fundamentally symbolic. This challenge might, however, be shrugged off by someone interested in cultural evolution either because the approach is in its infancy, or because treating connectionism as a challenge may be regarded by some as digging too deep into the nature of cultural replicators. It could be that when Hull says, in support of the claim that speech and script are not part of the cultural phenotype, that language is 'descriptive', he means that it is descriptive of cognitive entities and operations. If so, then an approach like PDP which claims to show that quasi-linguistic entities and operations do not *drive* cognition, is no problem as long as cognition can be accurately *described* in those terms.

However, the problem does not end there. For the last decade or more a battle has been raging *within* cognitive psychology between the supporters of single-code and multiple-code versions of the symbolic model of thought (see, for example, Kosslyn 1980, 1981; Pylyshyn 1973, 1984; Simon 1972; Shepard 1975). Those in the latter camp are presenting arguments and evidence suggesting, at minimum, that not all cognition is even describable in terms of propositions.

The supporters of the multiple-code view claim that, in addition to the propositional coding system, cognition involves visual, acoustic and motoric codes. Their research focuses on the visual code, and claims about its exact nature vary widely. The strongest suggest that the visual code is a spatial or analogue form of information representation, while others are more modest in suggesting that it is a type of cognition which shares symbolic and neurophysiological apparatus with the visual system itself.

The evidence presented in support of these claims is of two basic kinds. The first type suggests an equivalence on some dimension of performance between cognitive tasks assumed to involve mental images, and analogous physical operations on the world. For example, in a classic study, Shepard and Metzler (1971) presented subjects with pairs of line drawings of 3-dimensional objects and asked them whether the two were the same or different, that is, whether they could be made congruent by physical rotation. When the objects were the same, subjects' reaction times varied as a linear function of the angular difference in portrayed orientation; a fact not readily explicable in terms of the manipulation of mental propositions. What the subjects seemed to be working with were mental spatial analogues of the objects.

The second type of evidence supporting the operation of a visual, imaginal coding system focuses on its quasi-perceptual rather than on its alleged analogue properties. This evidence comes from studies showing that people sometimes confuse conscious mental images with percepts (Perky 1910; Segal and Fusella 1970), and from interference experiments. The latter, which rely much less on the validity of introspective reports, are based on the assumption that if the brain mechanisms representing and operating on images overlap with those that are perceiving and manipulating real objects, then performance on an imagery-dependent task should be especially susceptible to interference from a concurrent spatial perceptuo-motor task. For example, Brooks (1968) showed that when people are asked to make judgements about sentences and pictures that they have memorised, people take longer to report on pictures than on sentences if they have to indicate their judgements by pointing at a chart, and longer to report on sentences than on pictures if they have to make their judgements verbally.

Experiments of this kind may not be sufficient to support some of the more extreme claims made by 'pictorialists', but they are more than enough to sustain the weaker claims regarding the nature of visual coding (Goldman 1986). They thereby make it unlikely that speech and script, propositions, 'retain the structure of' or 'describe' all cultural replicators.

A philosopher of science might be inclined to sweep this objection aside on the grounds that all significant components of a conceptual system, all cultural replicators at least in the scientific domain, are propositional in nature. This philosopher, however, would have to be a traditionalist; one committed to the independence of discovery and justification. A post-Kuhnian naturalistic philosopher of science would find it difficult to turn his back on all the historical evidence attributing a key role to images, especially visual images, in scientific problem solving (see Wechsler (1978) for a small sample), and insist that the products of this process can be characterised and judged in an entirely different way, according to their propositional content.

Most philosophers interested in cultural evolution are, almost by definition, of the naturalistic variety. They might be relieved if it turns out that speech and script are not fundamentally different from other behaviour; that they are all interpreted products of cultural replicators, i.e. phenotypic expressions of memes. Part of this relief would be due to the fact that, even if one recognises the relative context-independence of language, regarding all behaviour as part of the cultural phenotype increases the chances that one model of cultural evolution will cover both the science and non-science domains.

The point of all this is: if one accepts that a fair proportion of cultural replicators do not have a quasi-linguistic structure, and that these sometimes replicate through language (just as some mental propositions sometimes replicate through the generation of non-linguistic behaviour), then one must acknowledge that cultural replicators have a heterocatalytic function; that memes are not merely manifested, but expressed. If, as Boyd and Richerson suggest, information in peoples heads is the cultural genotype, and behaviour, including linguistic behaviour, is the cultural phenotype, then cultural evolution is still neither literally nor metaphorically Lamarckian, but the metaphor does not break down where Hull thinks that it does.

Having made our main point, it is worth mentioning a couple of further consequences of attributing a heterocatalytic function to memes on the strength of evidence suggesting that thought is not exclusively propositional. First, as Hull points out, genetic heterocatalysis is characterised by massive information loss, and therefore recognition of memetic heterocatalysis leads to an expectation of lower fecundity on the part of cultural replicators. The sensation of having said less than one thought may be common and exasperating enough to make the wastefulness of cultural replication instantly plausible. However, those who are sceptical about introspective evidence might find it useful to consider Goldman's (1986) hypothetical example of the kind of error that may result from the linguistic expression of non-linguistic thought. He suggests that someone might deny the possibility of a chiliagon because they cannot form a visual image of one. Similarly, many insights may have been lost as a result of scientists, or their peers, finding that a 'mental vision' is either difficult to render coherently in words, or that in the course of imperfect translation it becomes a rather conventional idea. The history of science suggests that the double helix, benzene rings, and the polyphase system of electrical distribution, among many others, are exceptions.

Second, conceding that some important cultural replicators might be non-propositional creates problems in anticipating what their combined effects might be, whether or not subsequent segregation is possible. Furthermore, it makes blending, combination without subsequent segregation, intuitively more likely, and its effects on the availability of variants less predictable. As long as one defines cultural replicators in terms of either propositions or their phenotypic effects (and Boyd and Richerson in their discussion of blending appear to do both), one has access to frameworks for predicting the effects of combination. Propositional calculus or some other system of logic is available for propositions in general, and the 'intentional stance' (Dennett 1987) might help with those concerning social behaviour. However, while we have a good grasp of how propositions interact to produce other propositions, we have little to go on if we want to predict the products of interactions or combinations of other sorts of cognitive entities. The difficulty can be expressed analogically: what happens when two pictures are put together? Do they tend to overlap, or is one superimposed on the other? Does one mask the other, or do the features of both contribute to the product? Is each feature modified, or are various features of each picture maintained intact? Perhaps yet more difficult to fathom is how different sorts of cognitive entities (acoustic, visual, motoric, propositional) combine.

What does all this suggest? That blending inheritance of cultural replicators may sometimes result, not in rapid regression towards the mean, but in the generation of novel variants outside of the existing range. Even if blending would reduce variants, we needn't worry about it unduly. In view of the ways in which memes are doctored with idiosyncratic information as they are acquired, retained and expressed, that is, passed from head to head, the problem is not to find variants, but to find *replicable* variants.

Cultural interactors and selection

The general strategy offered at the start of this chapter, which is to attempt to map different kinds of evolution (biological and cultural in this instance) onto some mechanism-free characterization of evolution, has the advantage not only of illuminating the way evolution works at these different points in complex living systems, but in so doing may enrich and correct the abstract characterization. Almost everything in the last two main sections has dealt with replication. 'A little fidelity, not much longevity and a great deal of fecundity' might be the way one would want to sloganize the nature of cultural replicators. In that light one might want either to deny that cultural change occurs through evolution, or to adjust both ones model of cultural evolution *and* the abstract characterization of replicators in evolution. What of interactors?

In biological evolution, interactors are usually held to be phenotypes, the entities upon which selection forces act. What are they in cultural evolution? Boyd and Richerson are quite clear: the cultural analogues of phenotypes are behaviours. Therefore it is on behaviour that selection will act. Hull's position seem to be a little more complicated. He admits of both conceptual phenotypes – 'The conceptual phenotype is the *application* of theorems' (1982, p. 310; italics in the original) – and behavioural phenotypes – '... organisms are analogous to acts' (p. 302). On the basis of our previous arguments, it does seem necessary to extend the notion of cultural interactors to both overt behaviours and the entities upon which internal selection processes operate.

The case for behaviour as an interactor is clear. A model exhibits a certain behaviour and, however complex the intervening cognitive processing, the observer observes both the consequences of that behaviour for the model, and may come to exhibit that same or similar behaviour. If the consequences of the behaviour when it is performed by the observer are not to the observer's advantage, i.e. in the jargon of learning theory they are not reinforced, then the behaviour will be altered or eliminated. Here it is clearly behaviour upon which selection is being exerted and behaviour must therefore be an interactor in cultural evolution.

However, this particular kind of 'reality-testing' of culturally acquired information comes at the end of the sequence of cognitive processes described in previous sections of this paper. Information is cross-validated against, or filtered by, components of a dynamic cognitive environment at every stage in its acquisition, retention and expression, and these conscious and unconscious processes 'select' as surely as do any external world consequences of a behavioural act.

What does this tell us? Are cultural interactors and selection processes necessarily more complex than those of biological evolution? The answer is yes if one equates biological interactors with just organisms. The answer is no if one considers biological interactors to be more complex matters. G.C. Williams (1966) postulated three levels of selection, even though only one entity, the gene, is what is being selected. These three levels are the genetic, the somatic and the ecological. In this scheme, interactors are genes; developmental trajectories or, more simply, developing parts of the organism during ontogeny; as well as whole organisms that act on and within their environments. So if we follow G.C. Williams, then at some point, there is identity between replicators and interactors, but not at others. More recent accounts, like Brandon (1988), posit hierarchies of both replicators and interactors.

Such complexity fits much better the kind of picture that a psychologist conjures up when cultural evolution is spoken of. Perhaps biological and cultural evolution are about equal in complexity, and it is only in our present understanding of their mechanisms that a difference lies between these forms of evolution. When we began writing this chapter, our only published source of Hull on conceptual evolution was his 'Naked meme' essay. Having now seen his most recent writings (Hull 1988a, 1988b), especially his book on *Science as a process*, we feel that this is a conclusion with which he would entirely agree.

References

Bandura A. (1977). *Social learning theory*. Englewood Cliffs, NJ: Prentice-Hall.

Bartlett F.C. (1932). *Remembering*. Cambridge: Cambridge University Press.

Boyd R., Richerson, P.J. (1985). *Culture and the evolutionary process*. Chicago, IL: University of Chicago Press.

Brandon R.N. (1988). The levels of selection: a hierarchy of interactors. In Plotkin H.C. (ed.) *The role of behaviour in evolution*; 51–71. Cambridge, MA: Bradford Books, MIT Press.

Campbell D.T. (1974). Evolutionary epistemology. In Schilpp P.A. (ed.) *The philsophy of Karl Popper*; 413–463. La Salle, IL: Open Court.

Cavalli-Sforza L.L., Feldman M.W. (1981). *Cultural transmission and evolution: a quantitative approach*. Princeton: Princeton University Press.

Churchland P.S. (1986). *Neurophilosophy: toward a unified theory of the mind-brain*. Cambridge, MA: Bradford Books, MIT Press.

Dawkins R. (1976). *The selfish gene*. New York: Oxford University Press.

Dawkins R. (1982a). Replicators and vehicles. In King's College Sociobiology Group's *Current Problems in Sociobiology*; pp. 45–64. Cambridge: Cambridge University Press.

Dawkins R. (1982b). *The extended phenotype*. Oxford: Freeman.

Dennett D.C. (1987). *The intentional stance*. Cambridge, MA: Bradford Books, MIT Press.

Fodor J. (1975). *The language of thought*. New York: Thomas Y. Crowell.

Galef B.G. (1976). Social transmission of acquired behavior: a discussion of tradition and social learning in vertebrates. *Advances in the Study of Behavior* 6: 77–100.

Galef B.G. (1988). Imitation in animals: history, definition, and interpretation of data from the psychological laboratory. In Zentall T., Galef B.G. (eds) *Social learning*; 6–28. Hillsdale, NJ: Earlbaum.

Goldman A.I. (1986). *Epistemology and cognition*. Cambridge, MA: Harvard University Press.

Hull D.L. (1980). Sociobiology: another new synthesis. In Barlow G.W., Silverberg J. (eds) *Sociobiology: beyond nature/nurture?*; 77–96. AAAS Selected Symposium 35. Boulder, CO: Westview Press.

Hull D.L. (1982). The naked meme. In Plotkin H.C. (ed.) *Learning, development and culture: essays in evolutionary epistemology*; 273–327. Chichester: Wiley.

Hull D.L. (1988). Interactors versus vehicles. In Plotkin H.C. (ed.) *The role of behaviour in evolution*; 19–50. Cambridge, MA: Bradford Books, MIT Press.

Hull D.L. (1988). *Science as a process: an evolutionary account of the social and conceptual development of science*. Chicago: Chicago University Press.

Kosslyn S.M. (1980). *Image and mind*. Cambridge, MA: Harvard University Press.

Kosslyn S.M. (1981). The medium and the message in mental imagery: a theory. *Psychological Review* 88: 46–66.

Lewontin R.C. (1970). The units of selection. *Annual Review of Ecology and Systematics* 1: 1–18.

Loftus E.F. (1979). *Eyewitness testimony*. Cambridge, MA: Harvard University Press.

Lumsden C., Wilson E.O. (1981). *Genes, mind, and culture*. Cambridge, MA: Harvard University Press.

Rumelhart D.E., McClelland J.L. (1986). *Parallel distributed processing: explorations in the microstructure of cognition*, 3 volumes. Cambridge, MA: MIT Press.

Miller N.E., Dollard J. (1941). *Social learning and imitation.* New York: McGraw-Hill.

Morgan C.L. (1900). *Animal behaviour.* London: Edward Arnold.

Nisbett R., Ross L. (1980). *Human inference: strategies and shortcomings of social judgment.* Englewood Cliffs, NJ: Prentice-Hall.

Perky C.N. (1910). An experimental study of imagination. *American Journal of Psychology* 21: 422–452.

Pulliam H.R., Dunford C. (1980). *Programmed to learn.* New York: Columbia University Press.

Pylyshyn Z. (1973). What the mind's eye tells the mind's brain: a critique of mental imagery. *Psychological Bulletin* 80: 1–24.

Pylyshyn Z. (1984). *Computation and cognition.* Cambridge, MA: MIT Press.

Romanes G.J. (1884). *Mental evolution in animals.* London: Kegan Paul, French and Co.

Rosenthal T., Zimmerman B. (1978). *Social learning and cognition.* New York: Academic Press.

Segal S.J., Fusella V. (1970). Influence of imaged pictures and sounds on detection of visual and auditory signals. *Journal of Experimental Psychology* 83: 458–464.

Shepard R.N., Metzler J. (1971). Mental rotation of three-dimensional objects. *Science* 171: 701–703.

Shepard R.N. (1975). Form, formation and transformation of internal representations. In Solso R.L. (ed.) *Information processing and cognition: the Loyola symposium.* Hillsdale, NJ: Erlbaum.

Simon H.A. (1972). What is visual imagery? An information processing interpretation. In Gregg L.W. (ed.) *Cognition in learning and memory;* 183–204. New York: Wiley.

Smolensky P. (1986). Cognition: from microstructure to macrostructure. Paper presented at the 12th annual meeting of the Society for Philosophy and Psychology, Johns Hopkins University, Baltimore.

Thompson K.S. (1986). The relationship between development and evolution. *Oxford Surveys in Evolutionary Biology,* 2, 220–233.

Thorndike E.L. (1898). Animal intelligence: an experimental study of the associative processes in animals. *Psychological Review Monograph Supplements* 2: 1–109.

Thorpe W.H. (1956). *Learning and instinct in animals.* London: Methuen.

Waddington C.H. (1969). Paradigm for an evolutionary process. In Waddington C.H. (ed.) *Towards a theoretical biology,* vol. 2; 106–124. Edinburgh: Edinburgh University Press.

Wechsler J. (ed.) (1978). *On aesthetics in science.* Cambridge, MA: MIT Press.

Williams G.C. (1966). *Adaptation and natural selection: a critique of some current evolutionary thought.* Princeton: Princeton University Press.

Wilson E.O. (1975). *Sociobiology: the new synthesis.* Cambridge, MA: Harvard University Press.

Darwin's Theory and Darwin's Argument

M.J.S. HODGE

Division of History and Philosophy of Science, Department of Philosophy, University of Leeds, Leeds LS2 9JT, UK

1. Introduction: Darwin in New Jersey and elsewhere

One admirable characteristic of David Hull's work is that it is both genuinely interdisciplinary and consistently disciplined. His work presents therefore a splendid counterinstance to the usual tendency for discipline – in the sense of boundaries – hopping to be accompanied by discipline – in the sense of standards – dropping. In moving back and forth between history, philosophy, sociology and biology, David has always wanted to get things right by meeting, not dodging, the demands set for any good work on the subject in hand.

For this reason, he has been second to no one in our time in raising the level of interdisciplinary discussion of Darwin and his theory of natural selection. He has moreover contributed decisively to one special disciplinary element in this interdisciplinary achievement. Darwin and his theory have now been brought, in North America at least, within the domain of high, academic, professional philosophy of science, as can be seen by scanning recent *PSA* or *Philosophy of Science* volumes, or indeed by checking the index of such a prominent monograph on general philosophy of science as van Fraassen's *The Scientific Image* (1980). Others who have played major roles in advancing and securing this disciplinary development – Michael Ruse, Mary Williams, John Beatty, Philip Kitcher, Elliott Sober and Elizabeth Lloyd come at once to mind – owe a great deal to an author who had published both *Darwin and His Critics* (1973b) and *The Philosophy of Biological Science* (1974) before Nixon had had his comeuppance from Congress and the people.

My paper here offers to further this philosophical rehabilitation of Darwin's theorising; and it offers to do this in a way that is directly indebted to David's work. In his paper, 'Charles Darwin and Nineteenth-Century Philosophies of Science' (1973a), and in *Darwin and his critics*, David brought out for the first time how complex were the relations between Darwin's argumentation on behalf of natural selection, and diverse ideals of evidence and explanation then articulated by writers on philosophy of science such as Herschel, Whewell and Mill. Shortly afterwards, understanding of this topic was also enhanced by a dissertation by

Vince Kavaloski (1974), then a University of Chicago graduate student who had benefited from David's encouragement and counsel. What Kavaloski did was to show that one methodological ingredient in Darwin's theorising is indispensable to any attempt to grasp how that theorising stands in relation to the ideals of good science upheld by the philosophers of the time. This ingredient is Darwin's view of natural selection as a true cause, a *vera causa*, in the sense first adumbrated in Newton's first rule of philosophising as explicated by Thomas Reid in the eighteenth century.

Since Kavaloski made this contribution to the subject, others, including Michael Ruse (1976), Rachel Laudan (1982, 1987) and myself (1977, 1987), have extended his insight by applying it in more detail to the structuring and strategies of Lyell's *Principles of Geology* (1830–33) and of Darwin's *Origin of Species* (1859) itself. However, even more recently the subject has undergone a further transformation. For a succession of people – notably Paul Thagard (1978), Elizabeth Lloyd (1983), Philip Kitcher (1985a) and Doren Recker (1987) – have sought to relate the structure and strategies of the *Origin*'s argument to philosophical views unheard of in Victorian England; and unheard of there for the very good reason that they mostly trace to doctrines constructed in the last three or four decades, and prominently represented in what was in Darwin's day the College of New Jersey but is now Princeton University. For, in particular, Darwin's arguments on behalf of natural selection have been called in to illuminate, vindicate, and indeed sometimes to deprecate, Harmanian inference to best explanation, Hempelian explanatory unification and the Fraassenating semantics of empirical adequacy.

In these higher order, philosophical endeavors there has been recurrent if not invariable reference to Darwin's commitment to the *vera causa* ideal. So the new, often Princetonian, issues have been thoroughly entangled with a specific matter of historical scholarship. Not surprisingly, therefore, coming from the historical side as I do, I have been prompted to ask the two inescapable, Hullian questions: Are they getting the history right? And, if not, how can they get the philosophy right? By the end of this paper, I will, I trust, have shown two things. First, that often they have not been getting the history right and, second, that unless and until they do, they can not get the philosophy right either. Less belligerently, then, this paper seeks to advance the state of the philosophical arts as practised on Darwin's theorising, and to do so by establishing some minimal historical conditions for not going astray philosophically in such practice. More positively, this paper suggests some general considerations that we all, philosophers and historians alike, can do well to keep in mind whenever a particular scientific theory, such as Darwin's, is invoked in the evaluation of some general philosophical theory of theories.

The need to raise such general considerations is shown by a significant development concerning the philosophical study of the Darwin case. For a long time, there was a standard exercise involving this case. The logical empiricist tradition dominant in mid-century had elaborated what came to be called the 'received view' of scientific theories, the one familiar from the books of Carnap, Nagel, Braithwaite and others and examined so instructively by Suppe in the long introduction he wrote for *The Structure of Scientific Theories* (1974). The standard exercise was, therefore, to ask how well the received view fitted the Darwin case; or rather, perhaps, it should be how well the Darwin case fitted that view. The most sustained and informed answer to this question was given by our editor, Michael Ruse (1975). Michael knew his logical empiricists well, especially Braithwaite, and he knew his Darwin well; and he was indeed writing about Darwin's theory as set out in the *Origin*, not about any subsequent twentieth-century version of the theory. He concluded, did Michael, rather against his initial hopes, one suspects, that Darwin's theorising fits the received view only to a limited extent and that some of the fitting requires a fair bit of fudging. For a judicious critique of this fitting and fudging, one can now read Recker (1987).

However, the business has moved into a new phase since the days when comparing or contrasting Darwin's theorising with the received view was a dominant concern. For Kitcher (1985a) and Lloyd (1983) have broken with that concern. It is not merely that neither of them favors the received view. It is rather that they are claiming a much more positive outcome from their encounter with the *Origin*. Each is claiming that when the theory of theories he or she favors is brought to bear on the *Origin*'s theorising, then both the favored theory of theories and Darwin's theorising come out looking better than before. There is one obvious snag. Although Kitcher and Lloyd are both from the same small place in New Jersey and have been colleagues in San Diego, they disagree over what is the best theory of theories, the best metatheory, if this use of the word is allowable. Lloyd accepts that version of the so-called semantic theory of theories developed by van Fraassen. Kitcher, however, is developing his own theory of theories. It draws on Bromberger (the Why? man) and on Hempel; but not the Hempel of the covering law analysis of explanation so integral to the old received view, rather the Hempel of explanatory unification. To my knowledge, Kitcher has never elaborated his view of theories at any great length. But it is clear, from his books on mathematics, evolution and sociobiology and from related papers (1981, 1982, 1983, 1985a, 1985b) that his metatheoretical view has been designed from the start to fit two historical cases he has studied closely. One is the differential calculus as Leibniz and Newton launched it, the other is Darwin on natural selection. So, with Kitcher, we have what we have not had before, I think; and that is a philosophical theory of theories that has been deliberately constructed, not merely to illuminate, but actually to legitimate Darwin's theorising. For those of us working in the Hullian, hybrid zone of hypenated historico-philosophical Darwin studies, Kitcher's efforts on behalf of his theories must hold a special significance as a historical landmark if not a philosophical triumph.

2. Theories and metatheories in philosophy and in history

The need for a greater sense of the history of all such metatheoretical endeavors will form one conclusion to be drawn from the present paper. For we need a sense of history concerning what is a peculiarly twentieth century quest – the quest for a philosophical theory of theories, a general characterisation of how the structure distinctive of natural scientific theories allows them to function as they do. For, make no mistake, this quest does not go further back than the received view itself. Ask what it was the received view replaced, and one sees that it did not replace any comparable view or views. It was not a new set of answers to an old set of questions, for it addressed a need that had not previously been felt in that form. Even Duhem, who wrote on physical theory, its aim and structure, before the received view was invented, was not concerned with the common problems the received view and the semantic view have been designed to solve in divergent ways. For Duhem's quest (as Larry Laudan has pointed out to me) is to characterize any physical theory by appropriate contrast with metaphysical theorising, but he does not do this by offering a general account of the relationship, in scientific theories, between logical form and empirical content.

So what we have to keep in mind is that, in the eighteenth and nineteenth centuries, all sorts of things in all sorts of scientific domains were called theories, while no one thought to enunciate some general account of the structure and function of scientific theories such as the received and semantic views have offered to supply in our day. But surely all that we now know about people such as Lyell, Darwin, Herschel, Mill and Whewell, not to mention Comte and Mach, shows that they had some quite general views about theories as such. Indeed, they did. But it is just a mistake to presume that because those views were general they must have constituted, as the received and semantic views do, a general answer to the question, what is a scientific theory, in the relevant structural and functional senses.

This warning against this presumption is appropriate here, because we have to appreciate that in approaching Darwin's theorising through its place in the *vera causa* tradition, we are not placing it in relation to anything truly equivalent to the structural and functional views of scientific theory familiar in this century. This lack of equivalence will then prove to be one reason why fitting a twentieth-century style theory of theories to the *Origin* is a difficult exercise both to conduct and to evaluate. This difficulty and its historical source should be borne in mind, therefore, as we move from Darwin's understanding of his theory as a *vera causa* theory, to the issues raised by the various Princetonian explications of the *Origin*.

Before actually turning to the argument of the *Origin* itself, however, there are some preliminary matters to be mentioned. First, it will already be plain that I am focusing here only one among many clusters of philosophical issues surrounding Darwin's theory. In doing so, I do not mean to suggest that there is no philosophical interest in relating Darwin's theory to teleology, essentialism, creationism, probabilism, naturalism or capitalism and so on. Second, for our purposes here it is appropriate to focus on natural selection among Darwin's many theoretical proposals. Despite the claims of Michael Ghiselin (1969) and others, Darwin was no methodological unitarian. The structure and strategies of argumentation in play in the *Origin* are one thing, those in play in pangenesis (Darwin's theory of reproduction) another, those in play in his geological paper on Glen Roy another again. Third, our concern here is not with Darwin's original arrival at the theory of natural selection, but with his public case for it. There are instructive questions to be asked about the relation between the private inception in 1838–9 of Darwin's theory and its later presentation, but those relations do not form our principal business here, although some attention to them will be needed in sorting out Darwin's relations with Herschel and Whewell. Fourth, sometimes in historical case study work there is an explicit denial of any concern with the theorist's own intentions regarding his theorising. Indeed, it is sometimes implied that our interpretation of any theory should be kept strictly apart from the interpretation given it by its original author. This separatism may well be possible, even desirable, in other cases. But in the present instance it would be misleading. The most recent philosophical explicators of Darwin, notably Lloyd, have been insistent in claiming that their explications make sense not only of Darwin's theory but also of Darwin's own interpretation of its epistemic and systemic character. Any subsequent integration of historical scholarship and philosophical doctrines is thus invited to follow this precedent, and the present paper does so.

3. Darwin's one long argument and the *vera causa* ideal

Turning now to the structure and strategy of the 'one long argument' that the *Origin* presents, let me stress that the textual evidence for what follows is, much of it, available in an earlier piece of my own (1977); so, I will take various documentary details as given on the present occasion.

What can be shown, mainly by comparing the *Origin* with its two earlier unpublished, draft versions, the *Sketch* of 1842 and *Essay* of 1844, is that Darwin's text, in all three versions, was structured by him in accord with two distinct divisions or partitions, as the diagram for the *Origin* (Figure 1) shows.

The easiest partition to discern is a threefold one: an opening Part I concerns principles of variation and selection under domestication; a middle Part II concerns the application of those principles in the wild; and a final Part III exhibits the explanatory virtues of the theory of natural selection, in relation to various classes of facts about species. We may refer to this partitioning as the *three part* structuring of the argument.

Less easy to discern in the *Origin*, but not in the *Sketch* and *Essay*, is another threefold articulation that does not coincide totally with the *three part* one. For

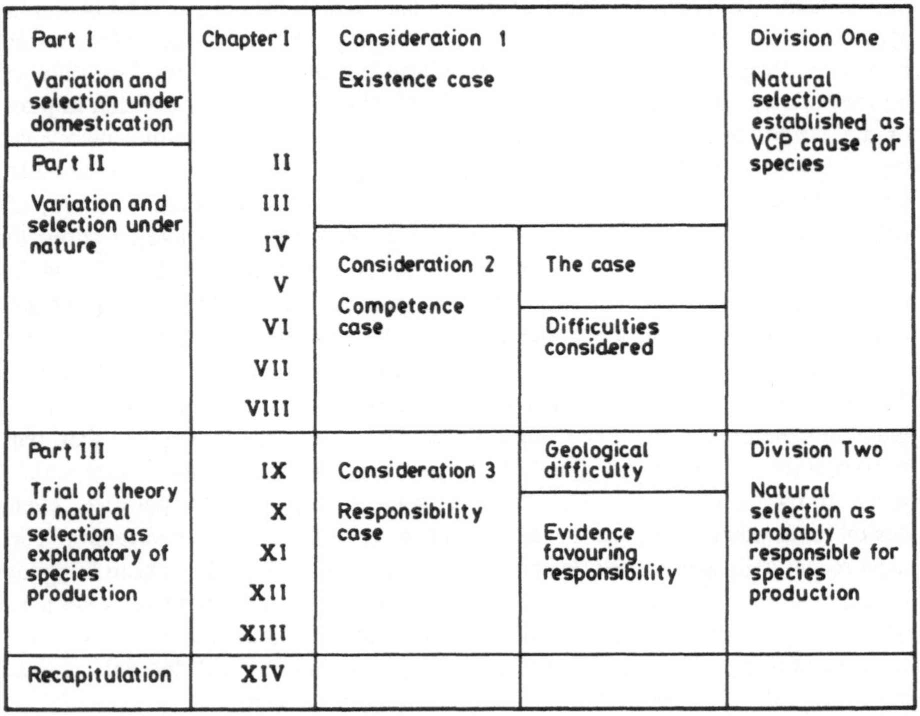

Part I	Chapter I	Consideration 1		Division One
Variation and selection under domestication		Existence case		Natural selection established as VCP cause for species
Part II	II			
Variation and selection under nature	III			
	IV	Consideration 2	The case	
	V	Competence case		
	VI		Difficulties considered	
	VII			
	VIII			
Part III	IX	Consideration 3	Geological difficulty	Division Two
Trial of theory of natural selection as explanatory of species production	X	Responsibility case		Natural selection as probably responsible for species production
	XI		Evidence favouring responsibility	
	XII			
	XIII			
Recapitulation	XIV			

Fig. 1. The structure of the *Origin* (1859).

Darwin's argument comprises, successively, a case (1) for the *existence* of natural selection; a case (2) for the *competence* or *adequacy* of natural selection to produce new species from old, and diversify them adaptively in the formation of new genera, orders and so on; and a case (3) for the *responsibility* of natural selection for the production of those species now extant and those extinct species commemorated in the rocks. This trichotomy we may call the *three case* structuring.

So, when we see that Part III in the *three part* structuring is the same as case (3) in the *three case* structuring, we see that this coincidence allows us to identify another articulation, this time a dichotomous one. For everything prior to case (3) – and so prior to Part III – constitutes a Division I, which sets out the theory of natural selection, so as to establish its *vera causa* credentials; then, in Division II, there is the verifying of the theory as an explanation for the origins of extant and extinct species.

Since the rationale for the *three case* and the *two division* articulations is the *vera causa* ideal, we need to get a clear view of that ideal without delay. Here it should be emphasized that the *vera causa* ideal is no esoteric historians' rarity. Anyone who has read in the philosophical literature surrounding science in the century from

Hume to Mill will have encountered the ideal in many more or less explicit invocations. Equally, specialist historians of the philosophy of science, such as Larry Laudan (1981) and Bob Butts (1970, 1973) – who have given us sophisticated exegeses of authors such as Thomas Reid and William Whewell on *verae causae* – have removed any excuse that the rest of us might have had for going astray on this subject.

In its original form, the *vera causa* ideal was given canonical explication by Reid in his interpretation of one of Newton's rules of philosophising. For Reid, the decisive element in the ideal was the contrast it drew between true, known, real or existing causes, on the one hand, and hypothetical, imaginary, unknown, conjectural or supposed causes, on the other. For the exemplary contrast was between Newton's gravitational force as a true cause, and Descartes' vortices of subtle matter as a hypothetical cause. Both causes, it was argued by Reid, are adequate causes. That is to say both would be sufficient or adequate to cause the effects, the orbiting of the planets around the sun, that are to be explained. As to their sufficiency, adequacy or competence, they stand equal. Where they differ is over the evidence for their existence. For the gravitational force there is independent evidence (from tides, for instance, and falling bodies on earth) for its existence; it is evidenced by facts other than the planetary orbits it is to explain; whereas for the existence of the Cartesian vortices there is no evidence except those orbits. The gravitational force, unlike the vortices, is therefore both a true, known, real or existing cause as well as a sufficient, adequate or competent cause. It meets, then, the requirement that one is not allowed to suppose, or infer, that one's explanatory cause exists merely because it is adequate to explain what it is to explain. Truth, existence and reality, on the one hand, and sufficiency, adequacy or competence, on the other, are independent considerations and their evidential demands must both be met for a cause to meet the contraints of the *vera causa* ideal.

So far, in this thumbnail epitome, the *vera causa* ideal may seem straightforward enough, especially as it has many echoes in twentieth-century discussions of, for example, what it is for a hypothesis to be *ad hoc*. But it should be plain, too, that there was always plenty of scope for complications and sophistication. Especially was this so concerning the kinds of evidencing that were thought appropriate in establishing either the existence of a cause or, conversely, its adequacy. Reid, for one, seems – sometimes at least – to insist on direct experiential acquaintance, generalisable by enumerative induction, as the only acceptable form of evidence for the known truth, reality or existence of a cause (L. Laudan 1981). More, much more, could be said, here, about such issues, and about the development of the *vera causa* ideal in the centuries since Newton, but our thumbnail epitome will allow us to bring out what we need to recognise in Darwin's argument.

Darwin's case for the existence of natural selection is more difficult to represent precisely than is often appreciated, as Ruse (1971) showed once and for all. But the gist of it is plain enough. Species in the wild are subject to changes in conditions of life; and domesticated species show that any animals and plants exposed to changed conditions vary and heritably so. There is hereditary variation in the wild, then. There is also a struggle for existence, for there is superfecundity and this entails a struggle for food, space and other limited requisites for life. In this struggle for existence, there is a differential survival and reproduction of hereditary variants, for some hereditary differences affect chances of survival and reproduction. There exists in nature, therefore, a process of selective breeding, a process analogous to the selective breeding practised by farmers and gardeners.

Darwin does not claim to establish the existence of natural selection by mere inspection of nature; that this process is going on in the wild is not given from experiental acquaintance with its presence there. However, his view is that its existence there is a reasonable conclusion from various premises, concerning heredity, geological change and superfecundity, that are either already accepted or are evidenced observationally.

Again, Darwin's main case for the adequacy of natural selection is far from straightforward in all steps, but its main outline – lucidly explicated recently by Kenneth Waters (1986) – is unmistakeable. Artificial selective breeding is known to be sufficient, adequate, competent to produce, within a species, distinct races – of dogs, for instance – adapted to distinct human ends. These races do not count, according to customary criteria, as distinct species. But natural selection has vastly longer to work and is much more comprehensive and discriminating. It can then produce races that do count as species, for they would be more permanently and more perfectly adapted and divergent in their organisation, and hence infertile with one another, true breeding and without intermediate varieties.

Finally, the case Darwin makes for the responsibility of natural selection for species formation, although tortuous in places, pursues one line throughout: namely, that the theory that natural selection did it is more probable, and so to be preferred over any rival, because it is better than any other at explaining facts of several general kinds or classes: biogeographical facts, embryological facts and so on.

4. Natural selection, true causes and Whewellian consilience of induction

Even this very brief consideration of how the *Origin* was structured so as to conform to the *vera causa* ideal, the old Reidian ideal, can put us right on many of the issues that have to be understood correctly, if we are not to go astray in attempts to derive lessons from this case for the philosophical theory of theories. One cluster of such issues concerns William Whewell and the wave theory of light. For Whewell and the wave theory clashed with that old ideal. What is more, after the *Origin* was published and in defending it, especially in correspondence, Darwin sometimes invoked the wave theory and Whewell. So, it may seem that we have to conclude that, after all, the old Reidian ideal is not as relevant as we thought, and that Darwin and natural selection really belong among the influences subverting that ideal. Indeed, this line has recently been taken in two valuable papers by Ronald Curtis (1986, 1987).

This is not the place for a full treatment of this line of historical interpretation (Hodge forthcoming). However it is necessary to insist here that the view Darwin took when constructing his theory in the 1830s, and when writing the *Sketch*, the *Essay* and the *Origin* in the 1840s and 50s, was one thing, and that the position he adopted in defending it later against objections was sometimes another.[1]

So the decisive precedent for Darwin's argument in the *Origin* was the old *vera causa* legitimation of gravitational astronomy, rather than the new epistemological rationale being given, in the 1840s, for the wave theory of light. Generally speaking, the wave theory was not being defended as *vera causa* legitimate (L. Laudan 1981). On the contrary, it was, especially by Whewell, being cited as not conforming to that ideal and, therefore, because it was good science, discrediting that ideal as a constraint on physical theory. Now, after he had finally published his theory of natural selection in the *Origin*, Darwin sometimes retreated to a similar style of defence, citing the wave theory in doing so. That is, he declared that even if the *vera causa* credentials of natural selection were disputable, it was not to be dismissed totally on that ground, because there were precedents, notably the wave theory, for taking seriously, even accepting as probably true, theories that did not meet the evidential requirements of that ideal. More particularly, it was agreed, Darwin reminded people, that although the existence of light waves was not evidenced independently of the many classes of optical facts the theory was able to explain, nonetheless that explanatory adequacy and achievement were grounds for thinking the theory at least probable rather than hopelessly conjectural.

However, although Darwin's move to this form of defence is apparent, it should not distract us from the fact that the *Origin* had not been written that way and that it was never rewritten that way either, and that the five revised editions of it that Darwin prepared from 1860 to 1872 kept the same structure and strategy of argument inherited from the *Sketch* of 1842. Moreover, if one reads carefully in those texts where the wave theory precedent is cited, one can see that that abandonment of the old ideal is often more apparent than real, and less than wholehearted.[2] For he will still insist that there is, after all, evidence for the existence and adequacy of natural selection independent of its explanatory virtues as exhibited in Division II of his book. But, be that as it may, the argument of the *Origin* in all its editions is a Reidian *vera causa* argument.

We can not, therefore, read the argument of the *Origin* as presenting a Whewellian consilience of inductions. The reason for this is quite simply that Whewell not only rejected the old Reidian *vera causa* ideal, he also offered his consilience ideal as an alternative to that older ideal. Most particularly and explicitly, he disputed the leading assumption made by the old ideal, the assumption that the truth, the existence of an explanatory cause is a separable issue from its adequacy (Whewell 1840, vol. 2, pp. 441–445). Often, the best we can do evidentially for an explanatory cause, Whewell argued, is establish its probable existence by establishing its explanatory adequacy for many distinct classes of facts, in a consilience of inductions as he dubbed it. For often we will not have independent – much less direct, independent – evidence of its existence. So one main rationale for Whewell's consilience doctrine was that the inductive credentials of a theory can be established without measuring it against that impossibly restrictive ideal of independent, direct evidencing of the existence of any causes it introduces.

So, to be sure, Darwin in his Division II, is doing very much the kind of thing that Whewell called consilience. But the old *vera causa* ideal also required one to do that, too, as can be seen by reading Herschel (1830) on Newton's gravitational theory (Good 1987). The decisive difference was that the old ideal laid down that this is not the only kind of evidencing one has to produce. One has also to produce independent argumentation for the existence of one's cause. So, the argument of the *Origin* taken as a whole is conformed to the ideal that Whewell sought to replace, not to the consilience of inductions that he offered as its replacement. It is, then, not correct to do as Ruse (1979) has done and propose that Darwin is being Herschelian (and so Reidian) in the first part of his argument and Whewellian, in the second. The whole argument is Reidian, and so, as a whole, un-Whewellian; and, as in the whole, so in the parts.

The attractive, persistent, but unsustainable, notion that Whewell's consilience doctrine is the inspiration for the *Origin*'s argument structure has no support from what we know of Darwin's reading of Whewell and other contacts with him. The consilience view is not presented explicitly until Whewell's *Philosophy of the Inductive Sciences* of 1840, and there is no sign that Darwin had read that book when he composed his *Sketch* in 1842. He did read Whewell's *History of the Inductive Sciences* of 1837, in the Autumn of 1838, and he may have read in it the year before. Also, the book emphasises the virtues of the wave theory as an explanation of many, diverse kinds of facts, but it does not set those virtues explicitly against the *vera causa* ideal. The book includes also a rejection of the old *vera causa* ideal as appropriate to geological science (Ruse 1976). All the signs are that this rejection left Darwin entirely unmoved. On the contrary, he shows every sign of remaining loyal to Lyell and Herschel's teaching that the old ideal is indispensable to the foundation of geology as an inductive science. Geology for Darwin, following Lyell, was the science that included theories about the origins of species.

Avoiding the Whewellian consilience reading of the *Origin* is very worthwhile, if only because we can then avoid mistakenly embracing the Princetonian proposal made by Thagard (1978). He holds that Whewellian consilience is tantamount to inference to the best explanation, as defended by Gilbert Harman, as the form of inference proper and typical for much natural science. And Thagard offers us Darwin in general, and the *Origin* in particular, as Whewellian and so, on his account, Harmanian. To this proposal, one must respond that even if the assimilating of Whewell to Harman is conceded, and there are reasons for resisting it, it does not make Darwin's book Harmanian because the overall argument of the *Origin* is not Whewellian.

Recker (1987) has rightly seen that Thagard's view of the *Origin* does not capture the way the whole argument is constructed; and he brings out well that there are difficulties for Thagard's further proposals, concerning simplicity, for instance. However, Recker himself has not succeeded fully in discerning how the whole argument proceeds. For Recker divides the *Origin*'s chapters into three blocks. The third and last block coincides, almost exactly, with Division II on the mapping I have given. But his first two blocks do not correspond either to the distinction I have given between Parts I and II or the distinction I have given of case (1) and (2). For Recker takes his first block of chapters to set out a general, positive case for the 'causal efficacy' of natural selection in relation to evolution; the second block to answer possible objections to the causal efficacy thesis; and the third to support that thesis by exhibiting its explanatory virtues regarding biogeography and so on. On Recker's account, then, it is a causal efficacy thesis, concerning natural selection, that is being argued for right through the whole book. Now, Recker is well aware that the positions of the Reidian Herschel and of the anti-Reidian Whewell on *verae causae* are indispensable in making sense of Darwin's argument. But because Recker sees a single thesis being argued for throughout the book, he has to see Darwin as deploying different *vera causa* strategies in different blocks of chapters, on behalf of this single thesis. Accordingly, he follows Ruse in seeing a Herschelian 'empiricist' *vera causa* strategy in play in the early chapters making up the first block, and a different, Whewellian 'rationalist' strategy in play in the third block. But this single thesis, three block, two strategy analysis, although it allows Recker to contribute many insights concerning Darwin and the *Origin*, rests ultimately on a mistake: the mistake of not seeing that the old Reidian *vera causa* ideal, as upheld by Lyell and Herschel, and learned from them by Darwin, required three evidential cases to be made on behalf of natural selection: an existence case, an adequacy case and a responsibility case. Recker holds that the unity of the *Origin* lies in its pursuing throughout a single causal efficacy thesis; but this is not so, rather it lies in Darwin's making three cases on behalf of his one cause. It is the cause, natural selection, itself, that provides the unity, together with the requirement that the case for the responsibility of natural selection for species origins requires that prior cases also be made for its existence and for its adequacy for such effects.

Although much more might be said, on the way the *Origin* reveals its descent from Lyell's and Herschel's reaffirmation of the Reidian tradition, enough has been done to show that correctly representing the structure and strategy of the book's argument is not always a straightforward task; but that it is an indispensable task if we are to correctly locate Darwin's work in relation to the epistemic and systemic ideals of evidence and explanation under consideration at the time.

5. Darwin and a semantic view of theories

It is a no less indispensable task if we are to assess fairly and fruitfully the suggestions, very different suggestions, made by Lloyd (1983) and by Kitcher (1985a), in their respective efforts to clarify 'the nature of Darwin's support for his theory of natural selection' and the nature of 'Darwin's achievement'.

Following van Fraassen, Lloyd adopts a semantic view of theories. She does not set out explicitly what this view is, but Ronald Giere (1983), another semanticist, proves a very good guide on this as on many other matters. Giere explains that on the van Fraassen version of the semantic view, that he shares with Lloyd, a theory is what is defined by a definition. The definition defines a kind of system: for instance, a Newtonian gravitational system is defined as a system with two or more masses moving in accord with Newton's three laws of motion and one law of gravitation. On one analysis of this view, moreover, the definition is taken to specify a state space in defining a kind of system. Thus for the Newtonian system, six state space variables, three for position, three for momentum, are assigned to each particle. So, a theory is what is defined by a definition of a kind of system, as specified in a specification of a state space. There is a sense, then, Giere explains, in which a theory is a model. A model, he says, is a set of objects satisfying a linguistic structure. The linguistic structure that a theory satisfies is the definition of the kind of system. Giere does not say so, but it would seem that a model, in this sense, need not be existentially instantiated; possible objects can satisfy linguistic structures.

Lloyd, although she does not say so, seems to accept all of this, as is shown by at least one of her other recent papers (1984). She is certainly in accord with the semantic view, as Giere expounds it, when she maintains that the theory of natural selection, in Darwin as elsewhere, is a group or set of model types. For a model type, she explains, is what one has if one takes a model that has values for some parameters and one leaves out the specification of those values. A model type has, then, parameters with unspecified values. Natural selection is a group of model types, rather than a single model type, because, Lloyd holds, the natural selection of bodily structures, say, is represented in one or more model types, while the natural selection of instincts, say, is represented by other, distinct model types. A particular model, of some particular instinct being modified, for instance, would have particular values specified for selective advantages, for variation and so on.

Now Lloyd's main proposals concern the support Darwin is giving his theory. Support, for Lloyd, following van Fraassen, is a semantic relation. And, semantic relations are to be distinguished from pragmatic relations, according to Lloyd, for pragmatic relations involve the uses of a definition or a model or whatever, by some user, for some purpose; as, for instance, when a theory is used by someone to explain some fact.

This invocation of the standard distinction, between what is semantic and what pragmatic, is decisive for Lloyd's analysis of Darwin's support for his theory, as Recker's (1987) appreciative critique of Lloyd brings out well. For Lloyd insists that Darwin's argument involves establishing semantic relations, more particularly semantic relations of empirical adequacy, relations that are independent of pragmatic relations of explanatory use. To see how Lloyd defends this conclusion concerning support and semantic relations of empirical adequacy, we have to see, therefore, how Lloyd understands the distinction between what I have designated Division I and Division II of Darwin's argument. For the business of Division I is, according to Lloyd, to support the group of model types that constitutes the theory of natural selection; while in Division II, by contrast, a host of particular models is being supported.

The support for the theory, in Division I, is provided, Lloyd holds, by showing how the assumptions of the theory, as a group of model types – assumptions about heredity, variation and differential survival and reproduction – are empirically adequate. Empirical adequacy is a semantic relation, an assumption being empirically adequate, on Lloyd's account, if its deductive consequences are instantiated by facts. In Division II, what is going on, then, according to Lloyd, is the construction of many particular models and an exhibition of their empirical adequacy. Thus a model for the origin of the Galapagos fauna is constructed and its empirical adequacy exhibited.

Now, for Lloyd, this exhibition of the theory is supportive because it is not a pragmatic achievement. Darwin, she insists, is not explaining how the Galapagos species come to be as they are. Rather, he is showing that such an explanation is possible. To show this is not to use the theory in explaining, it is to exhibit its empirical adequacy, a semantic not a pragmatic achievement. For what is shown is that the group of model types, that is the theory, includes a model type that can be specified in its parameters, so as to yield a particular model that is empirically adequate to the Galapagos facts.

There is more to Lloyd's suggestive and sophisticated discussion than the sampling given here presents, but we now have enough before us to begin assessing her main proposals. On the positive, favorable side, there are three manifest virtues in her account. First, the articulation of Darwin's long argument into two divisions is taken seriously, and the contrast between the generality of Division I and the particularity of Division II is appreciated. Second, the lack of direct confirmation for the conclusions of Division II, the lack that arises because there is no record – such as a vast film would provide – of the past history of life on earth, is very properly emphasised. Third, there is full appreciation of the way Darwin often provides sketches for possible explanations, sketches of possible sequences of events leading, say, to the present Galapagos species.

On the negative side, however, there are several fundamental difficulties. First, Division I of the *Origin* is not merely an attempt to support a definition of a kind of system. Lloyd seems to accept that a definition as such, apart from its applications and instantiations, has no empirical content. For she represents Darwin, in Division I, as providing empirical support not for the definition itself, but for the assumptions that must be made if the definition is to have any applications or instantiations. But, against this view, we must insist that Darwin is not concerned merely with empirical assumptions, in the sense of presuppositions, made in any application or instantiation of natural selection as a theory. Rather, he is concerned to show that natural selection, as he defines it, is existentially instantiated, in nature, in order that he can go on to show what kinds of consequences it is having and so what it is sufficient to produce. Lloyd's semantic view of the theory as a group of model types does not lead us to see, therefore, how the distinct existence and adequacy cases form the business of Division I.

Second, in insisting that Darwin's explanation sketches are exhibitions of semantic rather than pragmatic relations of support, Lloyd has to distinguish between Darwin's arguments in using his theory and his arguments in support of his theory in Division II. But this distinction has no foundation in Darwin's text as he himself understood it. It is a distinction required by the presuppositions of Lloyd's exegetical enterprise, not by the content of the case study itself. Lloyd's metatheory is itself not shown to be empirically adequate here, in that its consequences are not instantiated by the textual facts.

Third, Lloyd omits all consideration, even mention, of causation in general, and of natural selection as a causal process in particular. This omission is striking, to put it mildly, for in many of the texts Lloyd is quoting from there is abundant causal talk, including, in at least one case, explicit identification of natural selection as a *vera causa*.[3] But this omission is not surprising. We know that van Fraassen (1980) thinks that causation, like explanation, involves pragmatic not semantic considerations. For causation has to do, it is alleged, with a context of human actions and interests, and is not therefore to be confused with context independent semantic relations between statements, or with what makes statements empirically adequate. Lloyd is, then, being consistent in including in her entire paper not a phrase about causation or even any allusion to it. However, the consequence is that none of the features of Darwin's argument that have their rationale in his commitment to causal explanation, and to the *vera causa* ideal, are done proper justice in Lloyd's semantic explications.

The empiricism of van Fraassen's philosophy of empirical adequacy is an explicitly anti-realist and, especially, anti-causal-realist empiricism.[4] There are, then, deep reasons for Lloyd's failing in her attempt to exhibit Darwin as a fellow agnostic concerning causality and reality. For consider two traditions in the contrasting of mathematics and physics. One tradition – the one van Fraassen belongs to – says that mathematics and physics (including mathematical physics) differ ultimately in that mathematics, as such, has no empirical suport, while physics, even mathematical physics has some. Another tradition, however, holds that what marks off mathematics from physics is that the first has no concern for causal explanation. Darwin was obviously an heir to this tradition in its eighteenth-century forms. A historian has to read Lloyd's own paper as being, therefore, an incongruous conjunction of disparate traditions, the one Darwin belonged to and the one Lloyd herself descends from through van Fraassen. Lloyd's attempt to shift Darwin from his tradition to her own is, unknowingly, an attempt in effect to change the past, something even God is traditionally unable to do. Lloyd cannot be blamed, therefore, for failing.

6. Darwin and explanatory unification

Kitcher (1985a) starts from a view not of what theories consist of, but of what they do for us. A theory is a good one, he urges, if it unifies our beliefs by providing a few basic patterns of argument that can be used in the derivation of the many sentences that we accept. Accordingly, he holds that Darwin's theory is an explanatory device, a collection of problem-solving patterns, a collection of schemata, aimed at answering families of questions about organisms by describing the histories of those organisms.

For Kitcher, explanatory promise is not purely pragmatic because it is not irreducibly context dependent. Whether or not arguments can be applied to provide explanations, by providing answers to explanation-seeking questions, can be established independently of context, he says. The issue is whether those arguments can be part of a store of explanatory resources that provide explanations, by conforming to certain argument patterns. Explanatory promise is, then, a matter of potential unification.

Darwin's irregular branching tree of descent, and his process of natural selection, are therefore analysed by Kitcher as a very generalised explanatory resource. Darwin's exhibition of the resourcefulness of that resource store is achieved by using it in providing explanation sketches for a wide range of phenomena, sketches that exemplify a few patterns of argument. Most especially is this done in Division II of the *Origin*, where the common characters distinguishing some supraspecific taxonomic group of species – or their common confinement to some geographical region – are traced to a common ancestry for the group, and a subsequent history of limited adaptive divergence and migratory dispersal.

Kitcher urges that the genuine and controversial innovation in Darwin's *Origin* was not the quasi-deductive derivation of natural selection from heredity, variation and superfecundity, but nothing less than bringing historical questions about species origins within science for the first time, questions previously the subject of theological speculation. Darwin's main claim is thus, on Kitcher's view, that we can understand many biological phenomena in terms of what Kitcher calls Darwinian histories, with these histories conforming to a few basic patterns. The theory is simply the assertion that these phenomena are understandable in that way, for the theory consists of the demonstration that such unifying histories can be constructed.

On four counts, I submit, Kitcher's analysis is to be welcomed. First, he does not force on Darwin an inappropriate distinction between semantics and pragmatics. Second, Kitcher rightly refuses to identify Darwin's theory with the quasi-deductive existence case. Third, Kitcher is illuminating on the relation between Division I and II, and on the domination of Division II by a few recurrent lines of argument. Fourth, Kitcher brings out, better than anyone before him, the full force of those early criticisms of Darwin over the difficulty of getting testable consequences out of the theory and checking them against ascertainable facts. For, as Kitcher emphasises, there was considerable looseness over what were acceptable auxiliary assumptions to incorporate into Darwinian histories; and there was looseness over what were or were not relevant, independently-established facts, available for deciding whether the implications of those histories were confirmed by empirical findings.

Against these four welcome elements in Kitcher's achievement, there are some three less favorable considerations to be weighed. First, Kitcher underestimates Darwin's commitment to a positive causal adequacy case for natural selection. Before Darwin takes up difficulties, such as complex adaptations, which he says could possibly be due to natural selection, he works to establish under what conditions species will probably, indeed invariably, be formed by natural selection. Kitcher's theory of theories has inclined him, too much, to see Division I as merely preparatory to Division II, this last constituting Darwin's main achievement. He has not seen that Division II is often invoking the positive causal adequacy thesis of Division I. Second, although no aetiophobe, Kitcher never quite acknowledges Darwin's view that selection is explanatory because it is causal. Kitcher seems here still under Hempel's positivistic influence, in his unwillingness to make aetiological, rather than or as well as nomological, unification central to his account of explanatory unification. Third, Kitcher is largely mistaken in his claim that Darwin's achievement included – perhaps, for Kitcher, above all else – bringing the problem of the origin of species for the first time into science from theology. For a start, one needs to consider Lyell's explicit stand against those like Humboldt (no strong theist, by the way) who held that the origins of new species presented a mystery that lay beyond science (Hodge 1982). More generally, one needs to look at the discussions given of these issues by a writer such as Baden Powell (1855) writing in the wake of Robert Chambers' *Vestiges of the natural history of creation* (1844). Beyond that one needs to consider writers such as Buffon, in the previous century, who were not obviously excluding the explanation of species origins from science.

It is true that the problem of species origins, as Darwin found and engaged it in Lyell, was not seen as separable from theology. But then it is arguable that Darwin himself never separated it entirely from theology. For atheists there is no theology, granted. But for many theistic scientists – Darwin himself included, in the 1830s especially, but also still in the 1860s – the topic has remained partly a theological one. Indeed, the whole question of when and how theology and science became separated, institutionally as well as intellectually, is far from easy to assess and resolve, as a question of history. Fortunately, a critique of Kitcher's theory of scientific theories, as illuminating the Darwin case, does not need to take up this question, for Kitcher's unsatisfactory approach to the question is independent of his metatheoretical program. It stems more, one suspects from his campaigns (1982) against scientific creationism in our time, and from time spent at Harvard rather than at Princeton.

7. Epilogue: structure and function in scientific theories

One of Kitcher's most general metatheoretical theses (1985a) is a thesis he finds supported by two very dissimilar case studies: the development of the Leibniz and Newton differential calculus, and Darwin's natural selection. The thesis is that theories can be strong in two ways that may not covary. The calculus was, from the start, strong in axiomatisability; it could be given from birth an axiomatic structuring. It was also strong as a source of arguments applicable to geometrical, kinetic and dynamic problem solving. By contrast, Darwin's theory had low axiomatisability, but high explanatory promise, in that many unifying explanatory arguments could be drawn from it.

A historian cannot supply a verdict on the philosophical correctness and value of this thesis of Kitcher's. However, a historian may judge of its historiographical fruitfulness. Ever since Aristotle's *Posterior Analytics*, of course, if not before, there have been discussions of how the structural characteristics of explanatory arguments are related to their explanatory functioning. However, such concerns have not always been in the ascendant. As the Reidian *vera causa* tradition illustrates, in the eighteenth century, epistemological concerns about scientific theories were pursued often without sustained concern with structural issues. Familiarly enough, structural issues have come back to take the center of the philosophical stage in our own century, in the wake most obviously of the revival of formal logic late in the last century and the rise of logicism in the philosophy of mathematics.

It should not surprise us, then, when we look over the long run of history, to find that a theory, such as Darwin's, that draws its epistemic and systemic ideals from the eighteenth century, does not have a structure that is easily explicable by comparison and contrast with twentieth-century structural ideals. Moreover, and here is a stronger suggestion to end with, it may be worth giving a history a further role beyond that of merely taking the surprise out of such a finding. For, after all, Darwin's theory has a direct neo-Darwinian descendant in the synthetic theory that came of age in the middle of the twentieth century. The epistemic and systemic ideals of that theory may, then, be appropriately interpreted as continuous with the old eighteenth-century ideals (Hodge 1987). Perhaps, therefore, in so far as the synthetic theory is a canonical example of twentieth-century science, we should demand of a twentieth-century philosophical theory of scientific theories that it find a place for the concerns expressed in the articulation of those old ideals, no less than for the structural concerns so characteristic of our later age.

Notes

1. For a strong affirmation of Herschelian and Lyellian commitments, written in late October 1838, see Darwin (1839), pp. 615–625. For this dating, *see* Darwin (1986), p. 432.
2. The parenthetical parts of the following passage from a letter of Darwin (May 8, 1860) to Henslow are relevant here (Barlow 1967, p. 204): '... he [Adam Sedgwick] talks much about my departing from the spirit of inductive philosophy. – I wish, if you ever talk on subject to him, you would ask him whether it was not allowable (and a great step) to invent the undulatory theory of light – i.e. hypothetical undulations, in a hypothetical substance, the ether. And if this be so, why may I not invent a hypothesis of natural selection (which from analogy of domestic productions, and from what we know of the struggle for existence and of the variability of organic beings, is, in some slight degree, in itself probable) and try whether this hypothesis of natural selection does not explain (as I think it does) a large number of facts in geographical distribution – geological succession – classification – Morphology – embryology etc. etc. – I should really much like to know why such an hypothesis as the undulations of the either may be invented, and why I may not invent (not that I did *invent* it, for I was led to it by studying domestic varieties) any hypothesis, such as natural selection.'
3. Thus Lloyd (1983, p. 112) quotes the last three sentences only of the following postscript in a letter (22 May 1863) from Darwin to George Bentham (Darwin 1888, vol. 3, pp. 24–5): 'P.S. – In fact the belief in Natural Selection must at present be grounded on general considerations. (1) On its being a *vera causa*, from the struggle for existence; and the certain geological fact that species do somehow change. (2) From the analogy of change under domestication by man's selection (3) And chiefly from this view connecting under an intelligible pont of view a host of facts. When we descend to details, we can prove that no one species has changed [i.e. we cannot prove of any particular species that it has changed]; nor can we prove that the supposed changes are beneficial, which is the groundwork of the theory. Nor can we explain why some species have changed and others have not.'
4. Interestingly, Ian Hacking (1983, p. 277) reports:
   In an at present unpublished discussion note that I have just seen, Bas van Fraassen claims that causalism has its roots in Newton's search for *vera causa* (true causes) [sic] combined with the famous assertion, *hypotheses non fingo* (I do not make, or depend upon, hypotheses).

References

Barlow, N. (ed.) (1967). *Darwin and Henslow. The Growth of an Idea* London: John Murray.

Butts, R. (1970). Whewell on Newton's rules of philosophizing. In R.E. Butts and Davis, J.W. (eds), *The Methodological Heritage of Newton* pp. 132–149. Toronto: University of Toronto Press.

Butts, R. (1973). Reply to David Wilson: Was Whewell interested in true causes? *Philosophy of Science* 40: 125–128.

Curtis, R. (1986). Are methodologies theories of scientific rationality? *British Journal for the Philosophy of Science* 37: 135–161.

Curtis, R. (1987). Darwin as an epistemologist. *Annals of Science* 44: 379–408.

Darwin, C. (1839). *Journal of Researches into the Geology and Natural History of the various countries visited by H.M.S. Beagle.* London: Henry Colburn.

Darwin, C. (1859). *On the Origin of Species.* London: John Murray. Reprinted by Harvard University Press, 1975.

Darwin, C. (1888). *The Life and Letters of Charles Darwin.* (Edited by F. Darwin, 3 vols.) London: John Murray.

Darwin, C. (1986). *The Correspondence of Charles Darwin.* Volume 2: 1837–1843. Edited by F. Burkhardt and S. Smith. Cambridge: Cambridge University Press.

Ghiselin, M. (1969). *The Triumph of the Darwinian Method.* Berkeley: The University of California Press.

Giere, R.N. (1983). Testing theoretical hypotheses. In J. Earman (ed.), *Minnesota studies in the Philosophy of Science* 10: 269–298.

Good, G. (1987). John Herschel's optical researches and the development of his ideas on method and causality. *Studies in History and Philosophy of Science* 18: 1–41.

Hacking, I. (1983). *Representing and Intervening.* Cambridge: Cambridge University Press.

Herschel, J. (1830). *Preliminary Discourse on the Study of Natural Philosophy.* London: Longman.

Hodge, M. (1977). The structure and strategy of Darwin's 'long argument'. *British Journal for the History of Science* 10: 237–245.

Hodge, M. (1982). Darwin and the laws of the animate part of the terrestrial system (1835–1837): on the Lyellian origins of his zoonomical explanatory program. *Studies in the History of Biology* 7: 1–106.

Hodge, M. (1987). Natural selection as a causal, empirical and probabilistic theory. In L. Krüger, G. Gigerenzer and M. Morgan (eds), *The Probabilistic Revolution.* (2 vols,) Vol. 2, pp. 233–270. Cambridge, Mass.: MIT Press.

Hodge, M. (forthcoming). History, science, the earth and life. In M. Fisch and S. Schaffer (eds), *Essays on William Whewell.*

Hull, D.L. (1973a). Charles Darwin and Nineteenth-Century philosophies of science. In R.N. Giere and R.S. Westfall (eds), *Foundations of Scientific Method: The Nineteenth Century*, pp. 115–132. Bloomington, Indiana: Indiana University Press.

Hull, D.L. (1973b). *Darwin and his Critics.* Cambridge, Mass.: Harvard University Press.

Kavaloski, V. (1974). *The vera causa principle: a historico-philosophical study of a metatheoretical concept from Newton through Darwin.* Ph.D. dissertation, University of Chicago.

Kitcher, P. (1981). Explanatory unification. *Philosophy of Science* 48: 507–531.

Kitcher, P. (1982). *Abusing Science.* Cambridge, Mass.: MIT Press.

Kitcher, P. (1983). *The Nature of Mathematical Knowledge.* New York: Oxford University Press.

Kitcher, P. (1985a). Darwin's achievement. In N. Rescher (ed.), *Reason and Rationality in Science*, pp. 127–189. Washington D.C.: University Press of America.

Kitcher, P. (1985b). *Vaulting Ambition. Sociobiology and the Quest for Human Nature.* Cambridge, Mass.: MIT Press.

Laudan, L. (1981). *Science and Hypothesis. Historical Essays on Scientific Methodology*. Dordrecht: Reidel Publishing.

Laudan R. (1982). The role of methodology in Lyell's geology. *Studies in History and Philosophy of Science* 13: 215–250.

Laudan, R. (1987). *From Mineralogy to Geology. The Foundations of a Science, 1650–1830*. Chicago: The University of Chicago Press.

Lloyd, E. (1983). The nature of Darwin's support for the theory of natural selection. *Philosophy of Science* 50: 112–129.

Lloyd, E. (1984). A semantic approach to population genetics. *Philosophy of Science* 51: 242–264.

Lyell, C. (1830–1833). *The Principles of Geology*. (3 vols.) London: John Murray.

Powell, B. (1855). *Essays on the Spirit of the Inductive Philosophy, the Unity of Worlds and the Philosophy of Creation*. London: Longman.

Recker, D. (1987). Causal efficacy: the structure of Darwin's argument strategy in the *Origin of Species*. *Philosophy of Science* 54: 147–175.

Ruse, M. (1971). Natural selection in the *Origin of Species*. *Studies in History and Philosophy of Science* 1: 311–351.

Ruse, M. (1975). Charles Darwin's theory of evolution: an analysis. *Journal of the History of Biology* 8: 219–241.

Ruse, M. (1976). Charles Lyell and the philosophers of science. *British Journal for the History of Science* 9: 121–131.

Ruse, M. (1979). *The Darwinian Revolution: science red in tooth and claw*. Chicago: University of Chicago Press.

Suppe, F. (ed.) (1974). *The Structure of Scientific Theories*. Urbana: University of Illinois Press.

Thagard, P. (1978). The best explanation: criteria for theory choice. *Journal of Philosophy* 75: 76–92.

van Fraassen, B.C. (1980). *The Scientific Image*. Oxford: Clarendon Press.

Waters, C.K. (1986). Taking analogical inference seriously: Darwin's argument from artificial selection. In A. Fine and P. Kitcher (eds) *PSA 1986* 1: 502–513. East Lansing, Mich.: The Philosophy of Science Association.

Whewell, W. (1837). *History of the Inductive Sciences*. (3 vols) London: Parker.

Whewell, W. (1840). *Philosophy of the Inductive Sciences* (2 vols) London: Parker.

Some Puzzles About Species

PHILIP KITCHER

University of California/San Diego

In the Fall of 1974, my second year of teaching philosophy, I was giving a course in philosophy of science to a dozen or so bright undergraduates. After about three weeks, one of the students came to see me in my office. 'We find this material interesting,' he explained, 'but most of us are pre-meds, and the science we know best is biology. It would really help if you could give us some examples from biology, and not talk about physics quite so much.' The point was a good one. Like many philosophers of science of my generation, the standard examples came from physics – when I needed an illustration, I pointed to Newtonian dynamics, optics, electromagnetic theory, thermodynamics, and only occasionally ventured as far afield as chemistry. However, it was clear that the course would be improved if I honored my student's reasonable request, so I set off for the library in search of a key to reform.

I was lucky. There on the shelves was David Hull's *Philosophy of Biological Science*, relatively newly published in the Prentice-Hall series I knew and loved. I took it out and began to read. Almost immediately it was clear that this would not simply be a Useful Source of Improving Examples (although it did fulfil that function for my grateful students). Reading David's lucid discussions of reductionism and of the character of evolutionary theory, I realized that there were deep and important issues of which I had previously been ignorant, and a body of science that I would find difficult to integrate with the philosophical ideas I had absorbed in graduate school. It was clear that I needed re-education, and David's book pointed the way.

Other philosophers of biology of my generation probably have similar stories to tell. All of us owe Daivd Hull an enormous debt. For, at a time when biology was almost invisible in the graduate education of philosophers of science, he showed how exciting and significant the philosophy of biology could be. Moreover, the high scientific standards set in David's work made it clear that there could be no room for mere dabbling: biology, like physics, is serious, difficult, and demanding, and those who philosophize about it had better do their homework. David's example led many of us to the ever-hospitable Museum of Comparative Zoology at Harvard and to regular interchanges with the local population of biologists.

M. Ruse (editor), What the Philosophy of Biology is. pp. 183–208.
© 1989 *Kluwer Academic Publishers, Dordrecht*

As I have eradicated some of my initial innocence about biology, I have learned more and more from David's own work. On many topics, his discussions have influenced my own ways of thinking, probably beyond the extent to which I am aware. But there is one issue on which we are in deep disagreement. Following a provocative article by Michael Ghiselin, David has argued at considerable length for a view of species that seems to me to bypass the main questions that arise in this area of the philosophy of biology.[1] The aim of the present essay is to continue the debate between us. But it seemed to me wrong to launch into the arguments without some prefatory acknowledgement of my intellectual debts. And perhaps those who champion David's view of species may draw the obvious moral from my story: the re-education stopped too soon.

1. Individuality again

According to (Ghiselin 1974, Hull 1976, Hull 1978) biological species are not 'spatio-temporally unrestricted classes' but 'historical individuals.' What does this claim mean? And why does it matter?

I have argued (Kitcher 1984a, 1984b, 1987) that there are conceptual difficulties in the position that Ghiselin and Hull wish to oppose: they are stalking a broken-backed chimera. What is a spatio-temporally unrestricted class? The obvious response is to say that a class is spatio-temporally unrestricted just in case, for any finite region of space-time that one chooses, there are members of the class that lie outside the region. But this will not do, since no class of physical objects is spatio-temporally unrestricted in this sense. Hull recognizes the point and proposes that a class is spatio-temporally unrestricted if its definition allows for the presence of instances that lie outside a spatio-temporal boundary. But this, I suggest, is a confused hybrid notion. Classes (or sets) as I understand them are entities that have their properties independently of the particular ways in which we choose to talk about them. Set-theoretic identity is extensional: $a = b$ just in case a and b have the same members.

So what? Well, let $\{a_1,...a_n\}$ be any finite set of physical objects. Let B be some finite region of space-time that includes all the a_i. We can pick out the set in two different ways: as the extension of the predicate '$x = a_1 \vee x = a_2 \vee ... \vee x = a_n$' or as the extension of the predicate '$(x = a_1 \vee x = a_2 \vee ... \vee x = a_n)$ & x lies within B'. Here I assume that the names a_i do not pick out their referents in ways that restrict those referents to particular regions of space-time. (If they do, choose different names). Now we ask if the set $\{a_1, ..., a_n\}$ is spatio-temporally unrestricted. Answer: yes, because the first way of specifying it sets no spatio-temporal boundary within which its members must lie. Answer: no, because the second way of picking it out does set a spatio-temporal boundary, *viz. B*, within which its members must lie. Both definitions identify sets with exactly the same members, and therefore, by the extensionality of set-theoretic identity, they pick out the same set. So, the set we have identified is both spatio-temporally unrestricted and spatio-temporally restricted. But that set was an arbitrary finite set of physical objects. Thus we can conclude that any finite set of physical objects is both spatio-temporally unrestricted and spatio-temporally restricted.

The contradiction arises because the notion of spatio-temporally unrestricted class with which we have been working mixes properties of entities with properties of their definitions. The *first* issue about the ontology of species is whether species are sets (with organisms as members) or whether they are mereological wholes (with organisms as parts). In his (1976) and (1978), Hull offers a number of arguments for thinking that species are individuals. He appeals to the character of evolutionary theory, the nature of natural selection, and the absence of laws about individual species. If these arguments are taken as directed at the conclusion that species are wholes rather than sets, then I think they fail to reach their target. As I have argued at length (Kitcher 1984a, 1984b), all of our discourse about evolution can be reconstructed equally well within set theory or within mereology. The moral that I draw from this – and that I shall develop in some detail below – is that the point Hull (and Ghiselin) really want to make has nothing to do with ontology. There is a *second* issue about the delineation of the species category on which Hull and Ghiselin offer a significant (though controversial) proposal, and this issue is orthogonal to the question whether species are individuals or sets.

Before presenting that issue, I want to consider a line of argument that Hull has recently offered.[2] The kind of reasoning that leads us to think of a species as a set of organisms, he suggests, should also induce us to think of an organism as a set of cells. Because of our size and perceptual abilities, we are able to see the gaps that separate the parts of species from one another, and thence arises the temptation to view the species as a set of organisms. But the accidents of epistemological access should not lead us to attribute an ontological difference where there is none.

I find this argument interesting, challenging, and ultimately unsuccessful. First, let us ask why we do not think that organisms are sets of cells. One important, and fairly obvious, point is that an organism consists of cells and extra-cellular matrix and the latter may play a crucial role in its development and physiology. Another is that the organism (conceived as existing over time) would be better viewed set-theoretically as a function mapping any time at which it exists onto the set of space points occupied at that time. Since Carnap and Reichenbach, this has been a standard way of thinking about physical objects in general, and organisms can be treated as special cases.

But there is a deeper point that can be appreciated by recognizing that there are some organisms that we can easily conceive as collections of cells (or, more accurately, there are some *stages* of organisms that we can view in this way). In such organisms as *Hydra* and *Dictyostelium* cells can function with a high degree of independence, and we can think of the organism as continuing to survive (albeit in a different form) even when the cells are dissociated. But this is not the rule with organisms. The distinction between an organism and a set of cells is vividly brought home to us when we recognize that it is in principle possible for the organization of the cells that make up a complex organism to be destroyed while each cell persists. The set of cells remains but it is no longer an organism.[3]

Let us ask the analogous question about species. Does a species continue to exist when we disrupt the relations among the organisms that are (on the set-theoretic view) members of it? I believe that a case can be made for an affirmative answer. If an endangered species becomes scattered so that human intervention is required if its remaining members are to reproduce, then there remains a chance of preserving the species: that, of course, is what motivates efforts that people sometimes make. Provided that there is a set of organisms belonging to the species, the species persists. Here we have a clear disanalogy with the relationship between organisms and cells, and Hull's argument is blocked.

However, if it is suggested that species are as dependent on the interactions among organisms as organisms are on the relations among cells, it is possible to make a different reply to Hull. Waiving qualms about obligatorily asexual species, let us suppose that it is crucial to the persistence of a species that some of its member organisms be combining their genes in the production of progeny.[4] Now we can say that a species is a set-theoretic entity, to wit a set of organisms subject to a particular relation (or, more precisely, the ordered pair of a set and a relation) where the relation obtains just in case there is that kind of reproductive behavior that is supposed to be crucial to the persistence of species. Could we conceive of organisms after the same fashion, treating them as sets of cells and pieces of extra-cellular matrix subject to relational conditions? Perhaps. However, at the present state of our knowledge, we can only guess at the complexity of the relations that would have to be adduced. We have not the slightest idea how to define organisms as sets of cells and pieces of matrix (whereas the specification of the relational properties that are required in the case of species seems relatively straightforward). Two points follow. First, the organization of organisms appears much more intricate than that of species – another disanalogy between organisms and species. Second, there is no firm basis for saying that organisms could not be identified with sets subject to a complex of relations (a complex which encapsulated *all* the intricacy of organization), since we have no idea what the explicit specification of the organization of organisms would look like.

I conclude that Hull's argument does not tell against the claim that species are sets. For, depending on your views about what is essential for the persistence of species, it is possible either to find a relevant disanalogy or to find a defensible version of the conclusion that organisms (better: organism-stages) are sets.

On to issues of greater biological significance. The traditional species problem was to delimit the species category by saying which superorganismal entities count as species taxa. If we decide the first question by saying that species are sets, then we can formulate this second problem as that of explaining which sets whose members are organisms are species taxa. Alternatively, if the first question is answered by claiming that organisms are individuals then the second task is to specify which individuals with organisms as parts are species taxa. Notice that it is not a consequence of the set-theoretic view of the ontology of species that *any* set with organisms as members counts as a species. Nor is it a consequence of the mereological approach to species that *any* individual with organisms as parts counts as a species.[5] There are numerous sets with organisms as members and numerous individuals with organisms as parts, and the vast majority of these sets and individuals are of no biological interest whatsoever. To solve the traditional species problem, further specification is needed.

As I interpret them, both Hull and Ghiselin disguise an interesting answer to the second question as a thesis about the ontology of species. The significant point is that species are 'historical individuals', chunks of the genealogical nexus. What makes an individual historical? In general I think that this is a hard question to answer, but, in the case of interest, it seems fairly clear that historical connectedness is critical. So, talking in the mereological idiom, we conceive of an individual with organisms as parts to be historically connected just in case for any organismal parts x and y such that x precedes y and for any organism z, if z belongs to a population that descended from a population containing x and that is ancestral to a population containing y then z is also part of the same individual as x and y. Note that the criterion for historical connectedness can easily be reformulated as a condition on sets. A set of organisms is historically connected just in case it satisfies the following condition: for any organisms x, y and z, if x and y are in the set and if z belongs to a population that is descendant from a population which has x as a member and that is ancestral to a population that has y as a member then z is in the set. Hull and Ghiselin *might* have expressed their proposal by saying that species are historically connected entities and shown a studied neutrality on the question whether they are individuals or sets.[6] In my view, of course, this reformulation would have avoided considerable confusion and would have forestalled attempts to give a priori arguments for significant biological theses.[7]

In its neutral version the Hull-Ghiselin proposal is still at odds with Ernst Mayr's biological species concept. For Mayr's account allows for the possibility of species that are not historically connected. Imagine that a species A splits into two parts at t_0, one part consisting of almost all organisms in A and the other of a small isolated population. A (or the bulk of A) persists unmodified, but the peripheral isolate evolves so that, at t_1, it has descendants that are reproductively isolated from A and constitute a new species B. However, the evolutionary change consists in a small genetic modification that is reversed in an isolated population that descends from B, so that, at time t_2, there are descendants of B that make up a population C that is reproductively compatible with A. (See Figure 1). On Mayr's account, the organisms in C are conspecific with the organisms in A. But now it is clear that A is

188

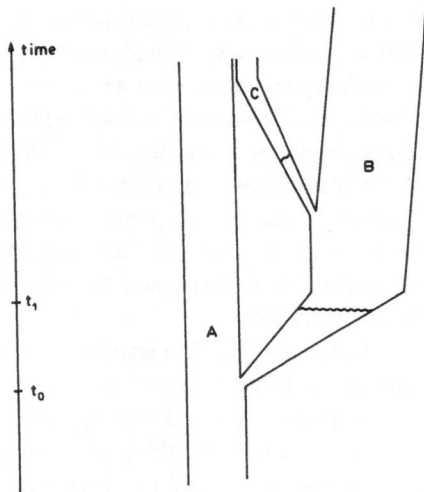

Fig. 1. Hull and Ghiselin *versus* Mayr.

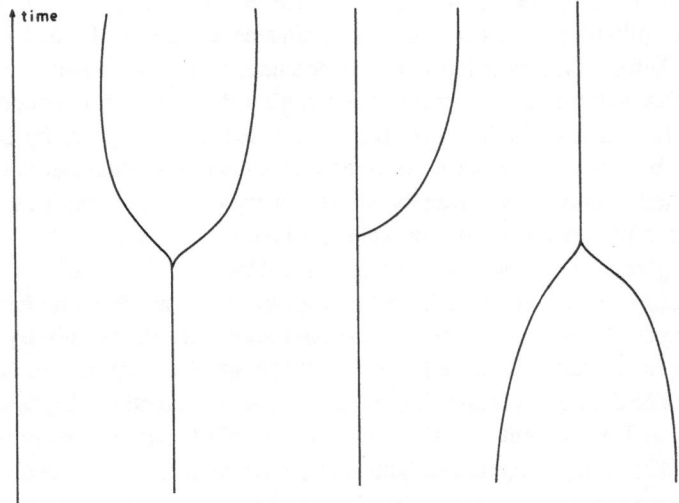

Fig. 2. Three modes of Phylogenetic Change (after (Hull 1978)).

not historically connected. For there are organisms – those in B – that belong to a population ancestral to a population of A and descendant from a population of A but that are not themselves included in A. Hence the biological species concept does not require species to be historically connected.

However, even though the Hull-Ghiselin proposal diverges from the most celebrated answer to the traditional version of the species problem, that proposal does not constitute a complete rival answer to the traditional question. Saying that species are historical entities narrows the range of candidate species taxa but still allows us different ways of splitting up the genealogical nexus. The whole of life – past, present, and future – is one very big historical entity, and, at the opposite extreme, timeslices of particular populations also count as historical entities. Somewhere between these extremes are the species, and, in his (1978) Hull canvasses some possibilities for delineating them. The diagrams that he presents (see Figure 2) are persuasive devices for leading us to think that the problem of breaking up the nexus has been solved – or can be solved relatively easily. But I want to urge that the diagrams conceal deep and important problems, that there are serious questions about what the lines and branch points actually mean.[8] The rest of this essay will be devoted to explaining what needs to be done to complete the Hull-Ghiselin account and why the task strikes me as formidable. I hope that the neutrality of the formulation of the ontological issue (sets versus individuals) will be apparent throughout.[9]

2. The trouble with populations

On the account of historical connectedness that I offered above, the historical connectedness of a species depends on the holding of certain relations among populations. This reference to populations in accounts of species is as necessary on the Ghiselin-Hull approach as it is to Mayr's well-known biological species concept. But what is a population?

One way to define the standard biologist's notion of *local population* is to take a local population to comprise all the organisms of a chosen species that are present in a particular place at a particular time (see, for example, Futuyma 1979, p. 506). There is no objection to using this definition for purposes of exposition, supposing that the notion of species can be taken as already well-understood, but it is useless in a context in which we are trying to use the notion of population to analyse the concept of species. However, Mayr has attempted to do better. He offers the following characterization:

All members in a local population share in a single gene pool, and such a population may be defined also as 'a group of individuals so situated that any two of them have equal probability of mating with each other and producing offspring', provided, of course, that they are sexually mature, of opposite sex and equivalent with respect to sexual selection. The local population is by definition and ideally a panmictic (randomly interbreeding) unit. An actual local population will, of course, always deviate more or less from the stated ideal. (Mayr 1963, p. 136; 1970, p. 182).

This passage offers a compelling picture that seems to encapsulate the ways in which many naturalists and theoretical biologists think about populations. Start with a particular sexual organism a. Consider all the organisms in the same region as a (where 'region' is defined as a specified function of the distance that a can be expected to travel to mate). Call the totality of these organisms T. Within T we are going to pick out those organisms b such that for any organism c in T, the probability that a mates with b is greater than or equal to the probability that a mates with c. (Here mating requires both copulation and the production of viable offspring). Let S_1 consist of all the b's that meet the condition just stated; intuitively, S_1 comprises the opposite sex of a within the local population. We now assume that, for any b belonging to S_1 there is a unique totality S_2 within T consisting of b's most probable mates (i.e. of those organisms c such that for any d in T the probability that b mates with c is greater than or equal to the probability that b mates with d), that S_2 is the same for each b in S_1, and that a belongs to S_2. Subject to these assumptions, the total local population to which a belongs consists of the organisms in S_1 and S_2.

There are a number of obvious worries that we might have about this picture. In some cases there will be organisms that are not among the most probable mates of their most probable mates. If a male bird of paradise has dull plumage, his potential mates will include females who do not include him among their potential mates. Examples like this – and it is easy to see that they are legion – prompt Mayr's suggestion that we treat the notion of population as an ideal, abstracting from the actual differences in sexual selection. Of course, once we demand that mating must involve not only copulation but production of viable offspring, then we encounter troubles with those organisms carrying alleles that are not concordant with the alleles found in members of the opposite sex. If we do not make the demand, then we shall have trouble with populations in which males have the propensity to copulate with females of closely related species as well as with their conspecifics.

For present purposes, however, let us restrict our attention to the difficulty posed by sexual selection, and ask what is meant by claiming that various mating probabilities are equal or unequal. Imagine that a male organism a_1 actually mates with a female a_2 and does not actually mate with another female a_3. Assume that all these organisms are, from the naturalist's standpoint, members of the same population. If Mayr's account of population is to be accepted, then we need to defend the claim that the probability that a_1 mates with a_2 is the same as the probability that a_1 mates with a_3. In making this claim we are obviously expanding our horizons from contemplation of the actual situation alone. We envisage a range of possible situations and suggest that the proportion of situations in which a_1 mates with a_2 is the same as the proportion of situations in which a_1 mates with a_3. The actual world was, as it were, 'selected' from this range of situations, and the 'selection' produced a situation in which a_1 mates with a_2 and not with a_3. Now what determines the appropriate range of situations, the situations that we tacitly envisage in making our judgment of equiprobability? Or, to put the point another way, what features of the organisms do we allow to vary across this range of possibilities, and which do we hold constant?

Plainly if *all* the features of the actual situation are held fixed, then our consideration is limited to a unique situation, so that the proportion of cases in which a_1 mates with a_2 is 1, and the proportion of cases in which a_1 mates with a_3 is 0. If *no* features of the actual situation are held constant, then we shall be confronted with a range of possibilities so vast that it seems that the proportion of cases in which a_1 mates with any particular organism will be effectively 0. Our probability judgment rests on our striking just the right balance between these two extremes, in abstracting from some features of the actual situation and holding others fixed, so that the probability judgments made in applying Mayr's picture will identify 'local populations of conspecific organisms'.

A full account of what a population is must tell us how to strike this balance. It must explain how the conception of probability is to be applied here, specifying the class of possible situations that are to fall under our consideration. What properties of the organisms should be held fixed? Which features can be idealized? To see how difficult these questions are, let us consider some cases, which I describe from the perspective of a naturalist who uses the concept of species without analysis.

1. A local population of a social species with a dominance hierarchy in which smaller, weaker males rank lower, contains some males – the smallest and weakest – who do not mate at all. In judging that they have a nonzero probability of mating with high-ranking and low-ranking females, we abstract from the size of these males (i.e. from the characteristic on which their position in the dominance hierarchy depends).

2. Populations of two species, one of which is the dwarf form of the other, inhabit the same region. In judging that dwarf-dwarf matings are more probable than dwarf-normal matings, we do not abstract from considerations of size.

3. Two small populations of related species occur in a marginal habitat at the peripheries of the ranges of both species. In this region, hybridization occurs as frequently as mating between conspecifics. We avoid lumping the two populations by judging that each organism has a greater probability of mating with an organism from its own species. This judgment rests on abstracting from the composition of the fauna of the region. We distinguish the populations by considering what would happen if the region were not so sparsely

populated, which would raise the relative frequency of mating among conspecifics.

4. In a highly polytypic species, showing a continuously distributed range of morphological types, individuals of each type may have a greater propensity to mate with one another than with individuals of different types. Consider a region in which a small number of organisms of the species, exhibiting different types, meet and mate freely. We judge this group of organisms to be a single population, taking the probabilities of cross-type mating to be equal to those of intra-type mating, because, in this case, we do not abstract from the composition of the local fauna.

5. Two species may be reproductively isolated from one another by differences in the times at which they are active. (The differences can consist in differences between the daily cycles of activity and rest or in differences between breeding seasons.) If two such species occur in a given region, we judge the probabilities of various types of mating by holding fixed the times of activity of the organisms concerned. Were we to abstract from the differences in these times, the probability of interspecific matings would be as great as that of intraspecific matings.

6. In some cases, a species may include organisms with a broad range of times of activity. Extreme individuals may be debarred from mating because their times of activity do not overlap. Yet we may count these organisms as belonging to the same local population, by abstracting from the differences in times of activity, so that the probabilities of mating become equal across the species.

I claim that if 'species' is used as naturalists and theoretical biologists alike use it, then there are numerous examples answering to the descriptions 1–6. What these examples show is that properties of the organisms in question which are held constant in arriving at probability judgments in some cases are allowed to vary in other cases. In other words, the collection of possible situations, with respect to which the probabilities of mating are judged, cannot *obviously* be characterized in any uniform way.[10] If Hull and Ghiselin hope to deploy the concept of population to articulate the idea that species are historical individuals, then they need to articulate the principles we use in setting up the space of 'real possibilities' that underlies our probability judgments.

Since Hull has differed with Mayr's use of modal notions (*viz.* the *possibility* of gene exchange) and has insisted that our delineation of species should be based on the pattern of *actual* matings, it is worth exploring briefly whether there is any plausibility to the idea that we can avoid talk of possibilities and probabilities, either explicating the notion of a population in a nonmodal way or bypassing it and building up the concept of historical connectedness from the actual matings among organisms. One obvious trouble results from the fact that, in many species, vast numbers of organisms belonging to the same population do not mate at all. This difficulty could be overcome by supposing that organisms whose parents belong to the same population and that inhabit the same region belong to the same population. Unfortunately, that supposition would debar *by fiat* the possibility of instant speciation, and would yield counterintuitive results in the known cases in which polyploidy results from a single generation event.

Another worry stems from the fact that hybridization does occur in nature, and it is quite probable that there are some organisms that only mate with members of different species. Not only will such instances draw the boundaries of populations in the wrong ways, but, if they are accompanied by instances of relatives that engage in some matings with conspecifics, there is the obvious possibility that the transitivity of the relation *belonging to the same population* will lead to identifications of 'populations' that are assemblages of members of different species – perhaps even species that are quite distantly related but connected by a chain of close relatives.

Although both problems are serious, the most fundamental trouble for those who hope to avoid the modal intricacies of Mayr's concept of population seems to me to be a consequence of the fact that populations may have significant internal structure and may fall into groups that have been reproductively disconnected for a number of generations. In some instances in which this occurs there may be incipient speciation; in others not. I deny that we can distinguish the two types of case by appealing to the pattern of actual matings.

Let's consider two examples in which the conspecifics in a region are fragmented into reproductively disconnected groups. The first is an idealization of what actually occurs among the Serengeti lions. Imagine that the females of a species divide into small groups, that these females mate with one or two males who become associated with a group for short periods, and that each male only has one chance to become associated with a group. Under these conditions there is no chain of animals in the population such that *a* mates with *b* who also mates with *c* who also mates with *d* ..., so that ultimately every member of the species in the region is connected to every other member. Moreover, if there is a large number of groups, and if there is a strong tendency for males to take over groups including offspring of the females in their mothers' groups, then there are likely to be males and females 'in the same population' who have no common ancestor in recent generations.

The second example is focused on our own species. It is all too familiar that there have been groups with very strong taboos or laws against various kinds of miscegenation. There are probably some instances in which these taboos have been and are still effective, so that, within a given region, people with different phenotypes have been reproductively disconnected for many generations. I doubt that we want to classify these cases as examples of incipient speciation or to declare that the people concerned belong to different populations. Instead, we want to talk of an extreme of assortative mating *within* a single population.

The moral of this section should by now be apparent. If the Hull-Ghiselin account is to be developed as a reply to the traditional problem of delineating the species taxa, then there is a serious task of analysing the notion of population or of devising some surrogate. If we are even to *understand* the thesis that species are 'historical entities', the difficulties that I have indicated must be faced and overcome.

3. The idiosyncrasies of isolating mechanisms

The breaks in the genealogical nexus that are depicted in branching diagrams and that Hull uses to indicate the views about species he regards as serious contenders are typically connected with the attainment of reproductive isolation between populations. Two populations are said to be reproductively isolated from one another if there are mechanisms that prevent interbreeding between their members where they occur together in nature or that would prevent interbreeding between their members if they did occur together in nature. Of course, organisms from populations that are reproductively isolated from one another may produce hybrid progeny in captivity, in the laboratory, or even in places where disturbances of the habitat have produced a large disruption of the normal way of life.[11] Moreover, it is possible for there to be some gene flow between reproductively isolated populations, for example, across stable hybrid zones. Introgression is not precluded, but it must not proceed on so wide a scale that the evolutionary autonomy of either species is threatened.

There is an apparent tension within those accounts that make reproductive incompatibility central to speciation, whether they do so in the classic way of Mayr (the biological species concept) or whether they pursue the idea that species are 'historical entities' whose boundaries are marked by episodes of speciation that involve the attainment of reproductive incompatibility. The tension arises from ideas about evolutionary autonomy, specifically:

(a) A small amount of introgression is compatible with reproductive isolation between populations.

(b) A low rate of migration between spatially separated populations (of the same species) is sufficient to ensure that these populations are not (effectively) isolated from one another.

Simultaneous acceptance of (a) and (b) seems problematic. If limited migration between spatially separated populations serves as the 'glue' that binds those populations together, making them parts of a single (scattered) species, why does limited gene exchange between populations that are classified as belonging to different species not serve equally effectively to bind those populations into the same kind of genetic/evolutionary unit?[12]

Notice that it won't do to try to solve the problem by insisting that *whenever* there's limited gene exchange the populations in question belong to different species – for, as we saw in the last section, we want to allow for assortative mating within a single species and for island populations of a continental species. There may well be limited gene flow among some subgroups of *Homo sapiens*, among some subgroups of Serengeti lions, among some groups of anoline lizards in the Caribbean, and among oaks in California and Quebec.[13] What we need is a principle for drawing the species boundary, and, in defining the task, it is helpful to pose the issue of what exactly we are attempting to map by speaking of species in the first place.

Remarks by Dobzhansky, and subsequently by Mayr, make plain the motivation for insisting on reproductive isolation. In a famous passage (1937, pp. 311–312), Dobzhansky introduces the notion of reproductive isolation as the key to the understanding of local diversity. Without the attainment of reproductive isolation, he suggests, gene flow would be uninterrupted, so that, in any locale, the effects on one group of organisms would be felt by the rest of the living residents. Dobzhansky's case for the importance of reproductive isolation presupposes a thesis about the homogenizing effects of gene flow. Even small amounts of gene exchange are taken to threaten the obliteration of genetic differences. Hence the principled division that we sought one paragraph back should explain just how much gene flow can be tolerated without making one group's 'evolutionary tendencies and fate' felt by the other.

But it is possible to question the presupposition on which the connection between speciation and the attainment of reproductive isolation depends. As several empirical studies have shown, gene flow in some groups of organisms is far weaker than orthodox evolutionary theorists had supposed: for example, detailed research on dispersal of pollen by insects and by wind has supported the conclusion that '[p]ollen and seed dispersal are either exclusively local or highly leptokurtic' (Levin and Kerster 1974, p. 202). Given this result, it is not easy to see how reproductive community serves as an explanation for the genetic (morphological, ecological) uniformity found in some widely distributed species. As two of the most influential critics conclude: 'Our suspicion is that, eventually, we will find that, in some species, gene flow is an important factor in keeping populations of the same species relatively undifferentiated, but that in most it is not. As this becomes widely recognized we will see the disappearance of the idea that species, as groups of actually or potentially interbreeding populations are evolutionary units 'required' by theory' (Ehrlich and Raven 1969, p. 1231).

Theoretical considerations also reveal that reproductive isolation is not a *sine qua non* for the development and maintenance of diversity. It is at least theoretically possible for considerable differences to evolve within an interbreeding population: even in the absence of barriers to gene flow, sharp differences in the frequencies of alternative alleles can be maintained (see Endler 1977, Roughgarden 1979, pp. 240–254).[14] Combining the theoretical study of clines with empirical results about gene flow, it becomes hard to sustain the thesis that attainment of reproductive isolation is necessary and sufficient for two groups of organisms to be subject to distinct evolutionary 'fates'.

Hull's account of the historical individuals that count as species is, I have suggested, incomplete, but his remarks about species fission (and possible fusion) seem wedded to the notion that the genealogical nexus is broken into species at those points at which reproductive isolation is attained (see his 1978, pp. 344–349). Not only is this approach vulnerable to the familiar objections about the status of species in nonsexual organisms, but, given the considerations that I have been raising here, we need a serious defense of the view that reproductive isolation is necessary and sufficient for the integrity of historical individuals (or for those historical entities that constitute species). It is not just that Hull and Ghiselin have failed to say which among several proposals for splitting the genealogical nexus they are inclined to favor, but we are owed an account of why *any* proposal involving the interruption of gene flow among populations should be seen as theoretically crucial to species diversity.

Let me extend the point by taking note of a response that proponents of the biological species concept have offered to suggestions that gene flow may be insufficient to promote the cohesion of 'conspecific' populations. Mayr writes:

> Physiologists and embryologists, likewise, have published evidence for a remarkable uniformity of physiological constants through the range of most species. The essential genetic unity of species cannot be doubted. Yet the mechanisms by which this unity is maintained are still largely unexplored. Gene flow is not nearly strong enough to make these species anywhere nearly panmictic. It is far more likely that all the populations share a limited number of highly successful epigenetic systems and homeostatic devices which place a severe restraint on genetic and phenotypic change. (Mayr 1963, p. 523; see also Mayr 1970, pp. 300–301).

This response threatens the priority of the concept of reproductive isolation by hinting at a quite different approach to the delimitation of species taxa. Each species taxon is to be associated with an epigenetic system (or a small family of such systems). The persistence of uniform phenotypes across the broad range of a species is to be explained by the difficulty of introducing new alleles that perturb the phenotype, and, by the same token, the distinctness of species is grounded in their having distinct epigenetic systems. No mention need be made of reproductive isolation. It might turn out that the distinctness of epigenetic systems coincided with the possession of isolating mechanisms, or that it did so in most cases, but the division of organisms into species (on this approach) would not rest on the fact of reproductive isolation. What would make organisms belong to different species would be their possession of different epigenetic systems.

If Mayr's account of the persistence of uniformities in phenotype through the prevalence of imperturbable epigenetic systems were correct, then not only would the biological species concept fail to identify the crucial features on which species identity and species difference rest but, more to our present point, species would not need to be characterized as historical entities. Species taxa would be individuated by (families of) epigenetic systems. Of course, we could impose the *additional* requirement that organisms sharing epigenetic systems (of the same family) belong to the same species only if they belong to populations that are historically connected. However, if one believed that it is the presence of the epigenetic systems themselves that explains uniformities and differences, then it would be hard to see this additional requirement as anything other than an *ad hoc* salvaging of the Hull-Ghiselin thesis. Why should we care about reproductive connections if evolutionary fates are fixed by the (family of) epigenetic systems?

I shall conclude my worries about reliance on the notion of reproductive isolation, by considering a disturbing possibility. One of the intuitive, pre-theoretical, ideas that we might have about species is that organisms are either conspecific or not, and that, in either case, there is no relativization to any third factor. Given the organisms, their intrinsic properties and the relations between them, the answer to the question 'Are they conspecifics?' is fixed. I do not wish to claim that this pre-theoretical idea is entirely precise, or that it is sacrosanct. However, if we appeal to reproductive isolation as a criterion for species distinctness (or as a criterion for the occurrence of a speciation event that has split the genealogical nexus) then it seems quite possible that there will be a necessary relativization to the environment. This could occur in numerous cases where there are actual or potential disruptions of the habitat with consequences for the cycles of activity of organisms that do not normally overlap. However, I want to consider a pure example in which a mechanical barrier to gene flow might be breached by the environment.

Fertilization in sea urchins involves three fusions between sperm and egg. The first of these involves the acrosome (at the head of the sperm) and a jelly that surrounds the egg: a receptor molecule on the surface of the sperm responds to glycoproteins in the jelly and the result is a change in the pH of the acrosome, a change that allows for release of acting and (ultimately) for the penetration of the egg by the sperm. Two species of sea urchins are distinguished by different glycoproteins at the egg surfaces and by different molecules that bind the glycoproteins to the sperm. The result is that sperm of *Strongylocentrotus purpuratus* cannot fertilize eggs of *S. franciscanus* because the reaction is blocked at the first stage. However, in the presence of trypsin, the glycoproteins *will* bind to the sperm, and, in consequence, hybrid progeny are produced.

To the best of my knowledge, *S. franciscanus* and *S. purpuratus* are isolated only by the mechanism just described. But it is plain that the isolation is environment-relative. In a trypsin-rich environment, there would be no barrier to gene flow between the two species. Now it is doubtful that there are any such environments inhabited by sea urchins – at least outside the laboratory. However, the example[15] points to a general possibility: populations may be reproductively isolated simply because a particular reaction in the formation of a zygote is blocked; however, the presence of certain molecules in the environment – perhaps as a result of abiological features, perhaps because of the presence of further organisms – might allow the reaction to go forward; thus it is quite possible that there are organisms that are reproductively isolated in one environment and not isolated in another (slightly different) environment. If the *very same* organisms had been situated slightly differently, the question whether they are conspecifics would have received a different answer. But perhaps the appropriate moral to draw here is that our initial view about the non-relativity of species relationships is faulty, and that, in our normal speech, we tacitly relativize to the kinds of environments that actually occur.

4. Segmentation and serendipity

Imagine that the problems of previous sections have been overcome and that we have successfully made sense of the concept of a population and of a principled notion of reproductive isolation. I'll suppose that we have understood a *lineage* to consist of organisms in some original population (the *founding* population) plus all their descendants, and that our residual task is to segment lineages by using the notion of reproductive isolation to characterize *separation events*. When a separation event occurs, some stages of the lineage just after the event belong to a different species than stages of the lineage from which they descended. The problem is to articulate the idea, specifying exactly how reproductive isolation relates to segmentation.

One proposal is to allow for speciation by anagenesis. Two lineage stages belong to different species if, had they coexisted, they would have been reproductively isolated. This proposal, essentially Simpson's, faces certain obvious difficulties of application – especially within the context of a gradualistic approach to evolutionary change. Notoriously, it has inspired some systematists to express their gratitude for the incompleteness of the fossil record, on the grounds that the gaps allow the delimitation of species taxa![16]

For many contemporary systematists, there is no hope of finding a principled division of lineages while allowing for anagenesis. Instead, we should recognize that the genealogical nexus is broken at those points where speciation produces two contemporary populations that are reproductively isolated from one another. Cladogenetic speciation is completed when the post-speciation descendants of the stages of the lineage preceding the speciation event divide into two groups that are reproductively isolated from one another. For Hennig, a species comprises the organisms on a branch of a lineage bounded by consecutive speciation events: 'The limits of the species in a longitudinal section through time would consequently be determined by two processes of speciation: the one through which it arose as an independent reproductive community, and the other through which the descendants of this initial population ceased to exist as a homogeneous reproductive community' (1956, p. 58). Hennig is committed to two claims that distinguish his account from Simpson's: (1) speciation by anagenesis cannot occur; (2) ancestral species cannot survive the events in which they give rise to daughter species.

Wiley (1981) has amended Hennig's approach to avoid one source of controversy, and his formulation of an evolutionary conception of species is explicitly designed to wed Simpsonian and Hennigian insights. On Wiley's account each species comprises the organisms on a branch of a lineage bounded by speciation events (not necessarily consecutive). Thus Wiley takes over (1), but does not commit himself to (2). He writes: 'Ancestral species may become extinct during speciation events if they are subdivided in such a way that neither daughter species has the same fate and tendencies as the ancestral species' (1981, p. 25). It is fairly clear what Wiley has in mind. If speciation occurs by geographical isolation of a very small population of the ancestral species, so that the full range of antecedent genetic (behavioral, ecological, morphological) variation is retained in that portion of the ancestral species that is *not* isolated, then, in a very obvious sense, the evolutionary history of the branch of the lineage containing the unisolated moiety is unaffected by what occurs on the branch that contains the isolate. Had a cataclysm simply eliminated the organisms that were actually geographically isolated, the subsequent evolution of the unisolated organisms would (at least initially) have been no different. But in this case, there would have been no speciation event, and hence no principled splitting of the lineage into two 'sibling' species that succeed one another temporally. Wiley proposes that ancestral species can survive speciation events if their range of variation is not substantially depleted, and this eminently reasonable idea enables him to cope with cases that Hennig finds troublesome.

The differences among Simpson, Hennig, and Wiley are easily displayed diagrammatically (see Figure 3). Hull (1978) reproduces similar diagrams, and points out, quite correctly, that it is a significant and difficult issue to choose among the corresponding positions. In the remainder of this section, I want to underscore the difficulties. The Hull-Ghiselin thesis that species are historical entities is committed to the view that there is some principled way of segmenting the genealogical nexus. I hope to show how each of the available principles of segmentation is problematic.

Here is the strategy. In motivating Wiley's departure from Hennig, I developed an argument that contrasts the actual course of evolution with a slightly different possible situation. In the transition from the actual history of the world to this possible situation, the intrinsic properties of and direct relations among stages of one branch of the lineage were left unmodified. Yet Hennig's criterion for species delineation was found to yield different conclusions for the organisms on this branch in the two cases. What discredits the criterion is our acceptance of the following principle:

(*) A proposal to count lineage-stages as stages of the same species should depend only on the intrinsic properties of and direct relations among those stages. It should give the same results in cases which differ only in the existence or properties of organisms occupying a different branch of the lineage.

I shall now try to show how appeals to (*) cast doubt on some of the most basic features of the idea that species are segments of the genealogical nexus.[17]

Let us begin with the thesis that Wiley shares with Hennig, (2), the ban on anagenesis. There is an old worry about this thesis. It is apparently possible that a lineage should evolve quite dramatically without splitting: imagine a world in which the lineage is founded by a population of protists and then evolves into *Homo sapiens* by the sequence of genetic changes that actually link us to our protist ancestors. On the Hennig-Wiley criterion, all the organisms in this lineage would belong to a single species.[18] This strikes many people as counterintuitive (even insane). I shall defend the example and develop it so as to make clear the source of the trouble.

Notice first that the Hennig-Wiley criterion cannot be protected by dismissing the imagined possibility as unreal. It will not do to protest that, in any world in which there was an undivided lineage linking the protists to humans, the laws of nature would have to be very different so that the Hennig-Wiley criterion would be inapplicable. We can describe a world, like our own in certain critical respects, in which the lineage is realized. At each point corresponding to a speciation event in the actual world the same kind of thing happens. Part of the ancestral population takes the first step toward speciation, and, as it does so, the relict of the ancestral population is wiped out. Objection: the story cannot be quite parallel, because the organisms that were eliminated would have exerted selection pressures on the evolving lineage, and, in their absence, the course of evolution cannot be the same. Reply: the selection pressures have to be made up in other ways; one possibility is to suppose that another (distinct) group of protists gives rise to a branching lineage in which organisms evolve to exert the right kinds of pressures on the unbroken lineage.

The heart of the problem can be understood by beginning with the hypothetical situation of the last paragraph and tracing a continuous path back toward actuality. Choose any of the actual branching points along the protist-human lineage – say the event in which the first mammalian species originated from part of the ancestral population. In the hypothetical world, we assume that the first mammalian species

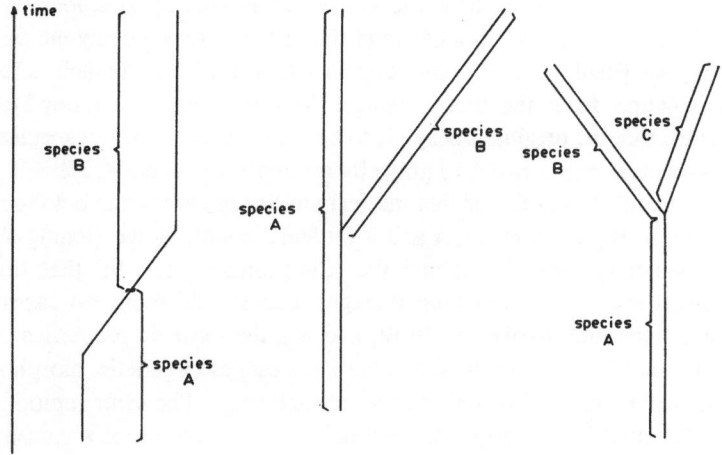

Allowed by: Simpson   Allowed by: Simpson, Wiley   Allowed by all
Banned by: Hennig, Wiley   Banned by: Hennig

Fig. 3. **Three Proposals for Splitting Lineages.**

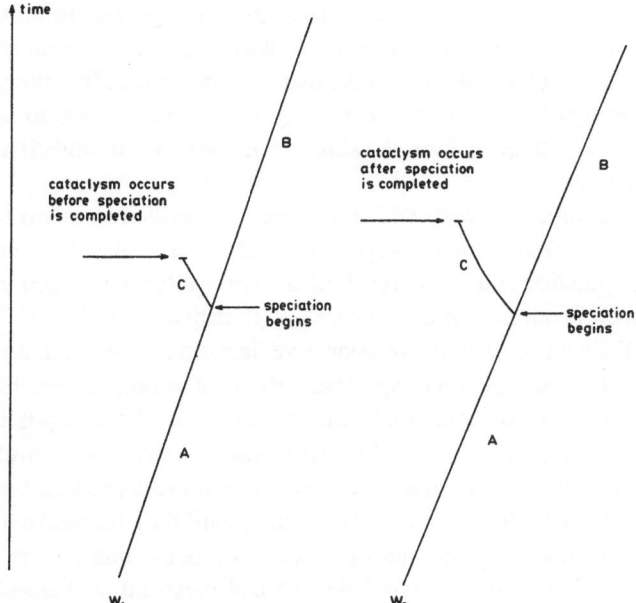

Fig. 4. **A Puzzle for the Hennig-Wiley Criterion.**

survived a cataclysm in which the rest of the ancestral population was wiped out. Now let us suppose that the time of the cataclysm is slightly postponed – the avalanche comes or the river floods a day later than before. As we delay the time of the catastrophe, we finally obtain a situation in which the relict branch achieves reproductive isolation from the main lineage. At this point, the Wiley-Hennig criterion demands that the original lineage is to be split into two distinct species.

The thought-experiment is easier to grasp by reference to Figure 4. Here W_1 is a world in which A and B (and C, for that matter) are lineage segments belonging to the same species. In W_2, by contrast, A and B (at least) count, by the Hennig-Wiley criterion, as distinct species. To defend the 'Simpsonian intuition' that lineage splitting is forced even in unbranching lineages, one should focus on cases like those contrasted here, and invoke (*). In W_1 and W_2, the intrinsic properties of the organisms in the $A+B$ lineage are the same: the same ranges of genetic, morphological, behavioral, and ecological variation occur at each stage. The same reproductive connections hold along the lineage. All that differs is the timing of a catastrophe that affects only organisms on a *different* branch. Appealing to (*), I claim that the difference is extraneous to the organisms in $A+B$, and that a proper division of the organisms of $A+B$ into species ought to yield the same result in each case.

Allowing for anagenesis would, of course, leave us with the puzzle of *how* to allow for anagenesis. That topic deserves a paper of its own, and I shall not pursue it here. Instead, I want to pose a problem, of the same general form, that strikes at all versions of the thesis that species are historical entities – including those that articulate the thesis along the lines indicated by Simpson. Unlike the argument just offered, we do not have to countenance any exotic possibilities to appreciate the force of the puzzle. It arises form the simple possibility of 'dumbbell allopatry' as a mode of speciation.

Imagine an evolving lineage which, at time t, is divided into two roughly equal halves by the interposition of a geographical barrier. Assume that, at t', the descendant populations on each branch of the lineage have diverged to a sufficient extent that each behaves as a good species with respect to the other. The criterion of species distinctness can be reproductive isolation – or something different, provided only that termination of speciation should conform to a familiar biological fact, to wit that speciation need not be instantaneous and that it is possible to talk of lineages as undergoing events of speciation (not necessarily at a uniform rate). Suppose, further, that from t' a condition of stasis prevails, so that the two lineage branches persist unmodified for a million years, until they become extinct. Finally, let us add the condition that the divergence of both branches is minimal for complete speciation. If the ancestral lineage had persisted unchanged beyond the point of geographical bifurcation, its subsequent stages would not be sufficiently distinct from the stages on either branch to count as a separate species. In other words, each incipient branch retained the full range of variation present in the ancestral lineage, there are evolutionary changes along both branches, and these

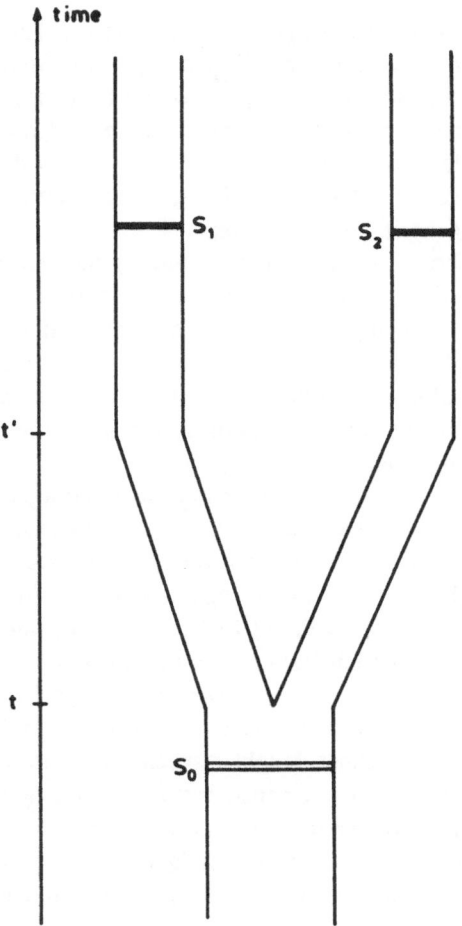

Fig. 5. Species Survival and Identity.

are, together but not separately, sufficient for speciation.

The envisaged situation is represented in Figure 5, where the horizontal axis represents whatever kind of change is taken to be relevant to speciation. By the criterion for speciation S_1 and S_2 are stages of different species. However, by the same criterion, S_1 and S_0 count as conspecific, and so do S_0 and S_2. Hence we face an apparent paradox: there is a species that embraces S_1 and S_0 and a species that embraces S_0 and S_2, but no species that embraces S_1 and S_2.

Formally, this is not a paradox. No contradiction arises unless one holds that for any organism there is at most one species to which it belongs. But if one retains that principle then one must decide which of the judgments about conspecificity to abandon. One approach (Hennig's) is to declare that the ancestral species becomes extinct at t, at which point two daughter species are born. This response falls foul of the argument given in motivating Wiley's departure from Hennig. Had either branch become extinct shortly after t, we would be happy to count the residual branch as a continuation of the ancestral species. Moreover, the situation is symmetrical. Worlds in which either branch survives and the other terminates are happily seen as worlds in which an evolving lineage gets about half way through what looks like anagenesis – and then stops.

Whatever condition is proposed for guaranteeing the persistence of the ancestral species – retention of full range of genetic variation, for example – can be built in to the scenario. Once again our judgments of conspecificity are grounded in (*). Because the time of extinction of one branch does not make a difference to the intrinsic properties of stages on the other branch, or to the direct relations among them, whether or not those stages belong to the same species cannot depend on whether or not the first branch persists.

If cases like this were to occur (and perhaps they do) a purely formal solution to the problem could be obtained by allowing the same lineage-stage to belong to two different species.[19] Just as two different roads can overlap on the same piece of tarmac, so, we might say, the same lineage-stage can be included in two different species. Biologists, I suspect, will not find this formal solution attractive. A more plausible way of treating such instances is to let one's judgments about division into species conform to the current needs of biological research. For someone investigating the acquisition of reproductive isolation, it might be appropriate to count both branches as distinct daughter species. On the other hand, a biologist concerned with the developmental constraints imposed upon organisms by the facts of their ancestry might prefer to view the branches and the ancestral lineage as constituting an (unusual) single species. Judgments about such cases seem to rest on whether one is more interested in the distinctness of the descendant branches or in their kinship with their common ancestor. I believe that there is no single, objectively right, way to segment the entire lineage into species. Various ways of proceeding offer partial solutions, emphasizing some biological features of the situation and downplaying others. I propose (once again) that we take a pluralistic view of species, allowing that there are equally legitimate alternative ways of segmenting lineages – and indeed legitimate ways of dividing organisms into species that do not treat species as historical entities at all.

5. Conclusions

To say more about pluralism and its virtues would change the focus of this paper. What I have been attempting to show is the extent of the work that needs to be done if Hull's conception of species as historical entities is to cope with the diversity of organisms. Pluralism enters the discussion only because it offers a way out of an apparent difficulty in segmenting lineages. Hull (1987) has complained that pluralism is the counsel of despair, and that monistic proposals for understanding species deserve a run for their money. There is surely a sound point here. Numerous instances from the history of science reinforce the judgment that theories need time to overcome apparently devastating objections. However, what concerns me about the proposal that species are historical entities is that the *difficult* problems about delimiting species taxa seem to have become invisible. As I read the recent literature – for example (Eldredge 1985, Mayr 1987, Ghiselin 1987, Hull 1987) – an unarticulated version of the proposal seems to be serving as the basis for suspiciously a-priori-looking arguments about evolutionary processes, while issues about the articulation of the proposal are ignored. My aim here has been to bring *some* of the problems back to center stage.

Of course, not all apparent puzzles deserve serious and sustained attention. Most philosophers are familiar with the dismaying degeneration that characterizes fields in which energy is lavished on counterexamples of no theoretical importance. One of the great merits of David Hull's approach to philosophy of biology has been his plea for the use of realistic examples and his dismissal of worries based on unconstrained philosophical fictions. I hope that the examples I have constructed are simply pure types of actual biological situations, so that they will strike him as the kinds of puzzles that his account of species will have to overcome. In this spirit I offer him, not a knockdown argument for pluralism, but just some puzzles about species.

Acknowledgement

Thanks to Michael Ruse for conceiving the idea of this volume and for inviting me to contribute to it.

Notes

1. The debate about 'the ontology of species' begins with (Ghiselin 1974), (Hull 1976, 1978). Criticisms are launched in (Kitts and Kitts 1979) and (Caplan 1981). Hull replies to these in his (1981). My own objections to the Hull-Ghiselin view are presented in my (1984a, 1984b, 1987), and some of my concerns are addressed in (Hull 1987). Sober (1984) is a reply to my (1984a) and my (1984b) attempts to rebut Sober's objections.
2. The following argument is briefly presented in (Hull 1987), but Hull has offered a more extensive version of it in conversation.
3. Talk of the persistence of sets all of whose members are physical objects is tricky. In one obvious sense, any set is an abstract object and therefore exists atemporally. But there is another notion of persistence that underwrites the intuition about the organization of organisms that I am attempting to articulate here. According to this notion, a set of physical object persists just so long as all of its members exist. When I speak of the

persistence of sets of physical objects, I shall be employing this latter notion.

4. This is the type of organization that Hull appears to emphasize in his (1978) – see for example p. 342. He has made the point even more explicitly in conversation.

5. This elementary logical point seems to have been very difficult to grasp – see for example Mayr's confession of bewilderment in his (1987) and Hull's (1987) acknowledgement that he shares Mayr's bewilderment. My (1987) tries to forestall the confusion, but perhaps I can make the point even more obvious by noting that the claim 'All A's are B's' ('All species are sets', 'All species are individuals') doesn't entail 'All B's are A's' ('All sets are species', 'All individuals are species').

6. It is worth noting that the concept of historical connection that I have just introduced isn't strong enough to generate Hull's 'conceptual point' (Hull 1978, p. 349) that species cannot re-evolve. To require that species can't become extinct and then re-appear, one needs a condition of *complete historical connection*: an entity with organisms as parts (or members) is completely historically connected just in case for any two organisms belonging to the entity there is a sequence of populations, all of the organisms in which belong to the entity, such that each population in the sequence is either an immediate ancestor or an immediate descendant of its predecessor and such that the organisms in question belong to the first and last members of the sequence, respectively. I don't hold Hull and Ghiselin to this requirement, because it seems to me to be incorrect. In my (1984a), I offered a hypothetical case based on what we know about species of the genus *Cnemidophorus* to suggest that the same species might have a discontinuous career. Examples of a similar kind are probably legion among microorganisms, and the fact that we are eukaryotes shouldn't prevent us from seeing the need for a species concept that can be applied to bacteria and viruses as well.

7. A paradigm example seems to me to be (Eldredge 1985), where I think that some very important ideas are obscured by developing and defending them in quite the wrong way.

8. A cautionary note: it is easy to draw branching diagrams and to canvass possibilities by appealing to them. But it is always worth asking how we link the organisms that the naturalist observes to the branching diagram. By this I do not simply mean to pose the cladists' central question of 'retrieving phylogeny' (in Elliott Sober's apposite phrase) but also to point out that we need to be told *precisely* what the phylogenetic branching diagram is supposed to represent. As Hull himself noted in a classic paper (1968), there is all the difference in the world between asking how we obtain evidence for a classificatory judgment and what the classificatory judgment means. So we should wonder not only how we are to find branch points and how we are to assign organisms to lineages, but what 'branch point' and 'lineage' *mean*.

9. Exercise for the reader; show that any statement made in subsequent pages that uses set-theoretic notions (e.g. *is a member of*) can be replaced without modification of empirical content by a mereological expression (e.g. *is a part of*) provided that the translation is done systematically, and that the converse is also true. Ambiguous expressions (*belongs to*) can be read either way.

10. The point that I have been making is an attempt to articulate what I regard as a deep insight in the critique of the biological species concept advanced by Sokal and Crovello (1970). The worry is not just that we cannot obtain *evidence* for reproductive community without introducing considerations of phenotypic similarity, but that the *concept* of reproductive community presupposes a projection from the events of actual mating, a projection itself defined by continuities and discontinuities in biological properties. Sokal and Crovello present their case in terms of the decision procedure of a field naturalist, but many of the observations they make can be freed from the emphasis on 'operational critria for application' and reformulated as points about the concept of reproductive community.

11. A classical example is that of the towhees (*Pipilo erythrophtalmus* and *Pipilo ocai*) which occur together in some places without interbreeding, but which hybridize freely where agricultural disturbance has disrupted their habitat. For concise discussion, see

Mayr 1970, pp. 73–74, 1963, p. 121.

12. Roughgarden (1979, Chapter 12) provides a review of some basic results about conditions under which migration of alleles among spatially separated populations is likely to be an evolutionarily significant force. He considers the relative strength of effects of migration and drift, and the interplay between migration and a spatially variable selection pressure. However, I have not seen any detailed discussion of the question when introgression in peripheral populations of a species becomes sufficient to break down the autonomy of the species. The conditions for introgression to be important can be expected to depend on (a) the spatial distribution of the populations of both species, (b) the migration rate among conspecific populations, (c) the selection pressures operating on different loci in different populations, and (d) the effective sizes of the populations in question. It seems possible that the phenomenon of introgression in hybrid zones should be asymmetrical: in other words, that genetic changes in one population should spread into the other population, but that there should be no significant flow in the opposite direction. Were this to occur, there would be a fundamental difficulty in talking about reproductive isolation as a symmetrical relation between populations.

13. The last example is due to Leigh van Valen. See his extremely interesting – and under-read – (1976).

14. My formulation here is deliberately conservative. Many writers would insist that morphological differences and isolating mechanisms can evolve without the interposition of a geographical barrier. See, for example, (Bush 1975), (White 1978), and, for a response, (Futuyma and Mayer 1980). If they are correct, then the case for making reproductive isolation crucial to the understanding of species diversity is even weaker than I have portrayed it as being.

15. For information and discussions relevant to this example, I am extremely grateful to William Loomis. For details on sea urchin fertilization, see (Vacquier 1979) and (Podell and Vacquier 1984).

16. See, for example, a first-rate textbook in paleontology, (Raup and Stanley 1978, p. 111). I should note that Raup and Stanley also lament the 'valuable information' that has been lost, so they do perceive the gappiness of the record as a *mixed* blessing.

17. Note that (*) is connected with the intuition that I canvassed in the last section in discussing the possible environment-relativity of species divisions. I don't rule out the possibility that we might ultimately want to reject (*), but I think that any such rejection would have to be based on a thorough scrutiny of the central ideas, aims and presuppositions of systematics, the type of investigation for which I have argued in my (1984b, pp. 628–630). In a line, philosophical discussion of species concepts ought to return to the issue of what we are after in devising a scheme for mapping organismic diversity, the issue that was originally posed by Mayr and Dobzhansky in the 1930s and 1940s, but that has become lost in subsequent discussions. The puzzles of this paper are intended as motivational preludes for the return.

18. One can't duck the issue by claiming that all the organisms belong to a single *taxon*, *Eucaryota*. For, if we believe that every organism belongs to a species, then trivially for each organism in the lineage there is a species to which it belongs. By the Wiley-Hennig criterion, any two organisms in the lineage are conspecific. Hence there is a unique *species* to which every organism in the lineage belongs. It makes little difference whether one calls this species *Eucaryota integra* or *Homo sapiens*.

19. Aficionados of the literature on personal identity will recognize the problem I have posed and the line of solution I indicate here. A similar puzzle and similar canvassing of possible solutions is given by Laurance Splitter in a forthcoming paper.

208

References

Bush, G.L. (1975) Modes of animal speciation. *Annual Review of Ecology and Systematics* 6: 339–364.

Caplan. A. (1981). Back to class: a note on the ontology of species. *Philosophy of Science* 48: 130–140.

Dobzhansky, T. (1937). *Genetics and the Origin of Species*. New York: Columbia University Press.

Ehrlich, P.R. and Raven, P.H. (1960). Differentiation of populations. *Science* 165: 1228–1232.

Eldredge, N. (1985). *Unfinished Synthesis*. New York: Oxford University Press.

Endler, J. (1977). *Geographic Variation, Speciation, and Clines*. Princeton: Princeton University Press.

Futuyma, D. (1979). *Evolutionary Biology*. Sunderland MA: Sinauer.

Futuyma, D. and Mayer, G. (1980). Non-allopatric speciation in animals. *Systematic Zoology* 29: 254–271.

Ghiselin, M. (1974). A radical solution to the species problem. *Systematic Zoology* 23: 536–544.

Ghiselin, M. (1987). Species concepts, individuality, and objectivity. *Biology and Philosophy* 2: 127–143.

Hennig, W. (1966). *Phylogenetic Systematics*. Urbana: University of Illinois Press.

Hull, D. (1968). The operational imperative – sense and nonsense in operationism. *Systematic Zoology* 17: 438–457.

Hull, D. (1976). Are species really individuals? *Systematic Zoology* 25: 174–191.

Hull, D. (1978). A matter of individuality'. *Philosophy of Science* 45: 335–360.

Hull, D. (1981). Kitts and Kitts and Caplan on species'. *Philosophy of Science* 48: 141–152.

Hull, D. (1987). Genealogical actors in historical roles. *Biology and Philosophy* 2: 168–184.

Kitcher, Ph. (1984a). Species. *Philosophy of Science* 51: 308–333.

Kitcher, Ph. (1984b). Against the monism of the moment: a reply to Elliott Sober. *Philosophy of Science* 51: 616–630.

Kitcher, Ph. (1987). Ghostly whispers: Mayr, Ghiselin and 'the Philosophers' on the ontology of species. *Biology and Philosophy* 2: 184–192.

Kitts, D.B. and Kitts, D.J. (1979). Biological species as natural kind. *Philosophy of Science* 46: 613–622.

Levin, D. and Kerster, H. (1974). Gene flow in seed plants. *Evolutionary Biology* 7: 139–220.

Mayr, E. (1963). *Animal Species and Evolution*. Cambridge MA: Harvard University Press.

Mayr, E. (1987). The ontological status of species: scientific progress and philosophical terminology. *Biology and Philosophy* 2: 145–166.

Podell, S. and Vacquier, V. (1984). Wheat germ agglutin blocks the acrosome reaction in *Strongylocentrus purpuratus* sperm by binding a 210,000 mol. wt. membrane protein. *Journal of Cell Biology* 99: 1598–1604.

Raup, D. and Stanley, S. (1978). *Principles of Paleontology*. San Francisco: Freeman.

Roughgarden, J. (1979). *Theory of population genetics and evolutionary ecology*. New York: Macmillan.

Sober, E. (1984). Sets, species and evolution: comments on Philip Kitcher's 'species'. *Philosophy of Science* 51: 334–341.

Sokal, R. and Crovello, T.J. (1970). The biological species concept: a critical evaluation. *American Naturalist* 104: 127–153.

Vacquier, V. (1979). The interactions of sea urchin gametes during fertilization. *American Zoologist* 19: 839–849.

Van Valen, L. (1976). Ecological species, multispecies, and oaks. *Taxon* 25: 233–239.

White, M.J.D. (1978). *Modes of Speciation*. San Francisco: Freeman.

Wiley, E.O. (1981). *Phylogenetics*. New York: Wiley.

The Rational Weight of the Scientific Past: Forging Fundamental Change in a Conservative Discipline

LARRY LAUDAN

University of Hawaii

Introduction

(One of the recurrent foci in David Hull's research has been the character of history, especially history of science. He has, indeed, been an eloquent defender of enlightened whiggism in history of science during decades when, for all the wrong reasons, whiggism has gone out of fashion. At the heart of the debates between anti-whigs and people like David Hull has been a disagreement about whether the past has a legitimate *justificatory* role in contemporary debates about science. I have written this essay for this volume as an effort to sketch out a picture of one important kind of justification which the history of science has often been expected to provide.)

For well over a century, scientists and philosophers of science have been struck by various apparent continuities in science. Thus, it has been commonly maintained (1) that earlier theories are limiting cases of later theories, (2) that successor theories explain all the successes of their predecessors, and (3) that later theories entail their predecessors. The conviction that there are pervasive temporal continuities in science was, of course, a cornerstone of both positivist and pre-positivist accounts of scientific knowledge. Within recent times, virtually all these claims have been seriously and (in my view) successfully challenged.[1]

However, as the case for wholesale retention of empirical content through theory transitions has been rapidly crumbling, many philosophers of science have begun to seek for continuities elsewhere – specifically in the aims and standards of the scientific enterprise. On this view, even if the *content* of our beliefs about the world shifts dramatically from epoch to epoch, there are nonetheless certain broad epistemic goals and methodological standards which have characterized science since its inception. Scientists, we are told, are and always have been seeking to find out the truth about the world and have utilized familiar methods of observation and experiment to ferret out those truths. This supposed constancy at the level of methods and aims avowedly gives science its coherence and individuates it from other, more ephemeral intellectual practices, lacking a fixed mooring in sound methods. Unfortunately, this view of the fixity of the aims and methods of science

M. Ruse (editor), What the Philosophy of Biology is. pp. 209–220.
© 1989 *Kluwer Academic Publishers, Dordrecht*

has recently come under such sustained criticism that it now looks about as unconvincing as its Parmenidean counterpart concerning the retention of content through theory change. The view of science which is now emerging in some quarters is thus wholly Heraclitean, insisting that science, diachronically viewed, changes its content, its methods and its aims.

Many philosophers are understandably troubled by the implications of this new picture of science. If there is no 'essence' which once and for all constitutes science as we know it, then what (they ask) would prevent science from evolving into something wholly other? These closet demarcationists rhetorically wonder whether science might turn into football or witchcraft if there is no permanent set of features which unambiguously marks it off from other forms of human cognitive activity. This particular worry is largely misplaced. Biologists long ago taught us how to identify and mark off classes and species of things without invoking eternal and unchanging characters. Hence the fact that the aims, methods and content of 'science' shift through time is by itself no obstacle to our finding ways of distinguishing science as *it presently exists* from other forms of human belief and activity, as they currently exist.

But there is something else which *is* genuinely disturbing about a thoroughly Heraclitean view of science, a kind of unstated apparent corollary of it. We might call it *the stochastic premise*. If the theories, methods, standards and aims of science are all in principle open to change, then it might seem that science could transform itself overnight into something radically different, even wholly unrecognizable. If all equally defeasible, then the aims, methods, standards and fundamental theories of a science could conceivably change all at once, so that there were no significant continuities whatever between what went before and what came after. A handful of scholars have maintained that precisely that kind of upheaval occurs from time to time; such, for instance, is what Foucault's 'ruptures of thought' evidently amount to.

I daresay that most of us believe there has never been such an abrupt and wholesale transformation of scientific thought (at least not since early antiquity). Quite the contrary, historians of science have taught us that scientific change is typically an extraordinarily gradual process. Even the great events which we honor with the title 'revolutions' exhibit startling continuities – sometimes of method and standards, sometimes of theory and experimental technique – with what went before.

By now our problem should be clear: if there are no traits which permanently constitute science, if science could in principle wholly reform itself every few years, why has scientific change been so (relatively) piecemeal and transitional? Why, to put it differently, is scientific change so locally conservative, when there appears, at least in the abstract, to be the possibility of abrupt, radical change across the board? That, in brief, is the question which motivates this essay.

One part, the less interesting part, of the answer to that question arises from the sheer intellectual difficulty of thorough-going, radical innovation. It is hard enough in any advanced science to produce plausible new theories *or* new methods *or* new standards. Concocting all at once may require the sort of Herculean effort which no mere mortal or even any single generation is capable of doing all at once. But there is another, and more interesting, reason for the drastically limited character of radical change in science. It has to do with what we might call *the rational weight of history* in directing and constraining far-reaching scientific change.

A general thesis about changing standards

To such first-generation modelers of scientific change as Thomas Kuhn, changes in standards and changes in core theories invariably went hand-in-hand. Rival paradigms always differed about both substantive matters and about standards. Subsequent research has led most of us to a rather more complex picture, according to which modifications in standards and in underlying ontology need not co-vary. Sometimes, of course, they do. Arguably, the mechanical philosophy of the 17th century sought to displace traditional views both about what is in the world and about how theories about that world should be legitimated. But there have been numerous scientific revolutions which involved mighty shifts in underlying ontology without any significant change in standards (e.g., the emergence of relativity theory or of benzene theory in chemistry). Equally, there are times when standards have shifted without a contemporaneous shift in the core ontology of a discipline (e.g., the emergence of statistical methods in 19th-century physics, the emergence of controlled experimental methods in 19th-century science generally, the rise of blinded experimental techniques in 20th-century pharmacology).

The focus of this essay will be on the general mechanics of standards change, whether those changes be accompanied by, or independent of, changes in fundamental theories. In an earlier work (*Science and Values*), I already addressed several aspects of this question. My concern there was chiefly with the logical and epistemic strategies open to scientists engaged in discussions about the respective merits of rival standards. The up-shot of that discussion was that any proposed new standard, before it can become acceptable, must satisfy certain demands of (a) logical coherence, (b) empirical realizability and (c) practical workability.[2] I want here to focus on a rather different, and decidedly more historicist (but no less epistemic), component of situations involving the proposed modification of standards.

Suppose that a scientist challenges the prevailing standards in his field by seeking to displace them with a quite different set of standards. What constraints (apart from those just mentioned) must the new standards satisfy? How can they be legitimated? Openly challenging the standards which have guided a discipline through significant parts of its development is, on the face of it, a fundamentally *revolutionary* activity. It involves the proposal that one should abandon the rules, conventions, procedures and understandings which (so it is widely supposed) have been responsible for much of the discipline's success – assuming, of course, that the discipline can lay claims to any notable successes. Put differently, the advocacy of new standards for a successful field potentially puts at risk the prospects of that field's continued flourishing. Clichés like 'if it ain't broke, don't fix it' or 'nothing succeeds like success' point to the kind of challenge faced by someone who would propose to transform the rules and standards by which a hitherto successful practice has been conducted.

Of course, none of these worries would apply if the discipline in question were notably unsuccessful. Chronic failure of a discipline to produce light rather than heat is an invitation to frequent and abrupt re-definitions of its basic standards. (Witness the ease and rapidity with which new standards are adopted in fields like literary theory or certain social sciences.) But that is not the sort of case we are dealing with here. Instead, we are attempting to understand how change of standards is negotiated in scientific disciplines which have noteworthy explanatory or predictive successes to their credit. Since it is natural to believe that the previous success of a discipline may have been the result of honing to the standards which guided research in that discipline, any proposal to replace those standards by new and untried ones runs the risk that the discipline will simply run out of heuristic steam. Again our question arises: how can one make the case for new standards for a discipline which gives the appearance of having flourished under the prevailing regime?

To put it simply, imposing new standards onto an already successful discipline rests on the ability of their advocates plausibly to *re-describe the history of the discipline*. That re-description will seek to show: (1) that the canonical achievements (or achievers) as recognized in the folklore of the field would still have been realized even if (counterfactually) the heroes of the discipline had been working within the guidelines associated with the new standards instead of the actual standards they were using; and (2) that although the older standards superficially appear to have been involved in producing the canonical achievements of the field, their actual role in producing those achievements was tangential and adventitious. I believe that historical research will bear out the claim that efforts to displace old standards by new ones has been successful in the natural sciences just to the extent that advocates of the latter have been able to mount a plausible case for *both* (1) and (2). The remainder of this essay will be devoted to fleshing out, and in minor respects qualifying, this strong claim.

Preserving the tradition

The picture I want to sketch about these matters has several key elements. To begin with, there is what I shall call the *Tradition*. It consists of certain historical achievements in the discipline which are regarded as landmarks and bench marks. Typically, they constitute both the paradigm cases of exemplary practice within the science and they (or elements or versions of them) continue to be used by practitioners of the science, either for pedagogical or research purposes. Obvious cases in point would be Newton's three laws of motion, Maxwell's equations, Hubble's techniques for estimating the size of the universe, or Pasteur's work on fermentation and spontaneous generation. In addition, before an historical achievement qualifies as part of the Tradition, it must address *central* issues in the field; it must offer solutions to absolutely core problems, problems which are close to heart of what the discipline is about.[3]

Who constructs the Tradition for a field and where it is written down? In a few cases, the Tradition is quite explicitly formulated as such and is initially the product of construction by a single individual or school in influential works explicitly devoted to history. Charles Lyell, for instance, in the long historical introduction to his *Principles of Geology* tells a tale about the history of geology which did much to define the Tradition of geological achievement for a century and more. Ernst Mach sought, with considerable success, to construct the Tradition for the history of mechanics; as Joseph Priestley earlier had for electricity, and before him as Johannes Kepler had done for astronomy. But it is not primarily explicit essays in the history of science which define the Tradition; the sequence is normally just the other way around. Histories of a science normally have their tables of contents dictated by prevailing perceptions about who and what makes up the Tradition. Those perceptions themselves arise out of textbooks on the subject, with their occasional vignettes devoted to the 'heroes' of the field. Those perceptions are yet more heavily shaped by the contents of the texts themselves. Physics students do not learn the law of inertia or the equations of electrodynamics; they learn 'Newton's laws' and 'Maxwell's equations'. The central laws and concepts of a science frequently come with the discoverer's name indelibly appended. The more central the law or concept in a contemporary science, the more likely that the work of its discoverer will find its way onto the Tradition. The Bohr and Rutherford models of the atom would probably be part of the Tradition for elementary particle physics. Numerous other atomic models from their era, however, including ones which were arguably as good, would not be.

Inclusion in the Tradition depends in part, too, on the pedagogical practices in the discipline: so long as 19th-century physicists learned their electrical theory by doing variants of Cavendish's classic work on charge-determination, that work certainly counted as part of the Tradition. It depends, as well, on whose work comes to be associated with the most-widely used laws and principle of the discipline. In sum, then, the Tradition is an artifact woven of threads made up of the stories that the practitioners of a discipline tell one another (and their students) about their past. But although sometimes mythic in fact, it is not deliberately mythic. The Tradition is meant to represent the best understanding about how the discipline in question came to be as it is; it is meant to be about the great turning points in its history and about the monumental contributions of its greatest practitioners.

We must be careful, however, not to suggest that the Tradition is more sharply defined that it often is in the natural sciences. In fact, there is much fuzziness around its borders and edges. One theoretical physicist's account of exactly what makes up the Tradition of physics would exhibit significant divergences from another's. Moreover, certain figures would be more squarely at the center of the Tradition, others towards its periphery. But in any advanced science, this much would certainly be so: there would be very broad agreement that certain specific individuals and certain achievements deserved a place of price within the Tradition, even if there might be disagreement about the centrality of other individuals and achievements.[4]

Anyone who has ever studied natural science will, I believe, recognize what I am describing as 'the Tradition' and its ubiquity in the pedagogy of science. But what is not so well understood is the role that the Tradition plays in debates about the standards, methods, and aims of the discipline. Indeed, I submit that the key, if often unintended, role of a discipline's Tradition is to qualify or disqualify proposed sets of standards. Specifically, any newly proposed standards for a science must, as a condition of their acceptability, be able to capture the (bulk of the) canonical achievements which make up the Tradition of that science. What 'capturing' amounts to is this: *it must be possible to show that, if the newly proposed standards had been in place in the past, they could have produced the achievements that make up the Tradition.* Thus, if someone proposes a set of standards for physics which entails (say) that Newtonian celestial mechanics is less 'scientific' than Cartesian mechanics or which suggests that Einstein's analysis of Brownian motion was bad physics, then those standards would have no chance of being accepted by physicists.

Two questions inevitably arise: Why must new standards satisfy this historical demand and why can they not be justified entirely on their abstract epistemic merits? Those are large questions to which we shall return repeatedly. But at this point in our analysis, the best answer I can give is this: confronted by a choice between a successful practice, on the one hand, and a plausible (*and* compatible) set of standards, on the other, scientists will happily accept both. But when it is impossible to do so (as, for instance, when the standards would make it impossible to have replicated the successful practice), then scientists express a preference for a proven practice over plausible standards incompatible with that practice. This, of course, is not yet to answer our questions, but it moves us one step in the right direction.

In any practice which is successful, there is a natural tendency to think that 'we must be doing something right'; otherwise how explain the success of the practice? But, or so one usually supposes, what codifies that successful practice – what makes it successful – is precisely the standards which guided experimental design, theory choice, etc., in the course of that practice. If someone now proposes to emend or drastically alter the standards associated with a successful practice, in ways which do not allow us to retain the most notable products of that practice, then it is wholly appropriate to resist the new standards – whatever their abstract epistemological merits might be.

To put the point thus, however, appears to beg a central question at issue; to wit, the apparent parasitic dependence of determinations of the success of a practice on the standards one has in mind. A critic might say: 'Look, whether a particular practice is successful in something we judge in light of a particular set of standards; hence one and the same practice may look successful or not depending upon what standards we have in mind. Insisting that any new standards for a putatively successful practice must allow the reconstruction of that practice *as successful* is to stack the deck hopelessly against the discovery of radically new, but perhaps significantly better standards.'

There is an important point here (to which we shall return below), although it is *not* the one it appears to be. The apparent thrust of this criticism (viz., that judgments of the success of a practice are relativised to the epistemic standards one is appraising) depends on a massive equivocation about what 'standards' are. Scientists' judgments as to the success of a scientific practice depend not on abstract epistemological and methodological matters but on palpably *pragmatic* ones. A science is successful just insofar as it manages to confer some measure of control of the subject matter under investigation on the practitioner who has mastered the practice. Thus, a medical practice is successful or not depending on the degree to which it gives its initiates the ability to predict and to alter the course of common diseases. An astronomical practice is successful to the extent that it enables one to anticipate future positions of planetary and celestial bodies. A theory of optics is successful if it can (say) predict the path of a light ray moving through various media and optical interfaces.[5]

These workaday assessments of the success of a practice are to be sharply contrasted with the often highly detailed epistemic and methodological judgments associated with 'standards'. Those latter standards will address questions like: when an experiment has been suitably designed, when an explanation of phenomena has been produced, when a theory enjoys a particular degree of credence, etc. Scientists generally do not invoke detailed standards of this sort to determine whether, viewed in the large, the theories of a discipline have been 'successful'.[6]

If my suggestion that there must be a pre-philosophical notion of empirical success – which is not itself beholden to elaborate epistemic or methodological doctrines – seems controversial, it is worth asking how it could be otherwise. Scientists, after all, generally do not learn confirmation theory or inductive logic or any of the other arcana of 'high epistemology'. If judgments of the success of science had to depend on familiarity with such esoterica, scientists would have no grounds for making 'success' judgments at all.[7] It therefore seems reasonable to conclude that we can generally ascertain whether a particular practice has been successful independently of subscribing to any particular model of scientific explanation or confirmation or inductive support.[8]

Having once determined that a particular practice is successful, we then judge a newly-proposed standard for the conduct of science by asking whether its utilization would have allowed that practice to arise and flourish.[9] Let me not be misread. I am not here insinuating that the judgments made by scientists about the success of theories are deeply inscrutable, that they are manifestations of Polanyi-esque 'tacit' knowledge, or that they are otherwise inexplicable or ineffable. In fact, the criteria guiding judgments of pragmatic success are pretty hum-drum and wholly accessible. They depend on answers to questions like: Does the theory work? Does it 'save the phenomena'? Can we use it for making effective interventions in the natural order? Does it enable us to foretell unusual events?

But, for the most part, the spelling out of such pragmatic criteria of success has not been the central task of epistemology or methodology, nor even of the epistemic standard-setting in which scientists often engage. Rather, the function of those activities has been (1) to understand why science is successful, (2) to determine the extent to which past pragmatic success betokens future success, and (3) to explore the relation between pragmatic success and the satisfaction of more demanding, esoteric epistemic requirements.[10]

Reconstituting the tradition

I have said that before a new standard can be effectively imposed on a successful discipline, it must first be shown that that standard will allow for the preservation of the Tradition in tact. To a first approximation, that is substantially correct. But things are not always so simple as that. Sometimes a proposed standard will fail to replicate an accepted Tradition fully and yet still win acceptance. Sometimes, certain of the elements which make up the Tradition itself will be dropped (either without substitution or to be replaced by others). The Tradition, in short, is not immutable.

Galileo, for instance, sought to impose new standards for natural philosophy; those standards implied, for instance, that most of Aristotle's astronomy and physics did not deserve the central place they had previously enjoyed at the core of the physics Tradition. Yet, as we know, something like Galileo's standards eventually prevailed and – at least for a couple of centuries – Aristotle was generally dropped from the Tradition which physicists proudly pointed to, largely displaced by the likes of Archimedes and Galileo himself. What is going on in such cases, and do processes like these show that the Tradition can be flouted at will when it conflicts with a nascent set of standards? I think not.

The Tradition cannot be ignored, but it can be partially reconstituted. The steps involved in removing a figure (or, more usually, the achievements of a figure) from prominence in the Tradition take a recurrent form. They involve, above all, presenting a compelling case that the achievement(s) in question, although presently figuring as part of the Tradition, do not rightfully belong there. The argument for excision of an item from the Tradition involves showing that the achievement in question fails to satisfy *prevailing* standards for scientific distinction. It is not enough for the advocates of new standards, when confronted by the failure of those standards to preserve the Tradition in tact, simply to dismiss the Tradition by virtue of its failure to comply with the newly proposed standards. That would simply be a tu quoque of major proportions. Rather, one must be able to show that an achievement, presently counted as part of the Tradition, fails to have earned a place there, *even by the standards now associated with the Tradition.*

Consider Galileo's strategy for de-throning Aristotle. It was quite clear that most of Aristotle's natural philosophy failed to satisfy the epistemic and methodological standards which Galileo was himself espousing. But, far from being an argument for de-canonizing Aristotle's physics, that is a presumptive argument for the rejection of Galileo's new standards (after all, they failed to preserve the Tradition). If Galileo is to justify his standards, he must show that they can capture a 'sanitized' version of the Tradition; and the disinfectant which one uses for the stable cleaning must not presuppose the new standards which are themselves at issue. Accordingly, Galileo goes to some pains to show that most of Aristotle's natural philosophy fails to satisfy those standards of rigor, proof and empirical support which had long been associated with the prevalent Tradition in natural philosophy. (As Galileo pointed out, most of those standards derived from Aristotle's own work in logic and epistemology). According to those standards, scientific knowledge is supposed (for instance) to be demonstrative and apodictically certain, and evidence cited was supposed to make the universals educed from them transparent. Yet – says Galileo over and again – Aristotle's physics and astronomy are a congeries of sloppy arguments, conceptual confusions, and sleights of hand with the evidence. The upshot of Galileo's analysis in the *Dialogue* and the *Discourses* is that Aristotle's physics fails to be good science even by Aristotle's own standards. By contrast, he insists, the physics of Archimedes[11] – all along a part of the Tradition, even if rather more peripheral than Aristotle – meets the very high standards for rigor in physical reasoning which Aristotle had insisted on.

This is not to suggest, of course, that Galileo wishes to endorse Aristotle's espoused methods and standards locks, stock and barrel. At important points, he has quibbles to make. But when it comes to Galileo's *grounds* for reconstituting the physics Tradition (so as, ultimately, to forge a yardstick for justifying his own standards), Galileo understood that any such reconstitution of the Tradition had to rest on an invocation of prevailing standards, rather than appraisal against new standards. Galileo's contemporary, Kepler, was similarly busying himself reconstituting traditions, although in Kepler's case it was the Tradition of astronomy rather than natural philosophy per se which drew his interest. We can see Kepler's reconstitution of astronomy by examining his lengthy historical treatise, *A Defense of Tycho against Ursus*.[12] Kepler, of course, wanted an astronomy which was causal, theoretical and ontologically committal – no mamby-pamby instrumentalism for him. But the prevailing standards in early 17th-century astronomy were decidedly instrumentalistic, and the then-accepted version of the astronomical Tradition included several apparent instrumentalists. With a view to eventually justifying his own standards, Kepler extensively re-writes the history of astronomy, shaping in the process a reconstituted Tradition which Kepler's standards can then be shown to recapture. Kepler's strategy involves a bit of excising instrumentalists from the Tradition as well as the reinterpretation of several putative instrumentalists to show that they were in fact realists. More generally, his history gives prominence to those astronomers who were causal theorists, who thought astronomy was a metaphysical probe for finding out what mechanisms really drove the heavens. What goes for Kepler and Galileo goes as well for Priestley, Mach, Duhem and a host of other scientist-philosophers who, in order to justify their proposals for transforming the standards of the sciences, have turned to history as the yard-stick for assessing the standards.

Conclusion

It is a tired cliché, both within science and outside it, that individuals re-write history to justify their current actions. That is often regarded as a wholly cynical process, as if any historical sequence could be put to virtually any use. I have tried to show here that scientists often turn to history to justify their standards, but that they are not permitted freedom to re-write it in any fashion they like. The history of a successful Tradition does not pick out a single set of standards which are the only standards which can be justified. That is why it is possible for scientific standards to change. But neither does the history of a successful Tradition provide a license for any standards whatever. And that is why the standards of science change much less often and rather less drastically than one might otherwise be inclined to expect.

In developing this account of the relations between scientific standards and scientific Traditions, I have been routinely tempted to draw on obvious similarities to changes in other areas of intellectual life far removed from science. Indeed, patterns of this sort might seem to govern shifts in philosophy, musicology, aesthetic theory generally, theology and historiography. But that is a story for another occasion so I shall continue to resist the temptation to explore the relevant parallels.

Notes

1. For a detailed discussion of the weaknesses of some of these claims, see Laudan (1984), ch. vi.
2. See Laudan, *op. cit.*, chs. iii-iv.
3. Many great achievements in the history of a discipline never get to the status of canonical precisely because, although brilliant, they nonetheless fail to tackle what are perceived as the fundamental problems.
4. Thus, I suppose every modern physicist would hold that Newton's planetary astronomy and Einstein's 1905 papers were indeed canonical achievements; whereas some might dispute whether Fourier's work on heat or Carnot's thermodynamics really deserved inclusion.
5. All these practices – medicine, astronomy and optics – have been 'successful' – albeit to varying degrees – since antiquity. By contrast, many sciences developed a successful practice relatively late in their careers. Matter theory, for instance, probably first became successful only in the 17th century.
6. There are, of course, certain theories in the natural sciences some of whose successes can apparently be judged *only* by the use of highly elaborate theories of empirical support. Thus, if we want to know whether statistical mechanics is 'successful' in accounting for Brownian motion, we need to rely on measures of success which go well beyond naive common sense. However, if the only successes which a theory could claim were those which required the mediation of an elaborate epistemic theory of empirical support, then such a theory could be fairly readily removed from the canon if there were shifts in methodological standards. This fact suggests the relative permanence of a theory on a discipline's canon will depend, in part, on how exotic a notion of success has to be invoked before that theory can be judged successful.
7. The proposal that I am making here is perfectly in line with the oft-expressed view of Reichenbach and Carnap that notions like 'degree of confirmation' or 'empirical support' are themselves technical notions, designed to be clarificatory 'explications' or 'rational reconstructions' of established pre-analytic judgments of empirical well-foundedness.
8. And a good thing too; for if we had to postpone a decision as to whether science was successful until we had an epistemically robust theory of empirical support, we should still be awaiting a verdict!
9. In other words, those who suggest that it is only by virtue of our epistemic standards that we are able to decide whether a practice is successful get things exactly back to front. We have criteria for the general success of a practice which are prior to, and ultimately adjudicatory of, our elaborated epistemic doctrines.
10. If pressed, I would be inclined to put it this way: the function of epistemology is to tell us how to determine which theories will be successful in the future. By and large, we do not need epistemology to tell us which theories have already been successful.
11. And, by implication, Galileo's own physics.
12. Available in English translation in a volume by Jardine (1984).

220

References

Jardine, H. (1984). *The Birth of History and Philosophy of Science*. Cambridge: Cambridge University Press.
Laudan, L. (1984). *Science and Values*. Berkeley: University of California Press.

Individuals, Species and the Development of Mineralogy and Geology

RACHEL LAUDAN

University of Hawaii at Manoa

Introduction

In recent years, philosophers and historians of science have belatedly begun to pay serious attention to taxonomy, largely as the result of work by David Hull and other philosophers of biology. Their interest has yet to spill over into inquiries into taxonomic practice in the other sciences. But since all sciences, even those with relatively sparse ontologies such as physics, must find some way of categorising the objects with which they deal, it seems reasonable to suppose that, just as our understanding of theories has been enriched by looking beyond physical theories, so too might our understanding of taxonomy be heightened by looking beyond biological systematics.

In this essay, I shall survey the development of mineral taxonomy from the early eighteenth to the early nineteenth centuries. Today mineralogy is a scientific subspecialty scarcely noted outside the small community actively engaged in it. Things were very different in the eighteenth century when mineralogy comprehended most of geology (as we now call it) and crystallography, as well as much of chemistry, and when mineralogists constituted a much larger proportion of the scientific community. Yet by the end of the second decade of the nineteenth century, this intellectual geography had been transformed. Mineralogy had been reduced to something like its present stature, crystallography had become largely a branch of chemistry, and geology had established its own way of looking at the world.

I shall argue that these changes in intellectual geography coincided with changes in taxonomic principles and in practice in mineralogy, changes that occurred as a result of the recalcitrance of the mineral kingdom to techniques that had been remarkably successful for the plant and animal kingdoms. I shall suggest that this coincidence is no accident but that reconsiderations of the ontology appropriate to understanding the earth's surface were one of the major driving forces in the emergence of different specialties from the umbrella discipline of mineralogy. In particular, debates about the mineralogical and geological analogues of the individual and the species within biology played a particularly important role in

M. Ruse (editor), What the Philosophy of Biology is. pp. 221–233.
© 1989 *Kluwer Academic Publishers, Dordrecht*

reshaping the organization of investigation into mineral substances in the late eighteenth and early nineteenth centuries.

The Linnean challenge to mineral taxonomy

Eighteenth-century mineralogists were preoccupied with taxonomy, new taxonomies appearing almost every year. They experimented with different taxonomic *principles* as well as with different substantive taxonomies and they employed technical taxonomic terms throughout their work, confident that their readership would understand the full significance of the terms. Hence like the other taxonomic sciences, mineralogy was given a great jolt when, in 1735, Linnaeus published his enormously successful taxonomy of the plant world. It offered mineralogists the hope that they could bring order to the chaotic state of mineral classification by applying Linnaeus's principles to the plant kingdom.

That minerals and rocks presented much greater problems to the taxonomist than did plants or animals had long been common wisdom amongst natural historians. True, botanists and zoologists had their problem cases – marine colonial organisms and plant varieties, for example – but these paled in comparison with the problems posed by the earthy, indistinct forms of most minerals. Half a century before the publication of Linnaeus' *Systema Natura*, John Woodward (1695, p. 170) had summed up the general opinion when he complained:

> To write of metals and minerals intelligibly and with tolerable perspicuity, is a task much more difficult than to write of either Animals or Vegetables. For these carry along with them such plain and evident Notes and Characters either of Disagreement, or of Affinity with one Another, that the several Kinds of them and the subordinate Species of each, are easily known and distinguish'd, even at first sight; the Eye alone being fully capable of judging and determining their mutual Relation, as well as their Differences.

The 'eye alone' usually failed to distinguish minerals since only the crystalline minerals – a small proportion of the total – had a definite form. Thus minerals and rocks, unlike most animals and plants, did not appear to fall into easily distinguishable groups or species, and therefore posed major problems for those who sought to systematize the mineral kingdom.

In the absence of individuals with readily recognizable forms, mineral taxonomists had traditionally relied on chemical and physical properties to differentiate minerals. This worked well at the level of classes, less well at the level of species. Agricola's classic *De natura fossilium* in 1543 (a work that traced its intellectual ancestry back to Aristotle's *Meteorologica*) set the precedent. Following its publication, mineralogists for the three succeeding centuries distinguished four mineral classes – the *earths* (resistant to both heat and water), the *metals* (resistant to water but melting in contact with heat and returning to their former shape on cooling), the *salts* (resistant to heat but soluble in water) and the *bitumens* (resistant to water but burning in contact with heat). They subdivided these classes according to different criteria, citing a variety of chemical and external characters.

Rocks – a technical term for aggregate bodies of one or more mineral that formed the significant features of the earth's surface – received scant attention by comparison. Most mineralogists simply assumed that, since rocks were, after all, masses of minerals, rock classification would follow naturally from mineral classification. Consequently they divided rocks too into earthy, metallic, salty and bituminous categories. But they did not accord rock classification separate status, regarding it as simply one more problem that they, unlike their botanist colleagues, had to confront.

Linnaeus, following the precedent set by Cesalpino, applied Aristotelian principles to the problems of taxonomy in a thorough-going fashion. There is no need to rehearse the success of his system which, in spite of sustained criticism, dominated taxonomic practice for the rest of the century. Using a few simple precepts, Linnaeus succeeded in bringing unprecedented order into botany, an order that scholars in a wide variety of disciplines naturally viewed with envy. As most eighteenth-century mineralogists understood Linnaeus' program, he was committed to the following propositions: that the world was made up of natural kinds, located at the species level; that these kinds had to be classified according to their 'essential characters;' that the different characters of the organism could be ranked or subordinated, so that taxonomists need consider only one or a small number of characters rather than the complete set; that, for plants, their external, visible, and countable sexual organs of plants could be regarded as their essential characters (or at least as mirroring their essential characters); and that the natural kinds in the world were so related that they could be grouped into a hierarchy of higher taxonomic units.

In the wake of Linnaeus' plant taxonomy, natural historians, including Linnaeus himself, considered whether the principles that governed the taxonomy of living beings could not be imported into their own domains, replacing the traditional chemical and physical methods. Linnaeus and his followers had several reasons for believing that methods successfully developed for the botanical 'kingdom' ought also to apply to the mineralogical. The widespread acceptance of the doctrine of the Great Chain of Being, which postulated that the species in the three kingdoms could be arranged in a series divided by only small steps, suggested strong analogies between plants and minerals. So too did the still widespread belief that minerals, like plants, grew in the earth.

Linnaeus himself initiated the effort to apply his principles to the taxonomy of the mineral kingdom, attempting to place mineralogy firmly in the sphere of natural history, not physics or chemistry. He argued that the methods by which minerals had formerly been classified, the 'Chemical' and the 'Physical,' should be abandoned in favor of the 'Natural [in the sense of natural historical] which considers their superficial and visible structure' (Linnaeus 1735/1802, pp. 7–9). He tried to show that mineral natural kinds had essential characters that could be mapped on to observable characters, preferably of a generative or sexual nature, and that the kinds could be arranged in a hierarchy. Because he was not sure that minerals could be sorted into natural kinds, Linnaeus admitted to some doubts about whether this program could be successfully accomplished. Unlike the species of plants that, borderline cases notwithstanding, he had in the main been able to distinguish by counting visible, external, sexual organs, minerals presented a more difficult case. The crucial problem, he conceded, the 'calamity' of mineralogy, was that it was 'scarcely ever, that the Species can be sufficiently determined, since in [minerals] the generation proceeds not from the egg; but [from] irregularly sportive nature' (1735/1802, pp. 7–9).

Nonetheless, determined to carry his system as far as he could, Linnaeus insisted that visible, countable, external characters be used to individuate minerals as they had been used to individuate plants. He used this method for the crystalline minerals and made some headway on the problem of sorting out the different crystal forms. Then, consistent as ever, he forced mineral species into a hierarchy of genera, orders and classes parallel to that he had employed for plants.

But Linnaeus's attempt to ground his mineral taxonomy in a sexual system foundered. He pursued some slight parallels between the generation of plants and the generation of minerals, at least during the early stages of the earth's history. Beginning from the quite conventional premise that the earth had once been composed primarily of the chemical element, water, Linnaeus (1735/1802, pp. 7–2) proposed that the water had generated a 'double offspring ... a saline male and a terrene female.' The saline male (salt or acid in modern times) was the form-giving principle, impressing shape on the terrene female (earth). Salts and earths were thus the 'fathers' and 'mothers' of all minerals. For most of the divisions at the level of species, however, Linnaeus resorted to the time-honored, essentially chemical, categories that had always been used by mineralogists. As to rocks, Linnaeus gave them little thought but would presumably have concurred with the accepted view that their classification would follow naturally once minerals had been sorted out.

During the succeeding decades, mineralogists considered, and eventually drastically modified or rejected, the Linnaean program for mineralogy. Although they differed about some points, all thought that the basic problem was that the disanalogies between plant and mineral individuals (which Linnaeus had recognized) outweighed the analogies (which he had hoped would prevail). The constitution of individual minerals was quite different from the constitution of individual plants and animals. Mineral individuals, unlike plant individuals, appeared to have no organization higher than the chemical that could serve to distinguish them as plants did. Consequently minerals did not reproduce sexually, as Linnaeus had recognized. Organization and reproduction marked the break between the living and the non-living, and efforts to use the same taxonomic principles for both seemed doomed to failure.

Late eighteenth-century mineralogists harped on these themes of lack of organization and lack of sexual reproduction in minerals. The great German mineralogist, Abraham Gottlob Werner (1774/1962, pp. xxvii-xxviii), insisted that:

> we immediately find a basic difference among [natural bodies] because they are divided into two principal kinds: in the first one, the essential features occur in their mode of association, whereas in the second, they occur in their composition. To the former belong the animals and plants; to the latter, the bodies of the mineral and meteoric kingdoms. Both, being natural bodies, are aggregated and their parts chemically associated; but the former consist of parts different from each other, which we call organs [*composita*], and in these bodies the essential features are represented by the manner in which these organs are associated to each other. The latter, on the contrary, are quite simple or consist of similar parts [*aggregata*], and their features do not correspond to the manner in which these parts are aggregated.

If a botanist were to divide a number of plants into the smallest possible parts, observed Werner, he would be unable to distinguish which parts belonged to which plant, because the essence of a plant lay in the association (or organization) of its components. But if a mineralogist were to divide a number of minerals into the smallest possible parts, each would still be recognizable because the essence of a mineral lay in its composition. In short, 'the natural succession of essential features in the [animal and vegetable kingdoms] follows their mode of association and in the [mineral kingdom], their composition' (Werner 1774/1962, p. xxviii).

The Swedish chemist-mineralogist, Torbern Bergman, stressed the connection between reproduction and organization in plants:

> there is a power implanted by the Creator in organized bodies, which, upon the acquisition of proper nutriment, unfolds and evolves the structure which before lay concealed in the fecundated egg or seed ... Hence it is that the leading features of the external parts agree with the internal properties, and when judiciously chosen, form sufficient characteristic distinctions (Bergman 1783, pp. 6–7).

By contrast, he explained (1783, p. i):

> the formation of fossils [minerals] is totally different. Here no system of vessels collects, distributes, secretes or changes the concurrent particles, but they run together by chance, and are solely connected by the power of attraction.

Daubenton, Buffon's protégé, pointed out that:

> among the minerals, there are no such thing as individuals and as a consequence

no such thing as species. We do not see minerals reproducing like plants and animals, by similar individuals, from generation to generation. A mineral is altered and destroyed by diverse accidents; its integrant parts are dispersed, are intermingled and are combined with minerals of different sorts, often very different from that which was decomposed. There are no such things as individuals, because there is no such thing as essential resemblance (1782, p. iii).

By the 1780s, mineralogists agreed that, given the disanalogies between the living and the non-living, the Linnaean plan for a grand natural history program that assimilated mineralogy to botany and zoology was in serious trouble. Consequently most mineralogists concluded that they had to return to the original principle of classifying minerals by chemical composition, with all the attendant problems at both the practical and the theoretical level that had been responsible for the chaotic state of mineral taxonomy in the first place. Practically, chemists were still struggling to develop methods of analysis capable of distinguishing different minerals. Chemical analysis was at best slow, laborious and difficult to carry out in the mines and mountains where it was most needed; at worst it was simply impossible. Theoretically, things were even worse. Chemists had no reason to believe that minerals (or other chemical substances) had fixed compositions. Indeed, according to the chemistry of the day, the number of basic elements or 'principles' was very small, of the order of three to five. These, it was thought, combined in various proportions to make all the objects in the animal, vegetable and mineral kingdoms. No one presumed that complex substances contained fixed proportions of the elements. Quite the reverse. Chemists and natural historians thought that the proportions could be varied indefinitely. Consequently they concluded that the number of chemical substances in the world was indefinitely large and that different chemicals or minerals could not be sharply distinguished. Botanists and zoologists had long since given up the hope of basing plant and animal classification on chemistry (Goodman 1971). The chances looked slim for mineralogy too, but unlike the botanists and zoologists, mineralogists seemed to have failed to find an alternative.

A straightforward application of Linnaean taxonomic principles to the mineral kingdom had foundered on empirical grounds, on the problem of finding mineral analogues for the plant individual and species. Mineralogists reacted in two quite distinct ways, though both ways were to involve redefining the investigation of the mineral kingdom. Some mineralogists, particularly in France, continued to subscribe to classical taxonomic constraints and hence to search for fixed mineral species, concentrating their efforts on the crystalline minerals. Other mineralogists decided that, at least so far as rocks and minerals were concerned, the traditional principles of taxonomy had served them badly and could therefore be treated quite cavalierly. They subordinated the search for fixed mineral species to the development of methods of identification, and replaced the priority of minerals by the priority of rocks.

Crystallographers and the mineral species

Rene-Just Haüy was the most important of those who continued to search for an analogue of the plant species (Mauskopf 1970, pp. 14–20). He retained the idea that some principle of organization higher than the chemical was required, but he separated this requirement from any considerations of reproduction. Instead he linked the organizing principle back to the chemical constitution of the mineral, thus achieving for minerals what botanists had not been able to achieve for plants. Haüy's key innovation was the postulation of a level of organization in the mineral below that of visible structure but above that of the ultimate constituents. He speculated that crystals were formed by building blocks larger than individual atoms though still too small to be observed. These he called the *integrant molecules*. Since an integrant molecule was formed of atoms of the chemical in question, it was ultimately chemical. But it also had a distinct form. This was not necessarily, or even usually the form of the crystal itself; at the visible level, this form was suggested by the cleavage pattern of a given crystal, a pattern that, although regular, differed from that of the crystal.

It was the integrant molecule that corresponded to the biological individual. A collection of the integrant molecules corresponded to the species. There was, Haüy insisted, a characteristic that:

> served by its invariability … as the rallying point for different bodies which belong to the same species. This is that which derives from the exact form of the integrant molecule, because that form subsists, without any sensible alteration, independently of all the causes which can make the other characteristics vary (Haüy 1801, p. 156).

He continued to clarify the analogy between integrant molecules and individual plants by explaining that:

> the species, in mineralogy, [is] a collection of bodies whose integrant molecules are similar and which are composed of the same elements united in the same proportion (Haüy 1801, p. 162).

Crystals grew by the stacking of integrant molecules on an original nucleus. The stacking pattern was specific to the mineral species.

Haüy was indebted to earlier writers for many of his ideas.[1] From Linnaeus, he took the idea that the external form of crystalline minerals was the key to their classification. From Bergman, Romé de l'Isle, and Daubenton (Hookyaas 1952), he took the germ of the idea of the integrant molecule. But it was Haüy who elaborated these ideas into a full-fledged theory of crystalline minerals. With great ingenuity, he managed to propose possible stacking patterns that corresponded to the known and constant interfacial angles of different crystalline forms, thus setting the theory on a firm empirical and mathematical footing.

Of course there were some snags. Most critically, Haüy's assumption that crystal form faithfully reflected chemical composition failed in certain key instances. Some chemical substances (calcium carbonate, for example) crystallized in more than one different form (calcspar and aragonite). On the other hand, diverse chemical substances could crystallize in the same form (the different alums). Eventually in the 1820s Mitscherlich resolved these problems with his concepts of isomorphism and polymorphism. But even before that, Haüy was so successful within his chosen subfield of mineralogy that Cuvier, himself an accomplished taxonomist, in recounting the history of the natural sciences, pronounced that taxonomy had reached its highest and most positive form in Haüy's systematization of the crystalline minerals. Indeed, completing the circle, in his *Théorie élémentaire de la botanique* (1813), Alphonse de Candolle acknowledged the influence of Haüy's crystal taxonomy on his own botanical taxonomy (Stevens 1984, p. 61).

Following Haüy's work, the study of crystals took on a life of its own, a life that was bound up more with the career of chemistry than with that of the earth sciences. Space prevents a detailed examination of the social and institutional consequences of this change in ontology and consequent change in taxonomy. Briefly, what happened was that chemists and crystallographers seized on the new conceptual tool that Haüy had provided and used it to explore the properties of a wide variety of inorganic and organic chemicals (Mauskopf 1970). But since only a small proportion of the known minerals had distinct crystal forms, most of mineral taxonomy was left floundering as it had been before. Even in Haüy's own taxonomy, he had to allow for categories of formless and massy crystals. As for rocks, although a few of them detectably crystalline by the naked eye, the vast proportion lay completely outside this means of analysis, at least until the introduction of microscopic examination of minerals in thin section. In short, the successful resolution of problems in the taxonomy of crystalline minerals spurred the independent development of crystallography while leaving most of mineralogy and geology unchanged.

Geologists and the individual 'formation'

Following the recognition of the difficulties confronting Linnaean taxonomy in mineralogy, the Germans and the French rejected the program more decisively than did the Swedes. Disillusioned, Werner declared that he would rather 'have a mineral ill classified and well described, than well classified and ill described (Werner 1774/1962, p. xxv). Although mineralogists did not (indeed could not) abandon attempts to create a classification of minerals, they viewed traditional theoretical constraints on taxonomy with a large pinch of salt. The old discipline of mineralogy limped along, making little significant progress over earlier work.

It was in the study of the rocks that the effects of this disillusionment were most evident. At just this time in the 1760s and 1770s, for a mixture of practical and theoretical reasons, mineralogists were demanding a satisfactory rock classification, not just relegating it to an anticipated future consequence of mineral classification (Laudan 1987, ch.3). Their failure to develop a satisfactory mineral classification made these demands even more pressing. Hence when Werner turned from the classification of minerals to the classification of rocks, he broke with the tradition of using mineralogy as the essential character of the rocks. Instead, drawing on some hints that were already present in the literature of the previous quarter century, he suggested an alternative. 'Mode and time of formation' – not mineral composition – constituted the 'essential differences' between rocks of various kinds, he declared (Werner manuscript, quoted by Ospovat 1971, p. 19). The basic systematic unit was the 'formation.' The formation, explained Cuvier and Brongniart to a French audience (1811, p. 8, my translation), was 'a group of beds of the same or different nature, but formed at the same epoch.'[2] Thus time of formation, not mineral composition, became the basic method of organizing rocks. Examples of formations were the 'Upper Freshwater Formation' which was 'marly' in France and 'calcareous' in southern England, and the 'Lower Marine Formation' which again had quite different lithologies in the two regions.

Geologists found the idea of the formation enormously exciting because it suggested a way of breaking the impasse in which mineralogy found itself. The story of the emergence of geology in the first third of the nineteenth century is the story of the enthusiasm with which those involved in what I term the Wernerian 'radiation' took up this question (Laudan 1987, ch.7). True, the problems of individuation and identification remained to be tackled, but geologists quickly developed workable solutions. Individuation, or 'independence' as it was called in the nineteenth century, was tricky for if time of formation defined formations it was not clear why there were (or should be) sharp breaks between one formation and the next. Werner himself had envisaged a gradually changing composition of the depositing ocean. Appreciating the problems of individuation that this created, first Cuvier and Brongniart, and later Élie de Beaumont argued that world-wide 'catastrophes' had occurred that marked the boundaries between the major formations. While this patently did not explain all the minor divisions, it at least suggested ways of approaching the problem. As to identification, the problem was how to determine the relative time of formation of a particular rock. Initially geologists tackled this by determining place in the succession, assuming that rocks lowest in the succession had been deposited first. Humboldt (1823, p. 9, my italics) stipulated that 'the *essential character* of the identity of an independent formation is its relative position, or the place which it occupies in the general series of formations'. But geologists quickly added an auxiliary technique. They assumed that fossils reliably indicated the age of a formation. Discussing the Chalk rocks around Paris, Cuvier and Brongniart (1811, pp. 10–11), my italics) explicitly stated that 'what *essentially characterizes* this formation is the fossils it contains, fossils completely different, not only the species, but often the genera from those contained in the calcaire grossier.' As with the postulation of catastrophes, this innovation raised new problems even as it resolved old ones. Some were largely practical: should one fossil be enough to identify a formation? or should whole fossil faunas and floras be used? and so on. Others were more theoretical, for throughout the eighteenth century, mineralogists had assumed that organic fossils indicated the environment, rather than the time, of the formation of the rock. They now had to sort out the relative influence of age *and* environment, something that still causes problems today.

Geologists realized that this flew in the face of all taxonomic practice and that it resolved the problems of individuals and species by making stratigraphy a science of individuals alone. A classification, after all, was supposed to be based on timeless essences, so that making time the essence was taxonomic lunacy. Adopting an organization of rocks that made the formation central meant abandoning the search for an analogue to the biological species. Unlike an organization of rocks by their mineralogy, which at least potentially identified natural kinds, formations by their very definition were simply individuals and the succession was simply a chronological arrangement of those individuals. A succession, said Humboldt (1823, p. 16) in a book devoted to extolling the virtues of the new system of organizing rocks, could 'in no respect ... be called a classification of rocks' however useful it might be in practice. But whatever qualms Humboldt and others might have about the niceties of whether a succession could count as a classification, the fact is that all mineralogists and geologists came to accept that this was a reputable, indeed enormously fruitful way of dividing up the rocks on the earth's surface.

Meanwhile, geologists continued to make mineralogical distinctions between rocks. Early nineteenth-century scientists clearly understood what was at stake. The Belgian geologist Omalius d'Halloy, who produced the first geognostic survey of the whole of France, distinguished in a paper read in 1813 (but not published until 1822):

> two principal points of view [which] seem equally to lead to the division of a country into physical regions determined by the nature of the soil [rocks]: in one it is considered geologically, i.e., according to the epoch of formation; in the other with respect to its mineralogical and chemical nature. [Omalius d'Halloy, (1822) translated in De la Beche 1824, pp. 296–97].

Mineralogists and geologists carefully distinguished the terminology of the two ways of organizing rocks. The leading nineteenth-century German petrologist, Karl von Leonhard worried that his countrymen tended to use *Gestein* (rock) and *Gebirgsart* (formation) as interchangeable terms. Alexandre Brongniart urged that the French distinguish the study of *'roches'* (petrological units that might or might not be the same age) from the study of *'terrains'* (formations).

Following, and, I would suggest, largely as the result of, the introduction of the concept of a formation, stratigraphy diverged from mineralogy and petrology, and all three were subsumed under the title 'geology' instead of 'mineralogy'. Again there is not the space here to follow out all these social and institutional shifts, but that they occurred is part of the common wisdom of the history of geology. On the one hand, geologists busily reconstructed the earth's chronology using formations as the basic unit. On the other hand, they struggled to understand the causal processes taking place on the earth – vulcanism, metamorphism, sedimentation and so on – utilising the mineralogical rock-type as the basic unit. Put another way, the problems geologists sought to tackle were crucially interconnected with the taxonomic structure they adopted. Those who asked causal questions used the more traditional classification for rocks were natural kinds that could function in general laws. Those who asked historical (or more accurately chronological) questions used the succession for formations were individuals that composed the historical sequence. The formation, was an individual, a historical entity. The central entity of stratigraphy was an individual, historical entity, not a natural kind. That is to say, the entities are historical not in the trivial sense of existing in time, but in the significant sense of being defined by the period of time in which they were formed. At least half a century before Darwin's work made most biologists re-conceptualise species as historical entities, the geological community was already struggling with this shift.

The problem was that, since these two activities utilised different basic concepts, cross fertilization of the two activities became extraordinarily difficult. The construction of causal theories about the earth did not draw on the chronological record that the stratigraphers were constructing (though the causal theories might be used to explain particular parts of that record). Conversely, the reconstruction of the earth's chronology did not draw on the principles established by the causal theories in geology.

Conclusion

By the end of the second decade of the nineteenth century, mineralogy in its old form had vanished for good. It was now simply the chemical and physical study of minerals proper. It was still closely connected to what was soon going to be called petrology, or the mineralogical study of rocks. But both these enterprises, despite their centrality to causal geology, failed to attract the majority of those who were interested in the earth sciences. Crystallography had much closer ties to chemistry than it did to mineralogy, let along geology. And geology was forging ahead, largely devoted to the development of a chronology of individual formations. Historians of geology have long debated the reasons for these changes. What I have argued here is that taxonomic problems were a major, perhaps the major factor, motivating the changes. Attempts to treat the mineral kingdom as if its ontology was closely parallel to that of the plant and animal kingdom failed on empirical grounds. One group of scientists (the crystallographers) held firm to their taxonomic principles and eventually achieved success with them by strictly delimiting the range of phenomena with which they dealt. Another group of scientists (the geologists) chose overwhelmingly to abandon the search for an analogue to the biological individual and species, thus transforming much of their science into a historical rather than a causal enterprise (though recognizing that for causal geology the traditional taxonomic approach was still necessary). This story should reinforce what philosophers and historians of biology have been arguing for the past couple of decades, namely that taxonomies cannot be dismissed as mere conventions but that they are an integral part of the empirical and conceptual foundations of a science.

Acknowledgements

I would like to thank David Hull, Ernst Mayr, and Fritz Rehbock for helpful comments on an earlier draft of this paper.

Notes

1. Interestingly the Chemical Revolution of the last decade of the eighteenth century appears not to have been the formative influence for Haüy.
2. In Werner's original scheme, time of formation determined mode of formation just because the mode of formation of a rock depended on the period it which it was formed from the ocean. Since this in turn determined its mineralogy, there was initially complete convergence between rock classifications based on mineralogy and rock classifications based on time. Within a matter of years though, belief that time determined mode of formation weakened, and the two kinds of classification diverged.

References

Bergman, T.O. (1783). *Outlines of Mineralogy*. Trans. W. Withering. Birmingham: T. Cadell.

Cuvier, G. and Brongniart, A. (1811). *Essai sur la géographie minéralogique des environs de Paris*. Paris: Bandouin.

Daubenton, L. (1782). Introduction à l'histoire naturelle. *Encyclopédie méthodique. Histoire naturelle des animaux*. Vol. 1. Paris: Panckoucke.

Goodman, D.C. (1971). The application of chemical criteria to biological classification in the eighteenth century. *Medical History* 15: 23–44.

Haüy, R.J. (1801). *Traité de minéralogie*. Paris: Louis.

Hookyaas, R. (1952). The species concept in eighteenth-century mineralogy. *Archives International d'Histoire des Sciences* 1: 45–55.

Hull, D. (1965). The effect of essentialism on taxonomy – two thousand years of stasis. *British Journal for the Philosophy of Science* 15: 314–326; 16: 1–18.

Humboldt, A. (1823). *A Geognostical Essay on the Superposition of Rocks in Both Hemispheres*. London: Longman.

Laudan, R. (1987). *From Mineralogy to Geology: The Foundations of a Science 1650–1830*. Chicago: Chicago Univers. Press.

Linnaeus, [1735] (1802). *A General System of Nature, Through the Three Grand Kingdoms of Animals, Vegetables, and Minerals*. Trans. W. Turton. London: Lackington Allen.

Mauskopf, S. (1970). Crystals and compounds: molecular structure and composition in nineteenth-century French science. *Transactions of the American Philosophical Society* 66: 1–82.

Ospovat, A. (1971). Introduction and notes. In edition of Werner, A.G., *Short Classification and Description of the Various Rocks*. New York: Hafner.

Stevens, P. (1984). Haüy and A.-P. Candolle: crystallography, botanical systematics, and comparative morphology, 1780–1840. *Journal of the History of Biology* 17: 49–82.

Werner, A.G. (1774). *On the External Characters of Minerals*. Trans. Albert Carozzi of *Von den äusserlichen Kennzeichen der Fossilien*. Urbana: University of Illinois Press.

Werner, A.G. (1786). *Kurze Klassifikation und Beschreibung der verschiedenen Gebirgsarten*. Trans. A. Ospovat (1971). *Short classification and description of the various rocks*. New York: Hafner.

Woodward, J. (1695). *An Essay Toward a Natural History of the Earth*. London: R. Wilkin.

Attaching Names to Objects

ERNST MAYR

Museum of Comparative Zoology, Harvard University, Cambridge, Mass., 02138

The process of attaching names (labels) to objects has been of great interest to philosophers since the days of the Greeks. However, in recent times other philosophical problems have seemed to be more interesting, and the subject of naming has been rather neglected. The few contemporary philosophers who have written about it have, on the whole, been quite unaware that the same problem of naming and fixing a name to a specified object has been one of the foremost concerns of taxonomy, one of the sciences. There is, however, one philosopher, David Hull, who did realize it and who attempted in a series of publications to show to what extent philosophy and taxonomy can help each other by reciprocal illumination. There are issues where the philosopher can help the taxonomist to improve his methods and definitions and others where the taxonomist has worked out methods that do the same for the philosophers.

Hull's interest in the problems of naming is part of a broader interest in all aspects of classification and is reflected in a series of papers and a recent book. It is he who directed the attention of the philosophers to the fact that taxonomists consider species to be particulars rather than classes. He also probed deeply into the questions whether the branching pattern of descent is all that is needed for a sound classification of animals and plants or whether the totality of characters in addition to genealogy should be considered. And it was he (Hull 1983) who examined the question whether the methods of zoological nomenclature for attaching labels (names) to definite objects would be suitable in philosophy. Could these methods also be used to attach names to ideologies, to research traditions, and evolving scientific concepts, he asked.

In this paper I am concerned with similarities and differences between philosophy and taxonomy in the methods and concepts of naming procedures. It is interesting that in a largely practical field, taxonomy, a highly elaborate system of rules developed, regulating the naming of objects and classes. How elaborate the rules are that govern the process of naming in taxonomy is best documented by the fact that their description requires 170 printed pages in the International Code of Zoological Nomenclature and that similarly elaborate rules exist in the codes for botanical and bacterial nomenclature.

M. Ruse (editor), What the Philosophy of Biology is. pp. 235–243.
© 1989 *Kluwer Academic Publishers, Dordrecht*

It is perhaps not surprising that taxonomists devote so much attention to problems of name-giving and to the relation between names and named objects. There are close to (or even more than) 5 million names (including synonyms) in existence for described living and fossil animals and plants. Complete chaos would surely prevail if highly specific rules had not been adopted by which the relation between these names and the named objects were specified. Much of this elaborate machinery deals with details that are of little interest to the philosopher. However, these codes of nomenclature also deal with important matters of principle; and it is these philosophical aspects of nomenclature which I shall now discuss.

What is the basic aim of an unequivocal system of namegiving? It is to have one unmistakable name for an unmistakably characterized object. Problems arise when several names exist for the same object (or are believed to refer to the same object), or the reverse, when only a single designation exists for what eventually turns out not to have been a single well characterized object.

Since the verbal descriptions of taxa are necessarily incomplete and frequently vague or ambiguous, it became obvious in the history of taxonomy that more secure standards were needed to tie scientific names unequivocally to objective taxonomic entities and these standards were referred to as *types*. Whenever there is the problem of what name refers to what taxon, the taxonomist solves it with the help of the so-called type method. The name of a heterogeneous taxon goes with that of its components that is represented by the type.

The type method, as it is articulated in the current codes of the taxonomists, achieves its objectives on the whole quite admirably. What is unfortunate, however, is that the meaning of the word type has undergone a remarkable and frequently overlooked change since the middle of the 18th century, when the type method was introduced. Linnaeus, as everyone knows, was a typologist (essentialist) and considered any specimen as typical that conformed to his concept of the essential characteristics of that species. Hence, a type by no means had to be unique. He and his contemporaries not only often recognized numerous types of a species, but, to make matters worse, did not hesitate to discard the original specimen on which the description of the new species had been based, when they received 'better' specimens of their species. The types of a species conformed for them to the Platonic concept of types.

Alas, this concept of the type was gradually being replaced by a very different one. When it was recognized, in the course of the 19th century, how variable many species are and that no single specimen could adequately represent the variation of a species, and when it was discovered that the so-called type series of many of the older names were composites of two or three similar species, the introduction of a fundamentally different type concept became necessary. In 1847 Whewell still described the type method in traditional Platonic terms: 'A typical representative of each species is chosen as its type and other members of the species are included in the species by means of their [similarity] to it' (1, p. 481).

Soon after this was written, however, a type was no longer considered as 'typical of the species' but rather as the name-bearer, or as Simpson (1961) called it, as the 'onomatophore' of the species. Even a totally atypical albino could be the onomatophore of a species, provided its other attributes clearly associated it with one particular species and with no other. Hull (1983, p. 484) has well described this fundamental shift in the meaning of the word type in the taxonomic literature. The change from one concept to the other took place gradually, and when it occurred it was not really noticed by the taxonomists. The result was great instability in the zoological nomenclature, just the opposite of what the type method had wanted to accomplish (Mayr *et al.*, 1953, pp. 212–245). As Hull has rightly pointed out, the type specimen has nothing to do with the descriptive characters of a species or an analysis of variation, its only function is to tie a name given by a taxonomist to a definite zoological entity. It is not its function to serve as the paradigm of the species, supplying an accurate catalog of its characters, although the type, being a member of a species, will normally have sufficient characteristics to associate it with one and only one species.

What in the type method of the taxonomist appealed to Hull is that it might solve a problem also encountered by the philosopher, that is, the availability of only a single name for several different objects.

The philosopher encounters this problem in a number of different guises. 'If conceptual systems are internally quite hetergeneous, if different conceptual systems contain instances of many of the 'same' concepts, if a conceptual system can undergo a total transformation of its elements while remaining the 'same' conceptual system, how are we to tell whether we have one conceptual system or two, either at any one time or through time? How can we name such slippery entities and continue to apply the same name to the same entity through time? When disagreements arise, how are we to reconcile them?' (Hull 1983, p. 479).

An apt example is the concept of *Darwinism*. Every historian knows that the concept Darwinism was composite from the very beginning (referring simply to evolutionary thinking, to Man's descent from the anthropoid apes, to the theory of natural selection, or to all of them), that its primary meaning changed through time (at present virtually restricted to selectionism), and that it was used by some authors only in a biological context and by others to designate the *Zeitgeist* of the post-Darwinian period (Greene 1986). Remembering that the type concept was introduced into taxonomy (at least in its modern version as name-bearer) owing to a dissatisfaction with descriptions and definitions, Hull felt that definitions or descriptions would never bring clarity in the case of changing or heterogeneous concepts, and that this can be achieved only by the fixation of a type, an exemplar. This might indeed be a successful method in certain cases, but I expressed my doubts that the type method could be applied in a majority of the relevant situations and illustrated the difficulties by attempting to select an exemplar for the concept Darwinism (Mayr 1983, pp. 505–507).

It might help to point out that Hull suggested applying the exemplar method to two major categories of names for heterogeneous objects. The one category consists of the cases where the object has been heterogeneous from the very beginning, and where clarification might be achieved by restricting the name to one of the components. This situation is well illustrated by the heterogeneity of the word teleological. As I showed previously (Mayr 1974, 1976) the word teleological has been used for four entirely different actual or hypothetical processes. When each of these is designated by a different name, considerable clarity is achieved. The four kinds of 'teleological' are now distinguished as (1) teleomatic, (2) teleonomic, (3) adapted, (4) cosmically teleological. My experience as a historian of science convinces me that a remarkably large proportion of the controversies in science are due to the fact that each opponent had a different concept in mind even though all of them used the same word for it. In the recent controversy on group selection in evolutionary biology, different authors had entirely different kinds of groups in mind. By distinguishing them as (1) Hamiltonian groups, (2) Sewall Wrightian groups, (3) Mayrian groups, and (4) cultural groups, one can greatly clarify the nature of the argument. It would seem to me that the clarification achieved in these and numerous similar cases was not due to a literal application of the type method, but rather due to the recognition that the controversial concept was heterogeneous, that it had to be partitioned, and that each of the components had to be distinguished by a separate name.

In the second category of nomenclatural problems described by Hull, the difficulty is caused by the concept itself changing in the course of time. This situation is well illustrated by the species concept in taxonomy, which changed from the typological species of Linnaeus to the biological species of the modern biologist, as particularly used by the students of behavior, ecology, and evolution. A gradual evolution of terms, sometimes resulting in drastically different end points, is characteristic for almost all research traditions. This is as true for the term Darwinism as for neo-Darwinism. When Romanes (1895) introduced this term he specified that it was a designation of Weismann's theory of evolution, in other words it was a term for Darwin's ideas of evolution, but with an explicit rejection of any inheritance of acquired characters. Forgetting the clear-cut designation of this term, recent opponents of neo-Darwinism understood under this term the paradigm of the mathematical population geneticists with its reductionist assumption of the gene as the target of selection. This is clearly not the neo-Darwinism of Romanes, and if it had to be labeled specifically, it might have to be called Fisherism, because R.A. Fisher had described the paradigm of the mathematical population geneticists most explicitly.

It will not have escaped the reader of this discussion that Hull's method of exemplars was in no case successful in straightening out the difficulties posed by evolving terms such as Darwinism. Indeed, as I pointed out (1983, p. 506), the choice of an exemplar for the term Darwinism would have surely created a major controversy. It would seem to me that description and definition, following careful analysis, are the only way to cope with the change in meaning of evolving concepts. Such changes must be coped with by continuing redefinitions of the terms. This is what happened in the case of terms like Darwinism, ethology, species, gene, fertilization, isolating mechanisms, and many others. Instead of coining a new term everytime the meaning changed a little or more profoundly, the definition was modified and the continuity of the term was thus preserved. There is no denying that this approach leads to occasional misunderstandings because among two users of the term one might still use an earlier, but the other a later usage and definition. Yet, most scientists consider this eventuality a lesser evil than the ever continuing introduction of new terms, as would be necessitated by the type method.

The type of a genus

The naming of objects in daily life and in philosophy usually follows a rule which we might designate as uninominalism, that is, one name per object. In the Linnaean method, by contrast, each species has two names, a generic and a specific one. It tells us for each individual not only to what species but also to what genus it belongs. For Linnaeus, owing to the essentialistic framework of his thinking, all species which he placed in a genus were 'typical'. They were typical because they conformed to the definition of the genus. And this was the concept adopted by his followers for the next 50 years or so. If it was found that some of the included species did not agree with either the original or a revised generic definition, they were no longer considered to be typical, and were placed into a different genus by a process called 'elimination'. Now only those species were considered typical that remained in the original genus. This was the Linnaean type concept of the genus. However, a profound change in the method of 'typifying' a genus eventually took place, when it became traditional to typify the genus by the selection of a single type species. This corresponded to the selection of a single onomatophore typifying a species. When no type species had been designated, a subsequent author was allowed to do so. Stability of nomenclature was maintained, in spite of the drastic conceptual change, whenever the subsequently designated type species was chosen from one of the species that had remained in the genus after others had been placed into different genera (under the old elimination method). However, it happened not infrequently that careless authors designated as the new generic type a species that had been previously removed to another genus. Inevitably such a thoughtless action caused utter nomenclatural confusion. This could have easily been prevented if the Code of Nomenclature had included a provision that the designated type species had to be selected from the species remaining in the genus if any species had in the meantime been removed to other genera. Unfortunately no such provision was included because no one at that time (in the 1890s) was aware of the drastic conceptual change of the concept 'typical'.

Several names for a single object

Another problem of nomenclature is likewise almost exclusive to taxonomy, the existence of several names for a single object. Which name is one to choose? I said 'almost exclusive', because an outsider approaching philosophy sometimes has the feeling that this is also a problem for philosophers. For instance, many philosophers use the terms 'natural kind' and 'class' as if they meant the same thing. And how does 'set' differ from 'class'? The reluctance of philosophers to provide unequivocal definitions for the terms they use makes decisions in such cases very difficult.

Such uncertainties no longer exist in taxonomy after it had become obligatory to typify taxa by concrete type specimens. If this were not the case we would have complete chaos, since many if not most species of animals and plants have been named repeatedly. The reasons for such duplication are manifold (Mayr *et al.* 1953). Sometimes, particularly in the immediate post-Linnaean period, it was philological pedantry that led to name changing; sometimes several authors described the same species in different journals or different countries; sometimes the first description was so incomplete or even misleading that (without examination of the type specimen) it was not recognized by later authors; sometimes the duplicated naming was merely due to a lack of diligence in searching through the previous literature. The rule that was ultimately adopted to settle the question which of several competing names should be considered the correct one, is the rule of priority. Among available, properly described species names that name should be adopted that had been published first.

The principle of priority, in spite of its simplicity and clarity, had unfortunately at first a most deleterious effect on the stability of zoological nomenclature. The reason is that it represented a drastic conceptual change which the framers of the Code of Nomenclature were either unaware of or had decided to ignore. Linnaeus himself, as much as he preached priority, was an inveterate name-changer, in most cases because he considered previously used names as not 'suitable'. For this he was upbraided by his friends. For instance, Peter Collinson wrote him in 1754: 'but my dear friend, we that admire you are much concerned that you should perplex the delightful science of botany with changing names that have been well received'.

Indeed, the 'rule of first revisor' seemed more important to Linnaeus than the rule of priority, and so it was for his followers. For instance, they all adopted the 12th edition (1766) of the *Systema Naturae* as the definitive basis of the Linnaean animal names rather than the 10th edition (1758) in which Linnaeus for the first time had used binominal nomenclature consistently. Authority, that of Linnaeus in botany or that of Fabricius in entomology, were considered more decisive rather than priority. Indeed priority in the hundred years after Linnaeus was almost consistently ignored whenever a name had achieved general usage.

The principle of strict priority was introduced in the first official code of nomenclature, the so-called Strickland Code (1843) of the British Association. When Darwin read it, he wrote to the author 'if I were to follow the strict rule of priority, more harm would be done than good' (LLD, Vol. I, p. 366), and this prophetic statement indeed became true for all branches of zoology. In some areas up to 50 per cent of well established names had to be changed for the sake of the strictly arbitrary principle of priority, thereby completely paralyzing the function of nomenclature as a key to scientific information.

The period when the Strickland Code was introduced was a period of great name-changing, and its primary purpose seems to have been to prevent all future arbitrary name changes. One might call this a principle of prospective priority. It is doubtful that the early promoters of strict priority realized that this principle would induce certain unscrupulous authors to search through the ancient literature for totally forgotten names, names that had not been used for 100 years or more. Actually, at the time when strict priority was introduced, both zoological and botanical literature were full of forgotten names that had been either overlooked or, more often, deliberately ignored for reasons too numerous to be specified in this context. Applying the principle of strict priority retroactively and digging up names from the old literature that had been forgotten for 50, 100, or more years, led to a most deplorable instability of nomenclature, exactly the opposite of what had been hoped for when the principle of priority had first been proposed.

In addition to strict priority there was a second principle of Linnaean nomenclature adverse to stability, the binominal system. Every time a species is transferred to a different genus, the binomen becomes different. Why did this lead so often to name changes? Linnaeus and his early followers recognized only about 500 genera of animals, while more than 100,000 are now recognized. As the genera grew in size owing to the discovery of new species, they were split again and again into smaller genera, and certain species transferred to these new genera. These species thus acquired a new generic epithet in their scientific name. Such changing of scientific names is of course entirely independent of the principle of priority.

Philosophy is lucky in having only relatively few situations where several names are used for the same concept and where a priority squabble might develop. Whenever doubts arise as to whether or not two terms refer to the same object, a clearcut definition might be able to avoid misunderstandings. I do not quite see how an exemplar could be used instead of a definition.

242

Conceptual change and stability of names

Perhaps the most important point that has emerged from my analysis is that zoological nomenclature could have escaped a great deal of trouble if the zoologists had worked more closely with the philosophers. As I showed, almost all concrete rules of nomenclature are based on certain concepts, but the zoologists failed to notice that some of these concepts changed drastically over time, resulting in conflicts between the methods of naming by the early post-Linnaeans and the taxonomists in the late 19th century who formulated the codes of nomenclature. A shift in some basic concept affecting nomenclature occurred on three occasions. This was not recognized in any of the three cases with the result that a retroactive application of the new concept resulted by necessity in great instability of zoological nomenclature. The three conceptual shifts were the following:

1. Replacement of the concept of a series of typical specimens by the concept of a single type specimen that has the function to serve as the onomatophore;
2. replacement, in the typification of a genus, of the concept of a series of typical species by that of a single type species that was designated either originally or subsequently, and
3. replacement of an optional use of the 'first reviser rule' to authority by usage by strict retroactive priority.

In all three cases it would have been easy to prevent the great instability of existing nomenclature by placing some constraint on retroactive application. However, in none of the three cases did the authors of the codes realize that a major conceptual change had occurred and that constraints on retroactivity were necessary. The unfortunate result was that the function of scientific names as keys to scientific information was severely damaged. Some measures of preventing such upsetting introductions of new changes have now been built into the currently valid Code of Zoological Nomenclature (3rd edition), but unfortunately long after most of the damage had been done.

References

Darwin, F. (1887). *The Life and Letters of Charles Darwin.* Vols. 1–3. London: Murray.
Greene, J.C. (1986). The history of ideas revisited. *Revue de Synthese* 4: 201–227.
Hull, D.L. (1983). Exemplars and scientific change. *PSA 1982*, Vol. 2: 479–503.
Hull, D.L. (1988). *Science as a Process.* Chicago: Chicago University Press.
International Code of Zoological Nomenclature. 3rd ed. 1985. London: International Trust for Zoological Nomenclature.
Kitcher, P. (1978). Theories, theorists and theoretical change. *Philosophical Review* 87: 389–406.
Kripke, S.A. (1972). Naming and necessity, pp. 253–355. In Davidson, D. and G. Harman (eds). *Semantics of Natural Language*, 2nd ed. 1972. Dordrecht: Reidel.

Mayr, E. (1974). Teleological and teleonomic: A new analysis. *Boston Studies in the Philosophy of Science* 14: 91–117. Proceedings Boston Colloquium for the Philosophy of Science, 1969/1972. Also, pp. 383–404 in Mayr, E. (1976). *Evolution and the Diversity of Life*. Cambridge, MA: Harvard University Press.

Mayr, E. (1983). Comments on David Hull's paper on exemplars and type specimens. *PSA 1982*, Vol. 2: 504–511.

Mayr, E. (1988). *Toward a New Philosophy of Biology. Observations of an Evolutionist*. Cambridge, MA: Harvard University Press.

Mayr, E., Linsley E.G., and Usinger R.L. (1953). *Methods and Principles of Systematic Zoology*. New York: McGraw-Hill.

Putnam, H. (1973). Meaning and reference. *J. Philosophy* 7: 699–711.

Putnam, H. (1975). The meaning of meaning, pp. 131–193 in Gunderson, K. (ed.) *Language, Mind, and Knowledge*. (Minnesota Studies in the Philosophy of Science, Vol. VII). Minneapolis, MN: University of Minnesota Press.

Romanes, G.J. (1895). *Darwin and After Darwin*. Vol. II. Chicago: Open Court Publ. Co.

Simpson, G.G. (1961). *Principles of Animal Taxonomy*. New York: Columbia University Press.

Strickland, H.E. (1842). Rules for zoological nomenclature. Report of the 12th meeting of British Association held at Manchester in 1842. *Brit. Assoc. Adv. Sci. Rpt.* 1842: 105–121.

Suppe, F. (1977). Exemplars, theories, and disciplinary matrixes, pp. 483–499 in Suppe, F. (ed.) *The Structure of Scientific Revolutions*, 2nd ed. Urbana, IL: University of Illinois Press.

Whewell, W. (1847). *Philosophy of the Inductive Sciences*. London: Parker.

From Reductionism to Instrumentalism?

ALEXANDER ROSENBERG

Philosophy University of California, Riverside, USA

Reductionism in biology never really recovered from chapter one of David Hull's *Philosophy of Biological Science*.[1]

After Watson and Crick but before that work, the prospects seemed bright that in biology we could find what all recognized to obtain in the physical sciences: relatively 'smooth' reductions of less fundamental theories to more fundamental ones, both across the history of science and down the hierarchy of completed scientific achievements. Of course everyone recognized that the simple account of reduction, in terms of derivability of laws and connectability of terms, was too coarse. Nevertheless, there was something about the relations between Kepler's laws and Galileo's, on the one hand, and Newton's on the other, that substantiated the reductionist's picture. And the more complex cases of theoretical subsumption that followed down to quantum electrodynamics and the general theory of relativity seem to lend equal support to the doctrine.[2] There were of course 'flies' in the ointment: problems about connecting terms with incompatible 'meanings' like Einsteinian mass and Newtonian mass,[3] but these were presumably problems in the philosophy of language, not the philosophy of science. Then of course, reduced theories need to be 'corrected'[4] and there were the problems associated with what counts as a correction of the reduced theory, as opposed to a new theory altogether. But these seemed to be problems about systematizing intuitions that were pretty clear.

1. The problem of the many and the many

But in chapter one of the *Philosophy of Biological Science*, Hull showed that there is another problem, characteristic of the relations between molecular biology and Mendelian genetics that effectively blocked any actual reduction. It is the problem of 'the many and the many'. And like the philosopher's ancient problem of 'the one and the many', the many-many problem has reared its head in other contexts as well, since first raised by Hull, notably in the philosophy of psychology, and the social sciences generally.

M. Ruse (editor), What the Philosophy of Biology is. pp. 245–262.
© 1989 *Kluwer Academic Publishers, Dordrecht*

The problem of the many and the many is simple enough to state:

> Even if all gross phenotypic traits are translated into molecularly characterized traits, the relation between Mendelian and molecular characterized predicate terms expresses prohibitively complex, many-many relations. Phenomena characterized by a single Mendelian predicate term can be produced by several different types of molecular mechanisms. Hence, any possible reduction will be complex. Conversely, the same type of molecular mechanism can produce phenomena that must be characterized by different Mendelian predicate terms. Hence reduction is impossible. (Hull, 1974, p. 39).

The consequences of this many-many relation between molecular and Mendelian predicates, and the properties they denote, is obvious: if there are nomological generalizations at the level of Mendelian, or population genetics, they will be autonomous from any generalizations at the molecular level. For they will not be linked to molecular generalizations by bridge-principles of the sort reduction requires. Or at least, such bridge-principles will be hideously complex, each side of which is a disjunction over a vast number of further disjunctions.

So that any derivation of a Mendelian phenomenon in its generality, from molecular biology, will implicate almost all of what theory there is about all the macromolecules known to biochemistry. This sort of reduction is not only methodologically useless, it is probably unattainable. Reductionism thus casts little light on inter-theoretical relations in biology.

In the *Philosophy of Biological Science* Hull briefly explores a possible line of reply to his argument. Reductionists agree that a given polynucleotide sequence may result in two or more phenotypical effects, or vice versa: a given phenotypical effect may be the result of two or more different biosynthetic pathways from two or more different polynucleotide sequences. When this occurs, it is because of variations in other causally relevant factors, especially in the molecular milieu of the genetic material in the nucleus and the cell in which gene-products are synthesized. Adding these factors to a biochemical-Mendelian bicondition will provide the bridge-law required to effect the reduction; it will convert the many-many relation to a one-one relation of the sort reduction requires. To this Hull responded:

> The conviction that this can always be done is actually a covert restatement of the principle of deterministic causality. The same cause always produces the same effect. If the effects are different, then the cause must be different. But one should notice how much the notion of molecular mechanisms has been expanded. We are no longer correlating Mendelian predicate terms with molecular mechanisms but with the entire molecular milieu. One possible difference between those biologists who consider themselves reductionists and those who consider themselves organicists might hinge on just this distinction. Reductionists think that a specification of just molecular structures and mechanisms is

adequate to explain all hereditary phenomena. If so, reduction would be straightforward, if somewhat complicated exercise. Organicists, on the other hand, emphasize that reference to the entire molecular milieu, including the environment, may well be necessary. If so, the entire feasibility of the reductinionist' program is called into question. (Hull, 1974, p. 42).

It's worth recalling that Hull was not led by his arguments to deny that Mendelian genetics is reducible to molecular biology. The relation between these two theories is, he held, 'a paradigm case' of reduction. Hull's conclusion was 'that the logical empiricist analysis of reduction is not very instructive in the case of genetics.' (p. 44).

In the years since Hull wrote these words, general opinion in the philosophy of biology shifted first towards his view, and then past it to more radical ones. Hull accepted that Mendelian genetics is reducible to molecular genetics; he doubted whether molecular theory could deductively explain Mendelian processes; and denied that the relation between reduced and reducing theories satisfies the criteria of derivability of laws and connectability of terms that the classical account of reduction demands. Nowadays, the ruling view embodies the last two of these theses, but denies the first. By and large, philosophers of biology deny that reduction of any kind characterizes the relation between these two theories.

I wish to examine two questions that emerge from this consensus. First, there is the substantive biological question why reduction of the sort envisioned by logical empiricism is impossible in biology. This question is a request for a causal explanation, one which cites biological facts that result in the many-many relation. It must be sharply distinguished from the methodological question of the same verbal form ('Why is the reduction impossible?') whose answer is the problem of the many-and-the-many. The question I want to address is what *causal* facts about biological systems result in the theoretical problem of the many and the many, when such difficulties do not seem to obtain in physical science. Second, there is the question of what view of the nature of biological science emerges from reflection on this biological fact and from the denial that reduction obtains at all between these two theories. Is there still anything to be said for David Hull's conviction that despite all the vicissitudes, there is a relation between these theories properly called a reduction?

2. Selection for function is blind to structure

What facts about biological systems make 'layer-cake' reductionism, as it has sometimes been called, impossible in the life-sciences? One answer seems obvious to me, though I have never seen it explicitly advanced before: biological systems are the result of natural selection over blind variation. The fact that they are the result of adaptational evolution is the causal fact that explains the impossibility of reduction. This claim should not be surprising, since evolutionary adaptation is what is distinctive about these systems, by contrast to purely physical one. So it seems the obvious first place to look for a causal explanation of the differences between theories about them and theories about purely physical systems. In particular, such an explanation seems vastly to be preferred to one in terms of 'paradigm shift' or Feyerabendian incommensurability, or differing social construc- tions of science. A biological explanation for the impossibility of reduction is simpler, is based on far more secure premises, and is far more precise in its explanatory focus, than any of these non-biological explanations could be.

What is it about natural selection over variation that precludes reduction by derivation between biological theories? Natural selection 'chooses' variants by some of their effects, those we identify as their functions. Processes that are 'random' with respect to adaptation, result in combinations of many sorts: thus quantum electrodynamic processes result in nuclear and atomic phenomena, chemical bonding processes result in molecules, thermodynamic processes acting together with these other processes result in larger arrays, etc. These combinations of matter have various properties, some of which enhance the physical stability and persistence of the combinations, others of which render the combinations unstable. Stability and persistence among these physical combinations are obviously enough 'selected for', by thermodynamics. Among those stable combinations ones that replicate are even more strongly selected for – molecular configurations that for example catalyze the processes that led to their own creation, or foster the ther- modynamic circumstances conducive to their appearance, or otherwise make the combinations they instantiate more likely by their presence will be reproduced in larger numbers. And the same goes for larger arrays of molecules, all the way up to macroscopic configurations, of the sort detectable by naked eyes. This much is tautology.

What is not tautology is that among such combinations, at apparently every level above the nucleic acid, there are frequently to be found *physically distinct* struc- tures with some *identical* or nearly identical functional properties, different combinations of different types of atoms and molecules, that are close enough to being equally stable, and equally likely, for purely physical causes, to foster the appearance of more instances of the kind they instantiate. Thus, so far as adaptation is concerned, there are frequently *ties* for first place in the race to be selected. And, as with many contests, in case of ties, duplicate prizes are awarded. For the prizes are increased representation of the selected types. Thus, if urea and methane are equally stable, and if the presence of a urea or a methane molecule makes for equal increases in the probability that more urea and methane molecules will be produced in their vicinity, both will be equally well adapted. And so on 'up the chemical ladder.' It is the nature of any mechanism that selects for effects, that *it cannot discriminate between differing structures with identical effects.* Functional equivalence combined with structural difference must in the nature of the case increase as physical combinations become larger and more differentiated from one another. This may all be obvious at the level of the observable phenotype: as Hull points out in *The Philosophy of Biological Science*, the steel-grey color of a heterozygous Andalusian fowl 'could result from a variety of molecular situations.'

> Perhaps a blue-grey pigment is being produced. Perhaps only a black pigment is being produced, but in reduced concentrations. Perhaps both black and white pigments are being produced in equal amounts. These pigments could be distributed evenly throughout the feathers, or gathered together in small patches. In the latter case, the grey appearance would be a function of the visual acuity of the observer. (pp. 40–41).

But the blindness of selection to structure is evident at the most elemental level of molecular biology as well.

Below the level of the nucleotide, and the polypeptide, intermolecular combinations do not seem to provide different structure with sufficiently similar effects to be selected by a mechanism blind with respect to structure. One reason to believe this is the ubiquity of DNA as the medium of hereditary information. The information controlling development and heredity for all organisms, with the minor exception of the RNA viruses, is carried in DNA, and only DNA. There is no other molecule that serves as an information carrier. That this crucial biological function should be subserved by only *one kind of physical* structure has been the object of a fair amount of mystery mongering among biologists. Every other biologically interesting function is subserved by two or more, usually many more physically distinct structures and mechanisms. Why should this particularly complicated function be subserved by only one?

There seem only two biologically or physically plausible answers to this question: either the nucleic acids are so much better at information storage than any other molecular configuration that in the long period of evolution it beat out all the competition. Thus only organisms bearing DNA as the genetic material survived. Or at the level of molecular information-carrying there was only one way nature could skin the cat: given purely physical constraints, the only configuration that was ever capable of bearing information with the fidelity required for transmission and regulation is the nucleic acid. Thus, on this implausible scenario, it was selected for in the degenerate case: there were no competitors to beat out. Both of these alternatives can solve the further mystery of why RNA viruses and reverse transcriptase arose, by claiming that the ubiquity of DNA as the genetic material constituted an environment that only later made the combination of RNA viruses and reverse transcriptase adaptive. But it did so only long after the physical or adaptational fixity of DNA. In any head-to-head competition RNA loses to DNA for selection as the original genetic material, because of its substantially lower fidelity, and propensity to point-mutations. Once nature selects DNA as the genetic material, RNA finds itself in a new adaptive environment – characterized by the presence of DNA, in which RNA's effects on the configurations embodying it are selected for.

Neither of these two just-so-stories is entirely satisfactory. And this has led at least some distinguished molecular biologists to suggest that maybe the Earth was seeded with DNA, by extraterrestrial visitors, in order to produce biological systems that might otherwise not arise.

Whether or not nuclear acid was selected for as the sole winner in an evolutionary competition, at every level of configuration that includes the DNA, there have been ties in the race for selection, and duplicate prizes have been awarded. Consider for instance, the 'slack' in the genetic code. Polynucleotide chains need to code sequences of twenty amino acids. The four different nucleic acid bases, thymine (T), cytosine (C), guanine (G), and adenine (A) can be combined in sixty-four different sequences of three bases each. Since there are only twenty amino acids, only twenty nucleotide signals are required (plus sequences that signal termination of amino acid chains which compose proteins). Thus, there is redundancy in the genetic code. Every distinct amino acid besides methionine and tryptophan is coded for by at least two different nucleotide sequences, some by as many as four. Thus, valine is coded for by sequences of the following form: GUU, GUC, GUA, GUG. One obvious explanation for the redundancy of the code, or as biologists put it, its 'degeneracy', is the absence of a selective difference, either for themselves, or for organisms containing them, between these four sequences. If they all code equally well for valine, and coding is their *only* functional role, then selection will be blind to their structural differences.

But it has doubtless not escaped the reader's notice that all four of these nucleotide sequences share the same initial two nucleotides, and differ only in the third position. This in fact is true for all of the types of nucleic acids synonymous for a given amino acid. But such structural regularities are always an invitation to function-finding. Perhaps the fact that redundancy appears only at the third position in a codon suggests that selection is not blind to these differences, and that the persistence of 'synonymous' codons has an adaptive explanation after all? Indeed, it does, or at least one such an adaptive explanation has been offered. If it is right, the redundancy of the DNA code is not the result of a tie in the selective race, in which duplicate prizes have been awarded. Rather it results from adaptation to structural differences among transfer RNAs.

Transfer RNAs are the vehicles which bring amino acids to the ribosome, where the messenger RNA, transcribed from the DNA, directs the order in which the amino acids are bonded together into a protein. Each transfer amino acid bears an 'anti-codon' triple that matches up with the amino acid it bears, and bonds to the codon triple transcribed from the nuclear DNA by the messenger RNA. But when the third nucleotide of the transfer RNA anticodon is Uracil (which substitutes for DNA's thymine molecules in the messenger RNA's sequence), the molecular bond linking the transfer RNA to the messenger RNA will be equally strong between guanine or adenine and uracil. Thus, a codon of GUA or GUU will affect the same amino acid bonding. And the same is true when the third molecule in the anticodon is guanine: it bonds equally well with either cytosine, or uracil, thus permitting either one to serve as the last member of the codon which codes for valine.

Apparently selection has permitted several different molecular sequences to serve as the transfer RNA for a given amino acid. This is what results in the 'degeneracy' of the DNA code. Thus, we end up explaining why selection *seems* blind to a structural difference between DNA sequences by showing that it is not in fact blind to them, but blind to a structural difference between transfer RNAs. It is this diversity of ribonucleic acids with the same functions that selects *for* a structural diversity of nucleic acids with the same function.

Here we have, at a very 'low level' of organization, a phenomenon widely cited by opponents of reductionism: the direction of mereological causation seems the reverse of what reductionism requires: transfer RNA structure and function determines DNA structure. Instead of explaining the character of gene-products, the molecular structure of the genetic material is explained by these products' organization and function. Nature selects for function in the light of environmental constraints. These constraints include the molecular environment of the genetic material in the nucleus, and the cellular environment, including the organelles operating within the cell. So, at least sometimes a given single molecular structure may be selected for, because it is optimal in the light of a diversity of physically different systems with the same function. But it is more likely that the physical diversity of these 'higher level' systems will call forth not one but a diversity of physically different molecular structures with the same function.

Furthermore, if selection can operate in ways that discriminate between very slight differences in functional efficiency among molecular configurations, then structural diversity among systems with some of the same functions will be encouraged. A given physical system may have indefinitely many effects on its environment. Only a subset of these are actually identified functions – i.e. selected for by the systems' environment. And among these functions only a smaller subset will be identified as such by scientific inquiry. Moreover, a physical system may functionally subserve several different 'needs' of the large system in which it operates. Thus, for example, thymine is present in DNA both to code for amino acids like valine, but also has the function of insuring maximum fidelity, since it is less prone to point mutation than the Uracil that replaces it, at lower energetic cost, in RNA.

All the diverse functions a physical system has must be reconciled to the extent it is successfully selected for. And it is clear that though two different physical structures can fulfill a bundle of functions with equal adaptive efficiency, they may do so, by fulfilling different members of these bundles differently. Thus, some parts of the genetic material may be better at fidelity and worse at energetic efficiency than others, and yet both sorts equally adaptive. These differences in function must be traced back to differences in structure, differences that cancel out so far as overall fitness differences are concerned.

So, not only does nature select diverse physical systems because of their identity of function, but then the diversity of these selected systems itself acts as a selective force producing further diversity of structure with identity of function. And of course it isn't *identity* of function that's required. *Similarity* will do. Indeed, it may be the case that no two physically diverse structures have exactly the same set of functions. For their diversity entails that they have diverse effects. Even if some or all of these diverse effects are inconsequential for selection in any given environment, there is always some possible environment and some length of time in which the smallest structural difference or its immediate effect can bear a selective advantage or disadvantage.

In fact, it is not necessary to come in first at any given point of selection in order to receive the prize – the opportunity to replicate further. There are sufficient many non-selective forces that impinge on configurations, and that affect their prospects of reproduction from occasion to occasion, that just coming a close second or third to being most well adapted is often enough to ensure persistence and replication. This after all is what the phenomenon of evolutionary *drift* is all about. Sometimes, as between two or more variants, the most well-adapted is eliminated through purely non-selective accidents. Sometimes the environment does not remain constant for long enough to ensure that the optimally adapted structure is fixed or the less optimal structures are extinguished. This gives configurations in second place another chance. And if the chances of an eliminated first place configuration recurring are low enough, these second place configurations will never have to face competition from it again. Or again, when the previously most well adapted configuration recurs, the environment will have changed, or the previously second place configurations' successors may have been improved by further selection.

Natural selection thus makes functional equivalence-cum-structural diversity the rule and not the exception. By contrast, nonevolutionary processes, mechanical, thermodynamic, electromagnetic, chemical ones, make functional equivalence with structural difference the exception, if they permit it at all. This difference, resulting from the operation of selection for effects, explains why reduction goes smoothly in the physical sciences, and apparently not at all, in biology, and in every other discipline in which phenomena are explained by their effects.

3. Methodological morals

Reductionism is in some large measures the reflection of methodological and metaphysical commitments to the simplicity of nature. It is the attempt to make precise the notion that under the buzzing, blooming confusion of nature, there is a small number of mechanisms or processes, and a small number of different types of things, which can systematize and explain the world. This notion of simplicity is the corner stone of reductionism.[5] If the world isn't simple after all, then reductionism's metaphysical rationale and methodological justification is undermined.

But this seems to be the moral of natural selection for scientific method and metaphysics: above the level of the molecule, nature is not simple any more, and it is not simple because of the blindness to structural differences of selection for functions. Thus, above the level where ties result from nature's selection for function, regularities, such as there are, will not be explained by a small number of laws about a limited number of physically distinct systems, no matter whether explanation is a matter of logical derivation or some quite different relation.

What is more, the regularities selection for function generates will be far more restricted, exception-ridden, and complex, than we have come to expect in other disciplines. Consider the class of objects selected because they fulfil some adaptive function, Fx. We seek some interesting generalization about members of this class, perhaps something of the form $(x)(Fx \rightarrow Gx)$. Thus, we seek another predicate, Gx, true of all members of the extension of Fx. We can seek no structural property of members of the extension of Fx, for the class of F's is physically heterogeneous, since they have all been selected by effects. Of course, we may find some structural feature shared by many or most of the members of F. But if we do, it will be a relatively uninteresting one, or if interesting, then the class of objects with property F is probably very narrow: First, an example of the former case: Consider the class of all predator-avoiders. Well, it is easy to identify a property they share in addition to being predator avoiders: they all experience gravitational attraction. Clearly what we want is not just a property necessary for being a predator avoider, but one distinctive of it. This brings us to the second example: For a relatively restricted function property, like 'codes for valine', we can produce a disjunctive class, composed of instances of three types of DNA sequences. But this generalization, though exceptionless and precise, is about the best we can do.

If we try to frame generalizations about functional systems much above the nucleic acids in complexity, the result is always falsity or vacuity. Consider Mendel's laws of segregation and assortment: each attributes a functional property to the Mendelian gene. That is, they claim that all members of the functionally identified class, Mendelian genes, have further functional properties, in virtue of being Mendelian genes: they assort independently and segregate from homologous alleles. Given that both the antecedent and the consequent classes of these generalizations are identified functionally, and therefore their extensions include structurally quite diverse physical systems, it is no surprise these 'laws' should be false. For it is highly unlikely that any pair of physical systems identical for any one function should be identical for another. And if the systems are merely quite similar in respect of fulfilling one function, the chances that they will be equally similar in fulfilling another are even lower.

So, at every level of physical organization where organization reflects numerous selective ties, there will be no very lawful regularities, and such general facts as obtain, will not be explainable by any small set of laws about selectivity or structurally simpler subsystems. But reduction requires such lower-level laws and derivations of them from small sets of more fundamental laws.

What are the consequences of the complexity of nature above the level of single-winner selective interactions for the character of biological theory? Its consequences seem to be thoroughly instrumentalist. Or so I shall argue. Complexity restricts humans and other agents, with limited computational and cognitive powers, to the construction of theories that are at most useful instruments; that theories succeed one another not because they are improvements in the direction of greater generality or truth, but because they reflect changes in the agenda of problems to which biologists apply differing instruments; that the aim of biological theorizing is not, as it is in physical science, the identification of natural laws of successive generality, precision, and power, but the sharpening of tools for interacting with the biosphere. If reductionism is wrong, instrumentalism is right. And, of course, by *Modus Tollens*, if we are unwilling to endorse instrumentalism, we shall have to find a way of preserving reductionism from Hull's problem of the many-and-the-many.

A number of current accounts of intertheoretical relations in biology substantiate this forced choice between reductionism and instrumentalism. Perhaps the most fully worked out of them is Phillip Kitcher's '1953 and All That: A tale of Two Sciences.'[6] The view expounded is Hull's analysis refined by a generation's further work in biology and its philosophy. Kitcher does not argue that his account encourages an instrumentalist view of biological theory. Indeed, at points in his paper, he explicitly rejects it. His arguments against it are unsatisfactory. If I am right, then to the degree Kitcher's analysis reflects the best anti-reductionist treatment of biological theory, it is after all an argument for instrumentalism.

4. What has molecuar biology done for us lately?

Kitcher's aim is to 'account for the almost universal idea that molecular biology has done something important for classical genetics' in spite of the fact that the absence of transmission laws, Molecular-Mendelian bridge principles, or derivational explanations makes reductionism (and not just the Logical Empiricist account of it, apparently) impossible. What is needed, on Kitcher's view, is a whole new appreciation of what theory is in biological science (and perhaps elsewhere as well).

To begin with we cannot view Mendelian, or transmission, or what Kitcher calls 'Classical genetics' as a body of nomological generalizations. Mendel's laws were known to be false almost from the outset of their rediscovery, and no steps were taken to repair them. Presumably because such patched up generalizations were irrelevant to subsequent research in genetics (p. 342).

Instead of a hypothetico-deductive theory, or for that matter a complicated predicate, classical genetic theory as to be viewed as a sequence or 'linked chain' of 'practices'.

> There is a common language used to talk about hereditary phenomena, a set of accepted statements in that language..., a set of questions taken to be the appropriate questions to ask about heretary phenomena, and a set of patterns of reasoning which are instantiated in answering some of the accepted questions. The practice of classical genetics at a time is completely specified by identifying each of the components just listed.

The most important of these is what Kitcher calls the set of patterns of reasoning or argument. These patterns were originally focused on solving 'pedigree problems', roughly those surrounding the distribution of phenotypes in successive generations. Starting with Mendelian recipes for answering questions about pedigrees assumed the single locus, two allele cases with complete dominance. This argument pattern was superceded by one which accommodates epistasis, but was blind to recombination and linkage. With Morgan, a stipulation is added that probabilities of linkage in transmission are functions of co-location on the same chromosome. The original Mendelian pattern included principles of reasoning we recognize as the 'laws' of assortment and segregation, but these arguments were surrendered with Morgan for more complex ones that allow for reasoning about phenomena that cytology reports: duplication, cross-over, segregation distortion and meiotoic drive, etc.

In addition to the pedigree problem, over time other soluable problems arise in the history of classical genetics, as a result of cytological location of the gene. The theory of gene-mapping, and of mutations constituted new problem-solving patterns that appealed to and extended the original patterns of the theory of gene transmission.

What has molecular genetics done for the connected set of reasoning patterns that constitute classical genetic theory? Classical genetics makes certain presuppositions. (A theory presupposes a proposition p, if every instantion of some problem-solving pattern of the theory implies, p. 361.) And Kitcher tells us, these presuppositions were problematical – classical geneticists had no idea how they could be true: among them is the claim that genes replicate, that mutant genes are viable. These presuppositions seemed impossible, Kitcher alleges, given premises classical geneticists accepted (p. 361). Molecular biology shows how these presuppositions could be true. It does so by making these conclusions the results of its own patterns of reasoning.

But it does not do so by explaining generalizations of classical genetics – for what molecular genetics explains are not laws in classical genetics. Molecular genetics explains how genes replicate, but 'the claim that genes can replicate does not have the status of a law of classical genetic theory ... Rather it is a claim that classical genetics took for granted, a claim presupposed by explanations, rather than an explicit part of them. Similarly for the problematical presupposition that mutant genes can replicate. It is only these presuppositions of classical genetics that molecular biology explains (by deductive systematization, as reduction requires), but they are not strictly part of explanations at all, so evidently molecular genetics explains nothing beyond classical demystifying problematical presuppositions. In the case of mutation, molecular genetics provides a 'conceptual refinement' of classical genetics. 'Later theories can be said to provide conceptual refinements of earlier theories when the later theory yields a specification of entities that belong to the extension of the predicates in the language of the earlier theory...' (p. 364). Molecular genetics provides an account of several different kinds of internal changes that result in mutant alleles, and it enables us to distinguish mutation from recombination. Here again, molecular genetics does something important for classical genetics, but not what reductionism suggests it should do. For that mutant alleles can replicate is neither a law of classical genetics nor part of the explanations which its patterns of reasoning provide. Or so Kitcher claims.

Only in what Kitcher calls the process of 'explanatory extension' does molecular genetics function in something like the reductionist's picture.

> [A] theory T' provides an *explanatory extension* of a theory T just in case there is some problem-solving pattern of T one of whose schematic premises (given in T's pattern of argument) can be generated (i.e. explained) as the conclusion of a problem solving pattern of T'. ... However it does not follow that the explanations provided by the old theory can be improved by replacing the premises in question with the pertinent derivations. (p. 365).

Explanatory extension in molecular genetics does not involve general molecular characterizations of all genes, but of particular ones. To cite his example, it enables us to derive the functional differences between sickle-cell hemoglobin cells and normal ones from a specification of the primary sequences of hemoglobin molecules, and the order of basis in normal and sickle-cell hemoglobin genes (together with certain boundary conditions). Explanatory extension in particular cases works because it focuses on differences from other cases, assuming a strong *ceteris paribus* clause. It does not aim at explaining generalizations about all classical genes, but only particular types, and it can afford to ignore the detailed biosynthetic pathways from genes to proteins, because in these cases the correlation between differences in gene sequences, protein primary structure, and phenotypic effects is relatively straightforward.

Explanatory extension is not reduction, claims Kitcher, for two reasons: First, most classical genetic phenomena can only (or can best) be explained by classical genetic patterns of reasoning, because the phenomena 'would look too heterogeneous from a molecular perspective. Intuitively, the cytological pattern makes connections which are lost at the molecular level, and it is thus to be preferred (for explanatory purposes). (pp. 371–2)' Secondly, sometimes the direction of explanation in genetics is from classical to molecular, or at least from cytology back to macromolecules: 'Understanding the phenotypic manifestation of a gene ... requires shifting back and forth across levels (of cellular organization) ... sometimes one uses descriptions at higher levels to explain what goes on at a more fundamental level.' (p. 371).

5. Instrumentalism

The reader familiar with the history of the philosophy of science will perhaps notice a similarity between Kitcher's picture of the structure of genetic theories with Stephen Toulmin's conception of scientific theories as 'inference licences entitling us to argue from known facts about a situation to the phenomena we may expect in that situation.' (Toulmin, 1953, p. 102) Following Ryle, Toulmin argued that laws of nature are not true or false, but 'inference-tickets,' valid in some regions but not in others: 'By making the journeys (inferences) so licenced, the physicist finds his way around phenomena. (Toulmin, 1953, p. 104).

As a view about the nature of physical laws, and an argument for instrumentalism, Toulmin's claims fell out of favor. One powerful argument against them was Nagel's point that the difference between 'material rules of inference' and substantive premises is arbitrary, in the context of scientific reasoning:

> ... questions can be raised about a theory when it is regarded as a leading principle that are substantially the same as those which arise when the theory is used as a premise. For whether or not a material leading principle happens to be a theory, the principle is a dependable one only if the conclusions inferred from true premises in accordance with the principle are in agreement with the facts of observation to some stipulated degree. In consequence, there is on the whole only a verbal difference between asking whether a theory is satisfactory (as a technique of inference) and asking whether a theory is true. (Nagel, 1961, p. 139).

If this view is correct, then there is an inevitable tension between Kitcher's claims that there are no law-like general statements in classical genetics (p. 340), and the claim that there are accepted patterns of reasoning in it.

What seems more reasonable is to hold that the laws about transmission, mapping and mutation are too complex, too disjunctive and too large in number for us to discover, express, and employ in arguments more precise than those we actually advance. Thus, the patterns of argument we employ are the best we can do, given our cognitive and computational powers. The law-like statements we can 'convert' patterns of argument into are known to be false, to have exceptions, undischarged *ceteris paribus* clauses, limited ranges of application, etc. They are not laws, but they are the best we can do, and given our needs and interests at any time, they are sufficiently good. Reductionism thus fails, because there are no *expressible* laws of classical genetic theory to reduce, and the patterns of reasoning classical genetics employs makes molecular details 'irrelevant'.

This suggestion gives the reductionist a rejoinder that Kitcher recognizes, and attempts to forestall: The claim, that molecular genetics is explanatorily irrelevant to the argument patterns characterizing classical genetics, will be held by reductionists to 'presuppose far too subjective a view of scientific explanation. After all, even if *we* become lost in the molecular details, beings who are cognitively more powerful than we could surely recognize the explanatory force of the envisaged molecular derivation.' There are classical laws, albeit complex ones, and we just are not smart enough to discover them or make use of them. Kitcher responds that this claim 'misses a crucial point. The molecular derivation forfeits something important.' (p. 348–9). Roughly, it is blind to important natural kinds of classical genetics, in particular, the class of gene-*pair* *separation* processes (which Kitcher calls PS-processes). Classical genetics appeals to cytology to describe and explain transmission as a PS process – it is by bringing transmission under the mechanism of PS processes that it is explained. A reduction of classical genetic theory to molecular genetics must preserve the explanatory power of the former theory, and to to this it must preserve the natural kinds of the theory. (p. 349).

> However, PS-processes are heterogeneous from the molecular point of view ... PS processes are realized in a motley of molecular ways.
>
> We thus obtain a reply to the reductionists charge that we reject the explanatory power of the molecular derivation simply because we anticipate our brains will prove too feeble to cope with its complexities. The molecular account *objectively* fails to explain because it cannot bring out that feature of the situation which is highlighted in the cytological story. (p. 350).

This argument hinges on two crucial unargued assumptions: first that PS processes are natural kinds, and second that explanatory success or failure is 'objective'. Both terms, 'natural kind' and 'objective' are undefined in Kitcher's argument. And on at least some accepted accounts of the meaning these terms, the claims Kitcher makes employing them are doubtful.

Since Mill's *System of Logic* 'natural kinds' have been explicated by appeal to laws. The simplest view is that a kind is natural, as opposed to artificial, only if one or more of the properties that characterize all its members figure in a small number of simple general laws. But on this criterion, or on any obvious complication of it, PS-processes cannot be 'natural kinds', because according to Kitcher, there are no laws in classical genetic theory. (Are there laws about PS-processes beyond genetics?) If nomological involvement is our criterion of natural kindhood, P-S processes do not constitute such a kind, and nomological explanation of instances of P-S processes need not preserve a role for the concept of a P-S process. Kitcher holds that explanation must objectively preserve reference to such processes. This suggests that his account of a natural kind ties it more closely to successful or acceptable or 'objective' explanations, than to laws of nature. If a kind is natural just in case adequate explanation subsumes expananda under it, then Kitcher's claim will be vindicated: Assume classical genetic explanations are 'objective.' Then any other 'objective' explanation of phenomena in the extension of 'P-S process' will have to advert to such processes, whether they are nomically grounded or not. Since 'P-S process' is not a natural kind in respect of molecular genetics, it follows that no molecular genetic explanation of phenomena in its extension will be 'objective.' QED.

But what could it mean to attribute 'objectivity' to explanations? One sense in which explanations are identified as objective is due to Carl Hempel. In 'Aspects of Scientific Explanation' he contrasts pragmatic and objective conceptions of explanation. A pragmatic conception is at least a three-term relation, among explanans, explanandum, and persons. By contrast Hempel seeks to propound an account of explanation which he describes as 'objective' in the sense that it 'does not require relativization with respect to questioning individuals anymore than does the concept of mathematical proof. It is this non-pragmatic conception of explanation which the covering-law models are meant to explicate.' Deductive nomological explanations 'are *objective* in the sense that their empirical implications and their evidential support are independent of what particular individuals happen ... to apply them, and the explanations ... based upon such laws and theories are meant to be *objective* in an analogous sense.' (Aspects, p. 426, emphasis added.) On this notion of explanatory objectivity, of course the analysis of natural kinds as essential parts of objective explanations, reduces to the treatment of them as *nomological* kinds. Accordingly, Kitcher can hardly embrace it.

Nevertheless Kitcher *seems* to embrace the same sense of explanatory 'objectivity' as Hempel, for he describes as 'subjective' those explanations that work because they are tailored to beings of a limited cognitive power (p. 349); he rejects the notion that molecular genetics fails to explain just because 'our brains will prove too feeble to cope with its complexities' (p. 350); he asserts that 'the commitment to several explanatory levels does not simply reflect our cognitive limitations. (p. 373) And, most important, classical 'pattern(s) of reasoning are *objectively* to be preferred to the molecular pattern' in explanations of transmission. (p. 371, emphasis added)

But while he seems to take seriously the importance of explanatory objectivity in this sense, Kitcher's arguments against the adequacy of reductionist explanations repeatedly appeal to pragmatic, inquirer-relative, 'subjective' considerations that *do* reflect our cognitive limitations. If the explanatory power of biological theory is ultimately 'pragmatic' however, then Kitcher must in the end tie *biological natural kinds* to these same *cognitive limitations*. It is in this respect that if correct, the antireductionistic picture of biological science is ultimately instrumentalist.

Kitcher writes that 'Classical patterns of reasoning (are) to be objectively preferred to ... molecular patterns ... (in the explanation of gamete distribution) ... because (they) can be applied across a range of cases which would *look heterogeneous* from a molecular perspective. (p. 370, emphasis added) This, as Hull first showed us, is true. But to whom will these cases *look* heterogeneous from a molecular perspective? The answer seems to be, to cognitive agents with our *interests* and our *powers*. And how much heterogeneity is an obstacle to explanatory success? Well, it depends ... on the cognitive powers of the agents offering and accepting explanations. After all, agents of our powers can tolerate a certain amount of heterogeneity; those who can keep more alternatives in mind, and follow out their implications more unerringly, can tolerate more heterogeneity in the explanation of gamete distribution, and other biological phenomena. Among such cognitive agents many classical patterns of reasoning might well be replaced by molecular patterns. The former patterns won't be reduced to them because they do not reflect laws. Among those with less tolerance for heterogeneity, neither reduction nor replacement is more than an abstract possibility.

Kitcher rightly notes that 'Intuitively, the cytological pattern makes connections which are *lost* at the molecular level, and is thus to be *preferred*'. Here 'lost' can only mean 'lost on us' or cognitive agents no more powerful than we are. It cannot mean that there are cytological sequences that are causally independent of molecular processes. As Kitcher notes, 'reductionists and antireductionists agree in a certain minimal physicalism. ... there are no major figures in contemporary biology who dispute the claim that each biological event, state or process is a complex physical event, state or process. The most intricate part of ontogeny or phylogeny involves countless changes of physical state.' (p. 369) Surely agents, who can tolerate vastly more heterogeneity in explanation than we, can discover and classify together molecular processes that have common effects in cytological connections. On them, such connections would not be lost at the molecular level. For us, however, as Kitcher says, the cytological pattern of reasoning is to be *preferred*. But preference as a criterion of explanatory adequacy can hardly be 'objective'.

From the claim that cytological connections are lost at the molecular level Kitcher infers that 'explanatory patterns that deploy the concepts of cytology will endure in *our* science because we would foreswear significant explanatory unifications ... by attempting to derive the conclusions to which they are applied using the vocabulary and reasoning patterns of molecular biology.' This seems to be correct, but only on the understanding that *our* science reflects our cognitive limitations. Yet this is not Kitcher's conclusion. Instead he draws the moral that 'the current divisions of biology (are) not simply ... a temporary feature of our science stemming from our cognitive imperfections, but ... the reflection of levels of organization in nature.' (p. 370–371)

(I do not challenge the second part of Kitcher's antireductionist argument, which he describes as stronger, but which seems to me to be weaker and less interesting. It is the claim that sometimes the direction of explanation goes from the less fundamental, the classical level, to the more fundamental, the molecular level. This will be a fairly obvious conclusion, if we accept the notion that often non-molecular principles have explanatory powers independent of any molecular derivation of them. For if we grant them such power independent of reduction, then applying them to explain molecular processes will be at least methodologically legitimate. And on the other hand, if it turns out that these principles have explanatory power because they are ultimately reducible to molecular ones, then their application in a molecular context will also be uncontroversial to a reductionist. After all, the reductionist is not an eliminativist about classical genetics. He does not deny the theory's claims, or the existence of the entities and processes the theory quantifies over. Kitcher's second antireductionist thesis thus seems weaker, rather than stronger.)

The complexity which selection for effects produces above the level of the molecule, makes Kitcher's account of the structure of classical genetics and its relation to molecular genetics a compelling one. For it explains why we can't discover classical laws, and must resign ourselves to patterns of argument. But if we accept Kitcher's treatment of classical genetics as characterized not by laws, but by principles of argument, and take seriously his denial that there are laws in classical genetics, then despite Kitcher's claims to the contrary, we are committed to treating biology as an *instrumental* science, that is, as a body of claims each of which is qualified by an implicit appeal to its usefulness for cognitive agents of our powers. For each material argument pattern in classical genetics there is a substantive statement which though quite false, is the most useful hypothesis for dealing with some problems of interest at some level of the development of our knowledge. Reduction is impossible because the generalizations about classical genetics it is supposed to reduce are unavailable. But this is no defect in classical genetics, for that theory is not a body of laws, but primarily a set of patterns of argument, patterns warranted by their usefulness in particular problems and at particular stages of biological development.

The reductionist's rejoinder, that there must in principle be such laws, though they will be too disjunctive, complex and heterogeneous to be of much use to *us*, is to be met with a shrug of the shoulders. Yes, what of it? Admitting this is of no moment to the methodology of biology; it has no consequences, practical or theoretical (though it might have some metaphysical consolation).

Thus it turns out that not only the content, but the character of biological theorizing is contingent on biological facts. Biology is at its best an instrumental science because of the operation of biological forces: first, the complexity of nature above the level of the molecule is the result of selection for function and its blindness to structure. Second, the biological fact that the sentient creatures who develop the subject have (through selection) come to have cognitive powers that limit their ability to deal systematically with this amount of complexity.

It is not clear to me whether this is a conclusion Kitcher will welcome, but it is certainly one Hull should be sympathetic to. For in addition to fathering the approach in the philosophy of biology which leads to this conclusion, he has also been at the forefront of attempts to provide biological explanations for the character of the scientific theories we accept and transmit.

On the other hand, this conclusion is likely to raise qualms, especially in erstwhile reductionists. For among our motives for reduction there is a commitment to realism about scientific theory, and to the continuity of scientific method from the physical to the life sciences.

Acknowledgements

I wish to thank Philip Kitcher for discussion and comments on this paper. My debts to David Hull are evident and enormous.

Notes

1. Hull, 1974. All page references in sections 1 and 2 are to this work.
2. For a vigorous and detailed exposition of this faith, see Yoshida, 1977.
3. Cf. for instances, Feyerabend, 1962.
4. Cf. Schaffner, 1967.
5. For a detailed account of simplicity that exposes its motivational force for reductionism see Sober, 1975, especially section 2.4 Explanations and Predictions, pp. 47–50.
6. Kitcher, 1984. Page references in sections 4 and 5 are from this paper.

References

Feyerabend, P.K. (1962). Explanation, Reduction, Empiricism. *Minnesota Studies on the Philosophy of Science*, v. 3, Minneapolis: University of Minnesota Press.

Hull, D. (1974). *Philosophy of Biological Science*. Englewood Cliffs: Prentice Hall.

Kitcher, P. (1984). 1953 And all that: a tale of two sciences. *Philosophical Review* 93: 335–373.

Nagel, E. (1961). *The Structure of Science*. New York: Harcourt, Brace, World. Now published by Hackett, Indianapolis, Indiana.

Schaffner, K. (1967). Approaches to reduction. *Philosophy of Science* 34: 137–147.

Sober, E. (1975). *Simplicity*. Oxford: Oxford University Press.

Toulmin, (1953). *The Philosophy of Science*. London: Hutchison University Library.

Yoshida, R. (1977). *Reduction in the Physical Sciences*. Halifax, Canada: Dalhousie University Press.

Systematics and Circularity

ELLIOTT SOBER

University of Wisconsin, Madison, USA

1. Introduction

A central theme in David Hull's writings about systematics has been the impossibility and the undesirability of theory neutrality (see, e.g., Hull 1967, 1970, 1979, 1988). Pheneticists at times defended their insistence on equally weighted characters by saying that the introduction of theoretical – in particular, evolutionary – considerations into the construction of classifications would somehow vitiate the taxonomic enterprise. More recently, some defenders of cladistic procedures have argued that phylogenetic reconstructions must be produced without exploiting nontrivial assumptions about the evolutionary process. Both these points of view Hull has greeted with considerable skepticism.

I have nothing much to add to Hull's critique of the demand for theory neutrality in classification, which I take to have the following general form: If no standards whatever are imposed on what counts as a character, then it seems inevitable that all pairs of taxa should emerge as equally similar.[1] In other words, if taxa are judged to differ in their pair-wise similarity, this must be because the systematist, either explicitly or implicitly, makes judgments about what counts as a character. If constraints on the individuation of characters are imposed, these presumably must issue from judgments about 'biological importance.' And if biological importance can take account of nonevolutionary theories, why is it barred from giving weight to evolutionary matters? So theory neutrality in classification is impossible. Nor is it desirable; although the use of a theory in classification makes the resulting classification vulnerable to revision if the theory is found wanting, the usefulness of classification inevitably depends on its reflecting theoretically important similarities and differences.

In this paper, I want to examine the issue of theory neutrality in a quite separate context – the reconstruction of phylogenetic relationships from the data of character similarities and differences. The three main taxonomic schools – pheneticism, evolutionary taxonomy, and cladistics – each have views about both classification and phylogenetic reconstruction. One school – cladistics – wishes to forge as close a link as possible between the branching pattern of the tree of life and the way life forms should be classified. The other two schools hold that branching patterns and classification need not coincide so closely. But regardless of one's position here, it is essential to see that classification and the reconstruction of genealogical relationships are separable undertakings. To see this is to grant only that a substantive argument is required to defend any of the three tendencies to which I just alluded.

My interest is not in the connection of phylogenetic reconstruction and classification, but in phylogenetic reconstruction for its own sake. Regardless of whether classification must reflect branching and branching alone, an evolutionist presumably wishes to reconstruct phylogenetic relationships. It is in this context that cladists at times have argued that some variety of theory neutrality is both possible and desirable. I want to get clear on the merits of this epistemological position.

The kind of theory neutrality that writers on cladistics have hankered after is different in kind from the sort that pheneticists have demanded. Some extreme statements not withstanding, the demand has mainly been for *relative* theory neutrality, not *absolute* theory neutrality. I'll explain what I mean by this contrast in due course. What is more, it is not difficult to see why relative theory neutrality can be very desirable in science. Here too I see a difference between the pheneticist argument in classification and the cladistic argument pertaining to the task of phylogenetic reconstruction. But what is desirable is not always possible. In conclusion, I'll briefly consider whether the kind of relative theory neutrality that cladists have demanded is likely to be attainable. But beginning at the beginning, I'll start with a sketch of what cladists propose as the proper procedure for phylogenetic reconstruction.

2. Parsimony

Cladistic theory denies that overall similarity is an appropriate method for reconstructing phylogenetic relationships. This may strike the uninitiated as an amazing claim: is not that fact that human beings and gorillas are much more similar to each other than either is to snakes one very good reason for thinking that humans and gorillas are more closely related to each other than either is to snakes? The cladistic reply is that the similarities just cited are of two kinds, and only one of them should be viewed as providing evidence about propinquity of descent.

To see what is involved here, consider Figure 1. We wish to infer the phylogenetic relationships that obtain among three species (*A, B,* and *C*). The (AB)C tree indicates that *A* and *B* are more closely related to each other than either is to *C* – note that the first two share an ancestor (represented by an interior node) that is not an ancestor of the third. The second tree represents the hypothesis that the phylogenetic relationship is A(BC); this means that it is *B* and *C* that are more closely related to each other than either is to *A*. How are we to tell which of these is the more plausible hypothesis of genealogy?

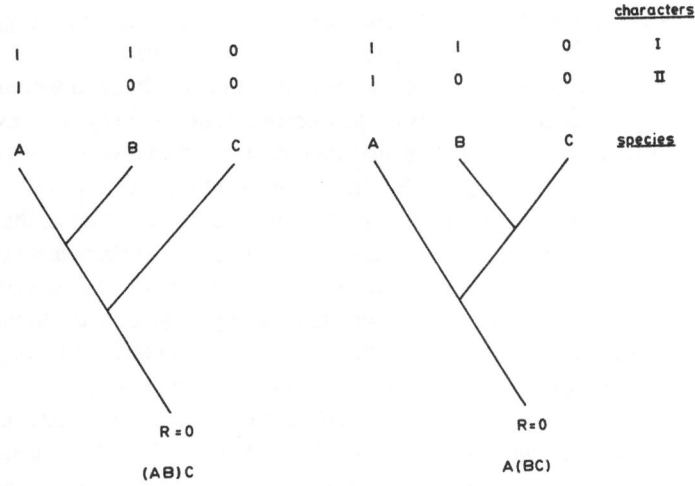

Fig. 1

We must infer this by examining the pattern of similarities and differences that obtains among the three taxa. Each of the two characters described in Figure 1 is assumed to come in two states, which are coded as '0' and '1'. Notice that the first character (I) is such that *A* and *B* are similar and that *C* is different, whereas the second character (II) is such that *B* and *C* constitute the similar pair. If we were to use the pheneticist measure of *overall similarity* to infer relationships, we would take these two characters as failing to discriminate between (AB)C and A(BC). On the other hand, if we had ten characters just like I and twenty just like II, overall similarity would say that A(BC) is better supported by this thirty character data set.

Notice that the two hypotheses agree that *A*, *B*, and *C* trace back to a common ancestor (*R*) at the root of the tree. This ancestor had many characteristics. With respect to character I, the ancestor was either in state 1 or in state 0; with respect to character II, it was either in state 1 or in state 0, and so on. We will say, as a matter of definition, that the character state exhibited by the ancestor *R* is *ancestral (plesiomorphic)* and that the alternative character state is *derived (apomorphic)*. Let us suppose that the ancestor possessed the 0 state for each character. In this case, '0' means ancestral and '1' derived.

I now can describe a difference between characters I and II. Character I involves a derived similarity, whereas character II involves an ancestral similarity. The fundamental tenet of cladistic methodology is that only derived similarities provide evidence of phylogenetic relationship. So if the data consisted of one character of type I and one of type II, cladistic methodology would single out (AB)C as the better supported hypothesis. Indeed, the same verdict would be reached if the data consisted of a single character of type I and any number of type II characters. Cladistic theory holds that not all similarities have evidential meaning; this is its basic disagreement with the phenetic idea of *overall* similarity.

The word 'parsimony' is used to describe this cladistic position: *the best supported phylogenetic hypothesis is the one that requires the fewest evolutionary changes in character state*. Notice that character I can evolve in the (AB)C with only a single 0-to-1 change occurring in the tree's interior, whereas the A(BC) tree requires at least two changes in state if character I is to be found at the tips of the tree. The derived similarity is thus more parsimoniously explained by (AB)C than by A(BC). However, character II – an ancestral similarity – can evolve on either tree with only a single change in state. Competing phylogenetic hypotheses can explain ancestral similarities equally parsimoniously, so cladistic theory concludes that ancestral similarities do not provide discriminating evidence.

Notice that the asymmetry just noted between character I and character II depends on knowing the character state of the ancestor R. If 1 were the ancestral state, all the judgments just made would have to be reversed. But how is it possible to tell of a given character which state is ancestral and which derived? This is the problem of ascertaining *character polarity*. A number of methods have been discussed, but the preeminent one, according to cladistic theory, is the method of outgroup comparison, illustrated in Figure 2.

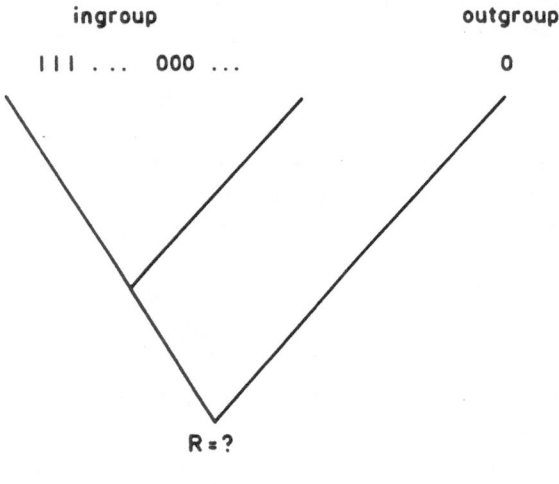

Fig. 2.

I will here ignore some of the complexities of this procedure, in order to focus on the main idea. The taxa whose phylogenetic relationships one wishes to infer comprise what is called the *ingroup*. Some ingroup species exhibit character state 1 while others exhibit character state 0. To polarize the character, one examines the character state of an *outgroup*. Ideally, this should be a species very closely related to those found in the ingroup, though more distantly related to them than any of them are to each other. The method of outgroup comparison says the following: *the character state found in the outgroup is the best estimate of the character state present at the root of the tree.* So in Figure 2, one would infer that $R=0$ – i.e., that 0 is the ancestral form.

The parsimony idea apparently underlies this procedure.[2] Note that if the root were assumed to be in state 1, then two changes would have to occur to produce the characters found at the top of the tree in Figure 2. At least one change would have to occur on the outgroup branch, and at least one would have to occur to generate the 0's found in the ingroup. On the other hand, assigning character state 0 to R has the result that only a single change – in the ingroup – is needed to account for the data.

It now is a matter of considerable controversy in the systematics community whether and to what extent cladistic methodology makes sense. Critics have alleged that the use of parsimony makes highly substantive (and often implausible) assumptions about the evolutionary process (see, e.g., Felsenstein 1983 and 1984). The standard version of this claim has been that the use of parsimony presupposes that multiple originations and reversions occur very rarely. By preferring hypotheses that minimize assumptions of change, cladists are assuming that the number of changes has actually been rather minimal. Cladists have tried to rebut this charge and to construct positive arguments showing that the use of parsimony in phylogenetic inference depends on few, if any, assumptions about evolution. The usual cladistic claim has been that parsimony requires only the assumption that the species under study are the product of descent with modification (see, e.g., Eldredge and Cracraft 1980, Wiley 1981, Farris 1983).

This large and central problem I will not attempt to solve here. However, I do want to make clear a few features of this dispute about the presuppositions of parsimony – ones that are relevant to the issue of theory neutrality in systematics.

3. The problem of circularity

Even the most ardent supporter of parsimony as a device for inferring genealogical relationships will grant that parsimony involves highly nontrivial theoretical assumptions. An example is provided by the method of outgroup comparison. To polarize a character found in the ingroup, I need to be able to assume that the outgroup really is an outgroup. This is not something given directly by unmediated observation, but is itself a historical claim about phylogenetic relationships.

Some critics of cladism have claimed that the outgroup method is thereby circular: one wishes to infer phylogenetic relationships by using parsimony, but to do so, one must already be prepared to make phylogenetic assumptions (Bock 1981, Cartmill 1981, Patterson 1982). This charge of circularity is misplaced: to infer relationships *within* the ingroup, one need not assume that the ingroup phylogeny is already known. Rather, what one needs is a phylogenetic assumption about the way the ingroup is related to the species selected as an outgroup. A phylogenetic assumption is required by the outgroup method, but not one that renders its use viciously circular.

A second instance of the theory-ladenness of cladistic procedures is provided by the process by which organisms in different species are said to be in 'the same' character state. Every organism is unique, if one only looks hard enough. And the traits found in one organism are different from those found in another, if one only describes each in a sufficiently fine-grained way. Yet, a description of taxonomic specimens as utterly unique would leave no similarities – no shared traits – on which to base phylogenetic inference.

So the systematist must have a way of determining when organisms share the same traits. The point I would make here is that this procedure is highly informed by biological and specifically evolutionary ideas. It is a familiar experience of fledgling systematists that they must learn to see the traits their teachers say are important. Again, this does not argue for the circularity of phylogenetic inference, but for what I hope is the uncontroversial point that theoretical considerations enter into this process in multiple ways.

Another theoretical assumption that seems to underlie the use of parsimony is suggested by the fact that both competing hypotheses in Figure 1 involve branches that split but never merge. Each species in such trees has precisely one immediate ancestor. But hybrid species are not like this. It is at present quite unclear whether the idea of cladistic parsimony can be extended to evaluating genealogical hypotheses that postulate reticulate (hybridizing) evolution. If extending parsimony into this domain proves to be impossible, then we will have to grant that the use of parsimony involves a highly theoretical background assumption – that hybridization has not occurred among the species considered.

Biologists should have no trouble granting these points, since they do not represent objections to what biologists wish to do. There seems to be no reason to reject the use of theoretical assumptions in inferring phylogenies, provided those assumptions are independently defensible. However, there is a standard problem that arises in scientific inference, which does make a kind of limited theory neutrality quite essential. This problem concerns how phylogenetic hypotheses, once they are inferred, can be used to infer other hypotheses.

Many cladists have wanted to see phylogenetic reconstructions as providing evidence that bears on substantive questions about the evolutionary process. An example is the hypothesis of an evolutionary clock. Is there an approximately constant rate of divergence among evolutionary lines, or do some evolve much faster than others? This question, it would appear, can be answered only by assuming a tree of the kind shown in Figure 1 and then determining whether rates of character turn over are the same in contemporaneous lines. If (AB)C is the true phylogeny, then the clock hypothesis predicts that the net similarity between A and C should be about the same as the net similarity between B and C. However, the clock hypothesis would have different implications, if the true phylogeny were A(BC).

The point here is that phylogenetic reconstructions provide evidence about the evolutionary clock only if those reconstructions were not based on the assumption of a clock. The use of parsimony to reconstruct phylogenies can help answer other evolutionary questions only if the method does not beg the very question at issue. The wish to avoid this kind of circularity is entirely legitimate.

A second and more general question is posed by the problem of saying why certain characters evolved by natural selection. If I am going to defend an evolutionary scenario that explains why a given character evolved, I must be prepared to describe the other traits already present in the population. A story explaining why a trait emerged may be plausible for one background of characters, but not plausible for another. A crude example is provided by the flightlessness of penguins. There are many organisms that cannot fly. The special feature of explaining this trait in penguins is that penguins *lost* the ability of fly. Within the birds, flight is an ancestral trait and flightlessness is derived.

So to judge the plausibility of evolutionary scenarios, one must be willing to make assumptions about phylogenetic relationships. If penguins had flying ancestors, one scenario may be plausible; if all their ancestors were flightless, quite a different one might be. Again, the constraint on inferring phylogenetic relationships is this: if you wish to use hypotheses of genealogy of test evolutionary scenarios, it had better not be true that the scenarios were themselves assumed in the process of inferring the genealogies.

To insist that a hypothesis be *absolutely* theory neutral is to claim that absolutely no theoretical considerations need enter into its justification. Phylogenetic hypotheses are clearly not theory neutral in this sense. Nor does there seem to be any good reason to fault them for failing to live up to some rarified philosophical ideal of absolute theory neutrality.

In contrast, I will say that a hypothesis is theory neutral, *relative to a set of assumptions T*, if T need not be assumed in justifying the hypothesis. Notice that many different questions about the relative theory neutrality of phylogenetic hypotheses arise; for each possible value of T, we may ask whether phylogenetic hypotheses are theory neutral relative to it.

The desire for relative theory neutrality is not a piece of misguided philosophy. A sensible systematist may concede that phylogenetic hypotheses are not absolutely theory neutral, but insist that they are relatively theory neutral in pertinent respects. There can be a very good reason for wishing to do this: given the use the systematist wishes to make of phylogenetic hypotheses, once they are inferred, it can be essential that those hypotheses be relatively theory neutral. Failure to insure this can result in an inference's being viciously circular.

4. Is relative theory neutrality possible?

Given the way I defined the idea of relative theory neutrality in the previous section, it should be clear that the title of the present section contains not one question, but a schema for indefinitely many questions. And given the gloss I've given to the idea of relative theory neutrality, I hope it is clear why this characteristic can be quite desirable. But what is desirable is not always possible. And so we reach the question: is it possible for phylogenetic inference to avoid the sorts of theoretical assumptions that cladists have frequently wished to avoid?

Some examples of this sort of problem will convey the kind of issue that now exercises systematists arguing about parsimony: Does the use of parsimony assume that change is very improbable? Does the use of parsimony assume that contemporaneous lineages have comparable rates of character evolution? I will not formulate answers to questions of this sort, because I think that the current level of understanding of the phylogenetic inference problem is very partial and preliminary. But I will formulate a general thesis about scientific inference that I think has some implications about this particular inference problem.

Statisticians standardly agree that hypothesis testing depends on adopting a model of the process that generates the data. That is, besides the hypotheses one wishes to test and the data one hopes will be useful in discriminating among competing hypotheses, there must in addition be a set of assumptions about how the data one obtains depends on which hypothesis is true.

Although there are many different contexts in which this meta-principle applies, it is perhaps simplest to see it at work in connection with the statistical idea of likelihood. One wishes to test two hypotheses H_1 and H_2 by seeing which confers the greater probability of the observations O. If H_1 says that O was to be expected, whereas H_2 says that O was very surprising, this counts in favor of H_1.[3] The point I want to emphasize here is that it will often be impossible to say what probabilities the hypotheses confer on the observations, unless one adopts auxiliary assumptions about the empirical processes at work.

In phylogenetic inference, this structure finds the following application. (AB)C and A(BC) are competing hypotheses. The observations are a set of character distributions. To use likelihood to discriminate between the hypotheses, one must be able to say which hypothesis makes the observations more probable. In the example discussed in Figure 1, we considered two characters. What probabilities do (AB)C and A(BC) each confer on this two character data set? This question is unanswerable, until we adopt assumptions about the evolutionary process that occurs in the trees depicted by the competing hypotheses.

Models can be described in whch the two member data set favors (AB)C – a result harmonious with the interpretation that parsimony sanctions of these data. However, other models can be described in which the two hypotheses are equally likely, while still others can be described in which A(BC) is more plausible. The meta-principle I want to stress is that *only in the context of a process model do observations have evidential meaning.*

As an example of the kind of thesis I want to defend, consider Figure 3.

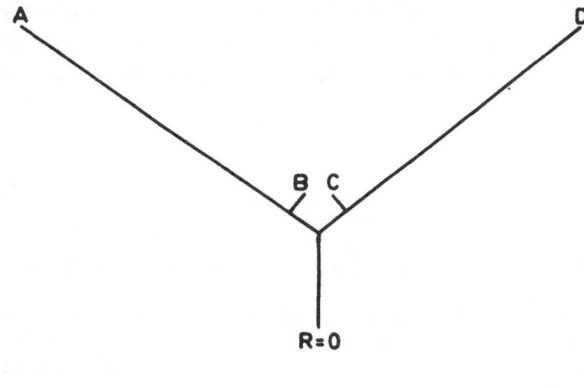

Fig. 3.

The length of a branch here represents the probability of a 0-to-1 transition occurring on it. Species *A* and *D* are more likely to exhibit a derived character state than either *B* or *C* are. I assume, further, that reversals – transitions from 1-to-0 – are impossible. Now suppose you sample three taxa at random from the tips of this tree and try to infer from their characteristics which pair forms a genealogical group apart from the third.

The extreme inequalities in branch transition probabilities generates the following curious result: If two taxa are found to be in the 1 state and one has the 0 form, the most likely hypothesis is that the taxa showing the derived state are *A* and *D*, whereas the species showing the ancestral form is either *B* or *C*. In this case, a shared derive character does *not* best support the hypothesis of close relatedness, contrary to what cladistic parsimony asserts.

I do not offer Figure 3 as a realistic model of the evolutionary process. It is a tinker toy example. Rather, my point is that the use of parsimony to infer phylogenetic relationships assumes that the model I have described is *not* realistic. But to assume that one model is false is just as much an assumption about the evolutionary process as assuming that another model is true.[4]

I therefore think it is inevitable that cladistic parsimony will be found to make substantive assumptions about the evolutionary process. The same holds true for the use of overall similarity as a device for inferring phylogenetic relationships. It is a reasonable research problem to see whether parsimony can be justified within the context of a model that makes only very weak assumptions. However, the meta-principle I described implies that it will not be possible to show that parsimony makes sense no matter what the evolutionary process has been like. Much hard work lies ahead before we can provide a reasonably detailed picture of the strengths and limitations of parsimony and of other procedures for inferring phylogenetic relationships. The hope that genealogical hypotheses should be relatively independent of assumptions about the evolutionary process is a coherent one; time will tell whether the kind of relative theory neutrality that systematists have wanted is in fact attainable.

Notes

1. This point can also be seen to flow from Nelson Goodman's (1970) discussion on the concept of similarity.
2. Space does not permit a detailed treatment of the question of whether using outgroup comparison to infer polarity requires different assumptions from the ones needed to justify parsimony as a device for inferring genealogies from already polarized characters. See Sober (1988), Chapter 6 for discussion.
3. Note that the likelihood of a hypothesis H relative to observations O is the probability of O, given H, not the probability of H, given O.
4. This example is inspired by the one presented in Felsenstein (1978), which was offered as a case in which parsimony will not be statistically consistent. His example and mine have features in common, but the points are different. Mine concerns likelihood, whereas Felsenstein's concerned the quite separate property of statistical consistency. See Sober (1988), Chapter 5 for discussion of Felsenstein's argument.

References

Bock, W. (1981). Functional-adaptive analysis in evolutionary classification. *American Zoologist* 21: 5–20.

Cartmill, M. (1981). Hypothesis testing and phylogenetic reconstruction. *Zeitschrift für Zoologische Systematik und Evolutionforschung* 19: 73–96.

Eldredge, N. and Cracraft, J. (1980). *Phylogenetic Patterns and the Evolutionary Process*. New York: Columbia University Press.

Farris, J. (1983). The logical basis of phylogenetic analysis. In N. Platnick and V. Funk (eds), *Advances in Cladistics*, vol. 2. New York: Columbia University Press, pp. 7–36. Reprinted in Sober (1984).

Felsenstein, J. (1978). Cases in which parsimony or compatibility methods will be positively misleading. *Systematic Zoology* 27: 401–410. Reprinted in Sober (1984).

Felsenstein, J. (1983). Parsimony in systematics: biological and statistical issues. *Annual Review of Ecology and Systematics* 14: 313–333.

Felsenstein, J. (1984). The statistical approach to inferring evolutionary trees and what it tells us about parsimony and compatibility. In T. Duncan and T. Stuessy (eds), *Cladistics: Perspectives on the Reconstruction of Evolutionary History*. New York: Columbia University Press, pp. 169–91.

Goodman, N. (1970). Seven strictures on similarity. In L. Foster and J. Swanson (eds), *Experience and Theory*. Boston: University of Massachusetts Press. Reprinted in N. Goodman (1972), 437–46.

Hull, D. (1967). Certainty and circularity in evolutionary taxonomy. *Evolution* 21: 174–89.

Hull, D. (1970). Contemporary systematic philosophies. *Annual Review of Ecology and Systematics* 1: 19–53. Reprinted in Sober (1984).

Hull, D. (1979). The limits of cladism. *Systematic Zoology* 28: 416–40.

Hull, D. (1988). *Science as a Process: An Evolutionary Account of the Social and Conceptual Development of Science*. Chicago: University of Chicago Press.

Patterson, C. (1982). Morphological characters and homology. In K. Joysey and A. Friday (eds), *Problems of Phylogenetic Reconstruction*. London: Academic Press, 21–74.

Sober, E. (1984). *Conceptual Issues in Evolutionary Biology: An Anthology*. Cambridge, Mass.: MIT Press.

Sober, E. (1988). *Reconstructing the Past: Parsimony, Evolution, and Inference*. Cambridge, Mass.: MIT Press.

Wiley, E. (1981). *Phylogenetics: The Theory and Practice of Phylogenetic Systematics*. New York: John Wiley.

David Hull's Conception of the Structure of Evolutionary Theory

PAUL THOMPSON

University of Toronto

David Hull is one of the founding fathers of modern philosophy of biology and he has written on most of the significant issues in the field. One area in which he has made extremely important contributions is sytematics. What has intrigued me for some time, however, is his conception of the structure of evolutionary theory and it is on this aspect of his work that I shall focus in this paper. The principal exposition of his conception is found in his book *The Philosophy of Biological Science* (1974) (page number references without source identification are to this book).

It is my contention that Hull's account of the nature of modern evolutionary theory differs in important respects from those of Ruse (see Ruse 1973) and Williams (see Williams 1970). And, I further maintain that most of the features of his account provide a more accurate representation of the theory that do the accounts of Ruse and Williams. One respect, however, in which Hull's account is similar to the accounts of Ruse and Williams is in its defence of the thesis that evolutionary theory conforms to the logical empiricist view that a scientific theory is an axiomatic-deductive structure which is partially interpreted by definitions called correspondence rules. I shall call this conception of theory structure the syntactic conception.[1] It is explicated and defended, in slightly different ways, by, among others, Hempel (1985, 1967), Braithwaite (1953), Nagel (1961) and Carnap (1936, 1937, 1956).

I argue that Hull's acceptance that evolutionary theory conforms to the syntactic conception of theory structure is inconsistent with his penetrating and perspicuous views on the nature of evolutionary theory. Hence, the central thesis of this paper is that there is an unresolved tension in Hull's treatment of the structure of evolutionary theory.[2]

I

In this section, I shall very briefly describe the main features of the syntactic conception. On this conception a scientific theory is an axiomatic deductive structure which is partially interpreted in terms of definitions called correspondence rules.

M. Ruse (editor), What the Philosophy of Biology is. pp. 275–287.

Both the formal language and the deductive apparatus of a scientific theory are that of first order predicate logic with identity (mathematical logic). The syntax of a theory is a formal system, the axioms and theorems of which when interpreted become laws of that scientific theory. Hence, a scientific theory is a deductively related set of laws such that from the axioms of the theory all other laws can, in principle, be deduced.

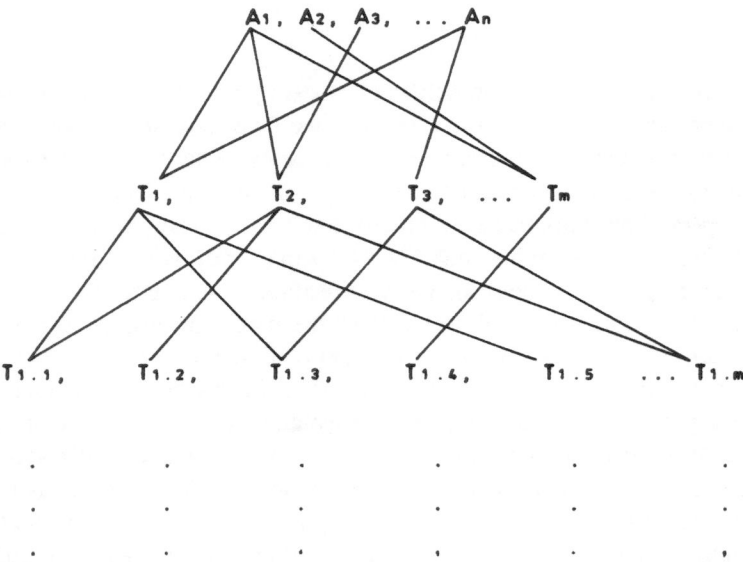

Lines indicate deductive connections. More than one line to a node indicates that the deduction of that theorem requires, as premises, all the theorems from which the lines come. The lines in this diagram illustrate a possible structure.

The paradigm example of this kind of structure is Newtonian mechanics. The axioms of Newtonian mechanics are the three laws of motion and the law of gravitational attraction. From these, it is held, all the laws of mechanics, in principle, can be deduced. For example, from them Galileo's law of free fall and Kepler's laws of planetary motion can be deduced. Laws on this view are *statements* of regularities in nature.

The semantics for a theory are provided by correspondence rules. In effect, correspondence rules provide a meaning structure for the formal system (syntax). Hence, by means of correspondence rules, one of the models (a meaning structure in which all of the axioms are true) of the syntax is specified. Correspondence rules define the theoretical terms of the theory by reference to observation terms. For example, a theoretical term like 'voltage' is partially defined by reference to readings on a calibrated meter such as a galvanometer. The definition is only partial because a term can be defined by reference to an open ended number of observational situations and procedures. New technologies, for example, will make possible new observations and new empirical operations and if the term was explicitly rather than only partially defined by a particular operation, the new technological procedure would, in effect, amount to a redefinition rather than simply an expansion of the definition of the term. For some theoretical terms empirical meaning is indirect. That is, some theoretical terms are defined by reference to one or more other theoretical terms. Ultimately, any chain of such definitions must end in theoretical terms that are defined by reference to observations. Because of this complex interconnection of theoretical terms, the meaning of any one term is seldom independent of the meaning of many if not all of the other terms of the theory. Theories, hence, have a global meaning structure – changes to the meaning of one term will have consequences for the meaning of many and usually all the other terms of the theory.

Hence, on the syntactic conception, a theory consists of a set of deductively related statements, the structure of which is provided by mathematical logic and the empirical meaning of which is provided by definitions in terms of correspondence rules which link theoretical terms like 'mass', 'electron', 'spin', 'gene', etc. to other theoretical terms or to observations. Ultimately these definitions provide direct or indirect partial empirical meaning to all the terms of the theory. In this way the theory is, as a whole, given empirical meaning. The statements of which a theory consists are generalizations (laws), a small subset of which are taken as the axioms of the theory. The axioms are laws of the highest generality within the theory. They constitute a consistent set no one of which can be derived from any subset of the others. All laws of a theory, including the axioms, describe the behaviour of phenomena. All laws except the axioms, in principle, can be derived from the axioms. Usually such deductions require numerous subsidiary assumptions. Explanation and prediction of phenomena consist in demonstrating that the phenomenon can be deduced (or that it follows logically with a high probability) from some subset of the laws of the theory conjoined with a relevant description of the prior state of the system.

II

As I indicated above, it is my contention that there is a tension in Hull's account of evolutionary theory: a tension between his acceptance that evolutionary theory conforms to the syntactic conception of theory structure and his account of the details of the theory. In this section I support my contention by providing two illustrations of what I argue are inconsistencies in his account: a tension between his multi-level, multi-field conception of the nature of the theory and the syntactic conception requirement of unity; a tension between his view that biology has very few, if any, genuine laws and the fact that theories, on a syntactic conception, *are* deductively organized sets of *laws* and that laws are of fundamental importance in explanations.

The first illustration focuses on what I take to be one of the insightful features of Hull's account of evolutionary theory. Hull is quite clear that the essential components of a Darwinian evolutionary theory are: non-random differential reproductive success of organisms, variation in populations, and heredity (pp. 50–51). He also seems to hold that theories about these different components are different than the modern synthetic theory of evolution (or neo-Darwinism) (p. 70). Indeed, he is quite explicit about the non-identity of the genetical component with an adequate theory of evolution:

> As might be expected, the genetical theory of evolution is mathematical in notation and deals with evolution in terms of gene and genotype frequencies. This single-minded concern in the genetical theory of evolution with genes has been intentionally emphasized (perhaps exaggerated) in the following pages in order to draw attention to the task confronting contemporary evolutionists – the need to synthesize Darwin's phenotypic version of evolutionary theory with the genetical version. Hence, this attempt is termed the synthetic theory of evolution (pp. 45–46).

This, I think, is a clear and accurate portrayal of the essence of the modern synthetic theory of evolution. There is, however, a tension between this characterization and Hull's attempt to argue that evolutionary theory conforms to the syntactic conception of theory structure.

Hull is correct that an adequate evolutionary theory has to encompass phenotypic and genetical aspects. That is it will include elements from ecology and population biology, and from genetics. The method of inclusion, according to Hull is to synthesize them. What exactly he means by 'synthesize' is not clear but it seems to be the bringing together of these domains into a unified theory. This is certainly what is required if an adequate theory of evolution is to conform to syntactic conception of theories. What is clear, however, is that no synthesis of this kind has been achieved. There is no theory with a single set of axioms from which all the laws of population genetic theory, ecology, selection theory, etc. can be deduced or induced. And, there is no common set of correspondence rules for these domains that provided a unified meaning structure.

Further evidence of this tension, I think, can be seen in several issues discussed by Hull. One example is Hull's ambivalence on this question of unification. While on the one hand he looks for a synthesis of phenotypic and genetic aspects in order to bring about a complete evolutionary theory, he espouses a non-unificationist position when discussing the notion of 'individual' in biological theory:

> Scientific theories do exist at various levels of analysis; they should be evaluated on their own merits rather than just in terms of the extent to which they can be reduced to the next lowest level. (p. 49)

In this latter case he is motivated by his views on reduction. The interesting question is: why should one attempt unification of the two levels of theorizing in the case of evolutionary theory but not in the case of other theories which exist at different levels? Reductionism of the kind with which Hull is uncomfortable is an integral part of the programme of what Hull calls 'the traditional analysis of scientific theories current in the literature' (p. 46) and to which, he appears to argue, evolutionary theory does, on the whole, conform.

Another example of this tension is Hull's espousal of Mary Willams' axiomatization of selection theory. He views her axiomatization as an example of the way in which evolutionary theory might be shown compatible with the syntactic conception:

> If one were to modify one's choice of the units that function in evolutionary theory and the type of prediction desired, it might be possible to organize evolutionary theory in a way compatible with the notion of a deductive scientific theory set out in the earlier pages of this volume. Mary B. Williams has done just this in her axiomatization of evolutionary theory (p. 64).

Although Williams' axiomatization is elegant and impressive – it is one of the best examples of the axiomatic method – it has been criticized on a number of counts, one of which follows directly from Hull's own view that evolutionary theory must encompass both genetic and phenotypic aspects. Elliot Sober (Sober 1984), for example, points out that the axiomatization: contains no source laws; does not mention mutation, migration, systems of mating and numerous other causes of evolution which evolutionary biology takes into account; excludes genetics; and is Malthusian – i.e., it incorrectly makes selection dependent on reproductive rates. Michael Ruse (Ruse 1977) also criticizes Williams' axiomatization for making no reference to genetics.

The important criticism of Ruse and Sober with respect to Hull's position is that Williams' axiomatization is inadequate as an axiomatization of evolutionary theory because it contains no hereditary mechanism. And, Ruse thinks that there is no satisfactory way to incorporate an account of heredity into her axiomatization. Therefore, it fails to produce an account which synthesizes genetic and phenotypic aspects.

An interesting feature of Ruse's own positive account (see Ruse 1973), on the other hand, as I have argued elsewhere (Thompson 1987) is that it simply mirrors Williams' account since his account of evolutionary theory identifies the theory with population genetic theory. Hence, phenotypic aspects play no significant role in his account of the theory. What is interesting about this is that both Ruse and Williams in their attempt to explicate the axiomatization in mathematical logic of evolutionary theory have produce accounts that far from synthesizing genetic and phenotypic aspects have identified the theory with only one or other of these aspects. This is a situation that one would expect Hull to find unsatisfactory.

A recent defence of Williams' account by Alexander Rosenberg (Rosenberg 1981, 1985) makes explicit the non-genetic character of her account. Rosenberg argues that it is a strength of Williams' account that it is neutral on the question of hereditary mechanisms and that it is a defect in accounts like Ruse's that they are so wedded to a particular theory of heredity. He argues that evolutionary theory is not dependent on any particular theory of heredity and that far from being the core of evolutionary theory, a theory of heredity is simply an assumption of the theory. The central mechanism of evolution is natural selection. Heredity is assumed as a background condition, albeit an important one. Hence, the theory of evolution is neutral on hereditary mechanisms. Since Ruse's account is not neutral with regard to the mechanism of heredity – it is entirely dependent on current population genetical theory – but, on the contrary, characterizes evolutionary theory entirely in terms of heredity, it misrepresents the character of the theory. Williams' account, on the other hand, has precisely the neutrality concerning hereditary mechanisms that is required as well as the correct emphasis on the centrality of natural selection.

Rosenberg is of course correct that no particular theory of heredity is required by evolutionary theory. All that is required is that characteristics of parents be transmittable to offspring. He is also correct that natural selection is a prime mechanism without which population genetics would predict a situation of almost complete stasis – the only changes being due to random drift and recombination. Further it is clear that natural selection usually acts on phenotypes and not directly on genotypes. It is, therefore, not accurately represented by selection coefficients in a genetic calculus. Despite all of this, however, his detaching of heredity from evolutionary theory is untenable and the reason follows from Hull's characterization of the theory as having phenotypic and genotypic components.

Hull's characterization makes clear that the problem with Rosenberg's position is that the fact that population genetics is not by itself sufficient to entail evolutionary change does not entail that it is not a central part of evolutionary theory. Natural selection by itself also does not entail evolutionary change since the effects of selection, without heredity, are limited to each generation and cannot be perpetuated or accumulated. What these arguments for individual insufficiency show is that both are essential. An adequate evolutionary theory, as Hull quite correctly maintained, is a composite of both heredity and natural selection among other things. And, despite Rosenberg's praise of neutrality on the question of hereditary mechanisms, this composite nature is not undermined by the unlikely possibility that our current accounts of the specific mechanisms of *either* selection or heredity are wrong.

What emerges from this discussion is that Williams' axiomatization identifies evolutionary theory with a theory of natural selection and that this identification misrepresent evolutionary theory as Hull understands it. It is a composite of heredity and selection as well as many other fields of biology.

In light of the above, it is a striking feature of Hull's account that he uses Williams' axiomatization as an example of the deductive structure of evolutionary theory when it is quite clear that her axiomatization is an axiomatization of only one of the component parts of a complete and adequate theory: selection theory. It appears that in his quest to demonstrate that evolutionary theory more or less conforms to the syntactic conception, Hull tacitly has abandoned his commitment to the need for a synthesis of the phenotypic features of the theory such as selection (which largely takes place at the phenotypic level) and the genetical features of the theory.

What is also striking about Hull's discussion of Williams is the extent to which the tension goes unnoticed. Although Williams' axiomatization of selection theory contains almost no genetics (and is certainly devoid of Mendelian genetics as Rosenberg makes clear), Hull, when elaborating on prediction as an important feature of Williams' treatment, uses a thoroughly population genetical example based on heterozygote superiority (the now famous sickle-cell anemia example).

> A second important feature of Williams' treatment is that she explicitly takes note of the types of predictions that can be made on the basis of evolutionary theory ... (p. 65).

Twelve lines later he writes:

> A traditional example of the type of predictions that can be made on the basis of evolutionary theory concerns superiority of heterozygote, for instance at the hemoglobin S locus in man in areas in which *Falciparum malaria* is endemic. Individuals homozygous for the normal hemoglobin allele are more prone to get malaria than those who are heterzygous, and individuals homzygous for the hemoglobin S allele are more likely to die of sickle cell anemia. The principles of Mendelian genetics, enshrined in the Hardy-Weinberg law, can be used to determine various gene and genotypes frequencies for a particular population, given the necessary information (p. 66).

From this discussion one can only conclude that Hull incorrectly assumes that Williams' axiomatization includes Mendelian genetics. Were this assumption correct there would be no tension in Hull's account on this particular point. The assumption, however, is incorrect and the tension is obvious.

The second of my illustrations of the tension in Hull's exposition is found in his extensive discussion of laws in biology. Hull holds that a genuine scientific law, unlike a descriptive generalization, is a generalization that actually is or eventually will be integrated into a theory – descriptive generalizations remain isolated statements (see p. 71). Employing this definition, Hull examines several different kinds of purported laws in biology. The overall conclusion of these examinations seems to be that biology has very few genuine laws. He concludes his examination of each of the types of putative laws with a claim about the paucity or replaceability of such laws. The following is the conclusion to the four types that he discusses. Each is discussed in detail by Hull.

His conclusion about developmental laws in ontogeny is;

> Such generalizations have in the past appeared to be merely descriptive both because they were formulated largely on the basis of observation and because they were not derivable in most instances form any well-formulated body of laws ... Some headway is being made in working out actual mechanisms in developmental embryology, but progress is proving to be slow and painful (p. 76).

His conclusion about phylogeny and cross-section laws is:

> Statements of the correlation of phenotypic characters can be counted as low-level biological laws but only when explained to include reference to the genetic makeup of the organisms and the biochemical reactions that produce them. In point of fact, very few cross-section laws in biology possess such backing. Most of them tend to be of the 'all swans are white' variety (p. 80).

His conclusion about developmental laws in phylogeny is:

> Currently these evolutionary laws are of limited scope and tend to exist in isolated, partially inconsistent clusters. They have yet to be integrated into a single unified system (p. 82).

His conclusion about historical laws is:

> I have argued that historical laws are in principle replaceable by laws concerning the present state of living organisms, including extensive knowledge of their genetic constitution. In actual practice, biologists do not have the knowledge necessary to do without their historical laws. They do not know the genetic makeup of the specimen before them. ... When historical laws are useful, they should be used. When they are the only laws available they must be used. In the preceding paragraphs I have argued that nothing except incomplete data stands in the way of replacing historical laws with non-historical laws (p. 87).

What emerges from these conclusions is that the only genuine laws in biology are those of genetics. It is the laws of genetics that one is led to believe will replace historical laws. And it is the laws of genetics with some laws of biochemical reactions that will provide an adequate underpinning for cross-section laws. The picture that emerges is that genuine scientific laws are derivable from or replaceable by laws of genetics and laws of biochemical reactions. If this is the case, then the laws underlying evolutionary change will be fundamentally genetic and, consequently, the axioms of the theory will be fundamentally the axioms of genetics. This is, of course, at odds with Hull's claim that evolutionary theory is a synthesis of both phenotypic and genotypic aspects. There is, however, another problem.

Hull devotes the latter part of the chapter in which he discusses laws to a discussion of explanation – specifically the 'covering-law' model of explanation which is part of the same 'traditional analysis of scientific theories' against which Hull is measuring evolutionary theory. Scientific laws play a fundamental role in the 'covering-law' model of explanation. While Hull is not explicit about his acceptance of the general features of the covering-law model, I think his discussion clearly points in that direction. Consider, for example, his critique of Goudge and Simpson on explanation in that chapter. He clearly thinks that the general form of the covering-law model is better than what Goudge and Simpson propose. And, he himself proposes no alternative. He simply refines the early statement of the model by Hempel and Oppenheim. Certainly nothing he says indicates that he does not think that scientific laws play a key role in explanation.

The problem is that Hull in his discussion of the possible range of laws in evolutionary biology has concluded that there are no genuine scientific laws which are not reducible to genetic laws or are not based upon them. Hence, all currently adequate explanations of evolutionary phenomena must be genetic or, in principle, replaceable by explanations employing genetic laws. There do not appear to be any genuine phenotypic level laws. Hence, there will be no genuine phenotypic explanations of evolutionary phenomena. And, as consequence, there cannot be a genuine phenotypic component to evolutionary theory (which is a deductively organized *set of laws*) or to evolutionary explanation.

I think that Hull's acceptance of the syntactic conception according to which laws in a scientific theory describe the behaviour of phenomena is at the heart of this tension in his view. If laws are statements about the behaviour of phenomena then explanation is quite naturally understood as a demonstration that the particular phenomenon in need of explanation is a phenomenon covered by some statement(s) of the theory.

To this point I have set out what I think is the tension in Hull's account. In so doing I have attempted to indicate my preferred solution: rejection of the syntactic conception. In the absence, however, of an alternative which resolves this tension my preferred solution is vacuous. I think this is the dilemma that Hull faced – perhaps unconsciously. At the time he was writing his book, the most powerful, elegant and widely accepted conception of theory structure was the syntactic conception. To accept that a tension existed between his account of the details of evolutionary theory and this conception of theory structure was to condemn his account of the details of evolutionary theory. What is remarkable about Hull's book is the extent to which he held firm to a position on the nature of evolutionary theory

– a position which I think is without a doubt the correct position – in the face of a conception of theory structure that did not accommodate his position. What I suggest in the next section is that an alternative conception of theory structure is available now and that adopting it resolves the tension I have identified. While I shall not set them out in this paper, there are reasons quite independent of the resolution of this tension for adopting this alternative conception (see, for example: Beatty 1980; Lloyd 1984, 1987, 1988; Suppe 1972, 1988; Suppes 1967; Thompson 1983, 1986, 1989; Van Fraassen 1970, 1980).

III

One of the central purposes of Hull's discussion of evolutionary theory was to assess how evolutionary theory measured up to the syntactic account:

> The purpose of this chapter will be to investigate the logical structure of evolutionary theory and to compare it with the traditional analysis of scientific theories current in the literature (p. 46).
>
> The question which will concern us in this chapter is whether evolutionary theory can be made to approach the deductive ideal with respect to the form of both its laws and its definitions – and if it cannot, whether this failure reflects inadequacies in evolutionary theory or in the currently accepted paradigms of scientific theories and definitions (p. 47).

And, he begins the next chapter with statements that make it clear that he believes that he has established in Chapter 2 that the laws and theories of biology are similar to those of physics:

> After all that has been said in the preceding chapters about genetic theories and evolutionary theories, it may seem somewhat gratuitous at this point to address ourselves to the question of whether or not there are any biological laws. Yet several features of the theories and laws that we have discussed thus far conspire to lead some philosophers, for example, J.J.C. Smart ..., to argue that 'there are no biological theories' analogous to the closely knit theories of physics; 'there are not even any biological laws.'

Hull correctly identified the choices to be made if it did not fit: accept that evolutionary theory is inadequate as a theory or reject 'the currently accepted paradigm of scientific theories and definitions.' Hull argued that evolutionary theory conformed to the syntactic view sufficiently closely (at least as closely as many theories of physics) and, hence, the choice did not have to be made. What I have been pressing is the view that it does not conform at all well and that this has resulted in a tension in Hull's account.

The alternative conception to which I have alluded is known as the semantic conception of theory structure. Since I and others have explicated this view at length in other writings (see, for example: Beatty 1980, 1980b, 1987; Lloyd 1984, 1987, 1988; Suppe 1972, 1988; Suppes 1967; Thompson 1983, 1986, 1989; Van Fraassen 1970, 1972, 1980) I shall not provide a detailed explication here. Instead I shall indicate very briefly two features of this conception that I think resolve the tension in Hull's account of the structure of evolutionary theory.

On the semantic conception, a theory is a specification, in mathematical English, of a system (a model theoretic structure). The relationship between a theory and phenomena is extra-theoretic. The relationship is held to be one of isomorphism or homomorphism. The establishment of this relationship requires the employment of numerous other theories including theories of experimental design, theories of goodness of fit, and other scientific theories. The two points about this conception relevant to the tension in Hull's account are: (1) that theories as model theoretic structures can interact (see Van Fraassen 1970 and Thompson 1986, 1989) with the nature of the interaction specified by laws of interaction which specify how the system behaves under inputs; (2) laws do not describe the behaviour of phenomena they specify the nature of the system (see Suppe 1972, 1976).

Since on the semantic conception theories can interact, it is possible to employ in a complex theoretical framework phenotypic and genotypic theories without requiring a formal synthesis or unification of the component theories: parameters of one theory become inputs in another. The behaviour of systems under input is specified by laws of interaction. These laws do not describe phenomena, they specify the behaviour of the system. This resolves one aspect of the tension in Hull's account.

Also, since on the semantic conception laws do not describe the behaviour of phenomena, phenomena are not explained by the covering law model. Instead, a phenomenon is explained if it is within the scope of a phenomenal system and is such that the occurrence of the phenomenon within that phenomenal system is consistent with there being an isomorphism between the phenomenal system and the system specified by the explaining theory. Laws on this conception play a very different role than on the syntactic conception and explanation is not a function of demonstrating that a phenomenon is described by a statement of the behaviour of a phenomena. Hence, Hull's view that there are few if any such phenotypic statements which fit his definition of a law does not imply that no specification of a phenotypic system is possible. Indeed, Mary Williams' axiomatization provides an excellent starting point for a formalization of a phenotypic theory in terms of a set of specifications. The differences will be that the formalization on a semantic conception will be in set theory with or without topology and not mathematical logic and, as a result, a phenotypic system formalized along the lines she has explored will be able to interact with a genotypic system. This resolves the other aspect of the tension in Hull's account.

In conclusion, Hull provided a remarkably insightful account of evolutionary theory but adherence to a syntactic conception of theory structure resulted in a tension in the account. This tension can be resolved while preserving his insights by adoption of a semantic conception of theory structure.

Notes

1. This conception has been called by different writers 'the received view', 'the hypothetico-deductive view' and 'axiomatic view'. While I have in a number of previous articles called it 'the received view', I now prefer the designation 'syntactic conception.' This designation is less pejorative and makes the appropriate contrast with the alternative conception which is widely known as the semantic conception.
2. In order to make clear the context within which I wrote this paper and my motives for writing it, let me be forthright. Although I agree with Hull on a wide range of issues, we disagree about the logical structure of theories. Indeed, he is strongly opposed to the view that I espouse which is known as the semantic conception of theories. He does not think that the semantic conception advances our understanding of scientific theories in biology. In this paper I intend to try (again) to demonstrate that on this point he is wrong. My strategy is to show that some of his most insightful contributions on the nature of evolutionary biology are at odds with his adherence to the syntactic view of theories. Indeed, I think this tension is unavoidable. The way to resolve the conflict is to reject the syntactic conception of theory structure and adopt the semantic conception.

References

Beatty, J. (1980a). Optimal-design models and the strategy of model building in evolutionary biology. *Philosophy of Science* 47: 532–561.

Beatty, J. (1980b). What's wrong with the received view of evolutionary theory? In P.D. Asquith and R.N. Giere (eds) *PSA 1980*, vol. 2. East Lansing: Philosophy of Science Association.

Beatty, J. (1987). On behalf of the semantic view. *Biology and Philosophy* 2: 17–23.

Braithwaite, R. (1953). *Scientific Explanation.* Cambridge: Cambridge University Press.

Carnap, R. (1936). Testability and meaning. *Philosophy of Science* 3: 420–468.

Carnap, R. (1937). Testability and meaning. *Philosophy of Science* 4: 1–40.

Carnap, R. (1956). The methodological character of theoretical concepts. In H. Feigl and M. Scriven (eds), *Minnesota Studies in the Philosophy of Science*, vol. 1. Minneapolis: University of Minnesota Press.

Hempel, C. (1965). *Aspects of Scientific Explanation.* New York: The Free Press.

Hempel, C.G. (1967). *Philosophy of Natural Science.* Englewood Cliffs: Prentice-Hall.

Hempel, C.G. and Oppenheim, P. (1948). Studies in the logic of explanation. *Philosophy of Science* 15: 135–175.

Hull, D.L. (1974). *Philosophy of Biological Science.* Englewood Cliffs: Prentice-Hall.

Lloyd, E. (1984). A semantic approach to the structure of population genetics. *Philosophy of Science* 51: 242–264.

Lloyd, E. (1987). Confirmation of ecological and evolutionary models. *Biology and Philosophy* 2: 277–293.

Lloyd, E. (1988). *The Structure of Population Genetics.*

Nagel, E. (1961). *The Structure of Science.* New York: Harcourt, Brace.

Rosenberg, A. (1981). The interaction of evolutionary and genetic theory. In L.W. Sumner, J.G. Slater and F.F. Wilson (eds), *Pragmatism and Purpose: Essays Presented to Thomas*

Goudge. Toronto: University of Toronto Press.

Rosenberg, A. (1985). *The Structure of Biological Science*. Cambridge: Cambridge University Press.

Ruse, M. (1977). Is biology different from physics? In R. Colodny (ed.), *Logic Laws and Life*. Pittsburg: University of Pittsburg Press.

Ruse, M. (1973). *The Philosophy of Biology*. London: Hutchinson.

Sober, E. (1984). Fact, fiction and fitness: a reply to Rosenberg. *Journal of Philosophy* 81: 372–384.

Suppe, F. (1972). What's wrong with the received view on the structure of scientific theories? *Philosophy of Science* 39: 1–19.

Suppe, F. (1988). *Scientific Realism and the Semantic Conception of the Structure of Scientific Theories*. Urbana: University of Illinois Press.

Suppe, F. (1976). Theoretical laws. In M. Prezlecki, K. Szaniawski and R. Wojcicki, *Formal Method in the Methodology of Empirical Science*. Wroclaw: Ossolineum.

Suppes, P. (1967). What is a scientific theory. In S. Morgenbesser (ed.) *Philosophy of Science Today*. New York: Basic Books, pp. 55–67.

Thompson, P. (1983). The structure of evolutionary theory: a semantic approach. *Studies in History and Philosophy of Science* 14: 215–229.

Thompson, P. (1986). The interaction of theories and the semantic conception of evolutionary theory. *Philosophica* 37: 73–86.

Thompson, P. (1989). *The Structure of Biological Theories*. New York: State University of New York Press.

Van Fraassen, B.C. (1970). On the extension of Beth's semantics of physical theories. *Philosophy of Science* 37: 325–339.

Van Fraassen, B.C. (1972). A formal approach to philosophy of science. In R.E. Colodny (ed.), *Paradigms and Paradoxes*. Pittsburgh: The University of Pittsburgh Press.

Van Fraassen, B.C. (1980). *The Scientific Image*. New York: Oxford University Press.

Williams, M.B. (1970). Deducing the consequences of evolution. *Journal of Theoretical Biology* 29: 343–385.

Kinds, Individuals and Theories

E.O. WILEY

Museum of Natural History, University of Kansas, Lawrence, KS 66045, USA

Those of the systematic community who have considered the nature of taxa have tended to treat them in one of two ways, as natural kinds or as individuals. Many members of my research community (phylogenetics) will probably wonder why it matters. After all, so long as we arrive at the same empirical conclusions given the same data, can this controversy be so important? Perhaps not, at least within the small confines of our community. However, I have concluded that this seemingly minute controversy has importance in illuminating some larger philosophical issues that are not confined to the phylogenetics community.

If taxa are natural kinds, then it is logically consistent to treat hypotheses of their relationships as hypotheses in universal form. If, however, taxa are (philosophical) individuals, then it is logically consistent to reject this notion and treat hypotheses of relationship as singular in form. Opting for one of these possibilities has two important consequences. The first is concerned with the strategy of hypothesis testing. If taxa are kinds and hypotheses of relationship are genuine universal hypotheses, then these hypotheses can be falsified but never verified. This produces an asymmetry between verification and falsification and Popper's (1968) philosophy has particular relevance (see, for example Wiley 1975, or Platnick and Rosen 1987). If, however, taxa are historical groups (philosophical individuals) and hypotheses of relationship are singular statements, then no asymmetry between falsification and verification exists (Hull 1983) and parsimony has particular relevance (Kluge 1984). The second consequence concerns the relationship between taxa and theories of process. If taxa are kinds, then we should expect them to function relative to evolutionary theory in a manner similar to that exhibited by other natural kinds relative to the theories of process that are associated with them. If, however, taxa are philosophical individuals, then we should not expect this relationship. Rather, we should expect taxa to be examples of natural kinds that are not, themselves, taxa.

In this paper I will argue the following theses: (1) monophyletic groups (natural taxa) are historical groups and are thus philosophical individuals, (2) phylogenetic hypotheses are singular statements, (3) taxa are defined by genealogy rather than by characters, (4) monophyletic taxa function significantly in evolutionary theory, but

M. Ruse (editor), What the Philosophy of Biology is. pp. 289–300.
© 1989 *Kluwer Academic Publishers, Dordrecht*

they do not function in the same logical manner as natural kinds nor do they have the same logical relationship as natural kinds to physical or biological theories, (5) the homologous characters (features) displayed by parts (individual organisms) of taxa do not function relative to evolutionary theories in the same manner that the characteristics that are displayed by members of natural kinds, and (6) characters are neither necessary nor sufficient, to recognize monophyletic taxa and thus several notions of natural process must be introduced as axioms to justify the use of characters in discovering natural taxa.

A taxon is not a kind, but 'monophyletic taxon' is

Natural kinds usually comprise a set of things, but they differ from most sets in that their definitions are ultimately justified by appeal to one or more fundamental underlying process theories (Rosenberg 1984). For example, the natural kind Helium comprises the set of atoms with the atomic number of 2. Helium is a natural kind, and not a mere set or class, because its definition is justified by process theories concerning atomic chemistry that explain how atoms with the atomic number 2 originate, why atoms with the atomic number 2 behave as they do, and how these atoms might change. In contrast, I might define, by extension, a set comprised of the Milky Way galaxy, three geese, and my left toe. No scientists is likely to mistake this set for a natural kind because this definition cannot be linked to any fundamental (or even trivial) process theory.

At least part of the controversy surrounding the nature of taxa concerns their definitions, or lack thereof. Do taxa have definitions analogous to those of natural kinds such as Helium? My answer is no, because (1) the definitions of natural kinds are universal (historically unconstrained) definitions whereas the 'definitions' (Platnick and Rosen 1987) of taxa are not definitions, but singular (historically constrained) diagnoses (Ghiselin 1984), and (2) the definitions of natural kinds have a different logical relationship to process theories than do the diagnoses of taxa.

Universal definitions versus singular diagnoses

When we say that the atomic number of 2 defines the natural kind Helium, we claim that this definition comprises a necessary and sufficient definition meant to apply to any and all atoms regardless of their particular origins. We see this immediately because Helium is a class based on what biologists would term 'convergence.' The property 'atomic number 2,' shared by all helium atoms, is a property that has orginated independently in each member of the kind. This property is not present because all helium atoms have descended from some common helium ancestor with the atomic number of 2. 'Convergence' is the mark of a truly universal definition. Members of the natural kind Star or the natural kind Galaxy are similar. There is no ancestral star which gave rise to all stars. The definition of this kind is based on a convergent character and such definitions are universal in form. I suggest that this is a general property of natural kinds. Indeed, I suggest that there is a direct relationship between the number of times a defining property has been convergently gained and the significance of the natural kind in question. *The power of natural kinds is vested in the ability of process theories to explain this convergence.* What makes natural kinds so important, and what makes the theories they are associated with so universal, is the fact that the definitions of these natural kinds are based on 'convergence' rather than on historical contingency (i.e., common properties shared because of common descent).

Just as there are many possible sets and only a few natural kinds, there are many possible taxa and only a few natural taxa. For example, we can contrast a taxon composed of humans, earthworms and ferns with a taxon composed of humans, chimpanzees, gorillas, and organgutans. The first would hardly be mistaken for a natural taxon because its relevance to evolutionary history is obscure. It is an arbitrary set. The second is hypothesized to be a natural taxon because the weight of evidence indicates that it is comprised of species that share a common ancestor to the exclusion of all other living species, and thus is a natural taxon (i.e., a monophyletic group sensu Hennig 1966). Some philosophers and at least a few phylogeneticists believe that natural taxa are natural kinds and that certain intrinsic characters (synapomorphies) define these taxa in a manner (I presume) analogous to the definitions of other natural kinds. However, if we examine this belief, we can arrive at a different conclusion. Consider, for example, the natural taxon Aves. One synapomorphy of this taxon is 'feathers present.' Does 'feathers present' have the same logical relationship to Aves as 'atomic number = 2' has to Helium? No, the property shared by birds is based on what biologists term homology. It is a historically contingent character. Birds share this character because they are descended from a common ancestor that had feathers. The property shared by birds is exactly opposite in its qualities to the character shared by helium atoms. No phylogeneticist would group taxa on the basis of convergence. It is precisely the historically contingent nature of synapomorphies like feathers that makes taxa like Aves natural. Aves is the logical opposite of Helium in terms of its 'definition.'

Kinds, taxa and process theories

If natural taxa are not analogous to natural kinds, then their relationship to process theories might also be different. One aspect of this difference lies in the observation that the defining characters of natural kinds are an integral part of the theories that explain them while the diagnostic characters of taxa are not. For example, the specific necessary and sufficient definition of Helium is embedded in the general theory of atomic chemistry (as are all definitions of all Elements). Thus we can predict the characteristics of unobserved and undiscovered elements. In contrast, the diagnosis of Aves is not embedded in any evolutionary theory (Darwinian or non-Darwinian). If Aves never evolved, our current theories (whatever their relative merits on other grounds) would not suffer. If helium atoms never existed, atomic physics would be in a great deal of trouble. The inability of various forms of evolutionary theory to provide the 'definitions' of taxa in the same way as atomic theory provides the definitions of elements does not present a problem when we realize that taxa are not kinds like Helium or Star, but philosophical individuals.

We should expect physics and chemistry to provide definitions for the natural kinds in their domains. We should also expect evolutionary biology to provide definitions for the natural kinds within its domain. These natural kinds are not taxa like *Homo sapiens* or Aves. Rather, they are kinds such as Codon, Mendelian Gene, Locus, Allele, Population, Evolutionary Species, and Monophyletic Group. Perhaps some of these kinds will turn out to be numerical universals rather than true universals. Some may turn out to be artificial classes and they will eventually be trashed along with their supporting theories. Some will turn out to be as sharply defined as Planet, if not so sharp as Helium. Others will be recognizable, function significantly in theory, but be rather fuzzy, like the natural kind BO Star. Some will continue to function in whatever current evolutionary theory is in fashion in a manner identical to the manner in which Helium functions in atomic chemistry, as the repository of examples that validate the process theory in question. For example, the whole idea of evolution would be seriously questioned if there were no monophyletic groups because a lack of monophyletic groups would imply a lack of common ancestry connections between species. If there were no monophyletic groups, then there would be modification, but no descent.

Taxa as individuals

The idea that taxa function as philosophical individuals has been discussed by many scientists and philosophers (Hennig 1966, Ghiselin 1974, 1981, Hull 1976, 1978, 1980, 1983, Wiley 1978, 1980, 1981a, Williams 1985, Rosenberg 1985, Ax 1987). *Homo sapiens* is an individual. It is an example of the natural kind Evolutionary Species (Wiley 1978, Ax 1987). Vertebrata and Primates are also philosophical individuals. They are restricted in time and space and the characteristics we use to identify *Homo sapiens* as both a part of Vertebrata and Primates are historically contingent homologies termed synapomorphies. However, not all philosophical individuals have the same qualities (Hennig 1966, Wiley 1980), and I believe that this fact should be recognized if we are to get at the problem of diagnosis versus definition. The problem can be stated thus: individuals are not supposed to be *definable* by their attributes, only by their origins. No necessary and sufficient attribute suffices to define an individual in the way that atomic number = 2 suffices to define a member of Helium. Yet, systematists use characters constantly to place individual organisms in groups and the groups recognized are 'justified' by these characters.

The problem is really two-fold, involving both a perceptual problem and a process problem. In a short paper (Wiley 1980), I attempted to distinguish between two sorts of philosophical individuals. The first sort comprises 'real' individuals, they are distinguished by the fact that they are parts of cohesive wholes. For example, I am a cohesive whole. Each of my cells recognize each other as 'self' and actively cohere with each other. Any cells that deviate from the party line are gobbled up. If they are not, they may cause my demise. *Homo sapiens* is also a 'real' individual. It has the populational analogy of cell cohesion, gene flow. Some parts may become isolated for a short time (geologically short, that is), some may commune only through intermediate populations. And, just like my cells, not all humans must look the same. *Homo sapiens* and me are 'real individuals' because our parts have (1) a common origin, and (2) actively cohere. We are not 'real' because we have defining characters like 'has lost most of his hair' or 'cranial capacity greater than 900cc.'

Vertebrata is a philosophical individual. It had an origin as a single species. At the time of its origin it had the same individual qualities as *Homo sapiens* and me. However, once the first speciation event split that ancestor into two species, Vertebrata ceased to be a cohesive whole. One part no longer communed with the other. That is what speciation is all about, making two (or more) individual lineages where only one existed before. Here is the problem: We can point at me, note my singular origin, and justify my existence because we can observe me. We can point to *Homo sapiens*, note various historically documented migrations, see intermediates between differentiated populations, and justify its existance because we can infer that *Homo sapiens* has cohesive properties. However, we cannot directly observe the Vertebrata or even (directly) the history of Vertebrata, and it is composed of different species, precluding our use of the cohesion criterion. How then, can we ever recognize Vertebrata as a natural taxon?

We recognize Vertebrata as a natural taxon because its members share several synapomorphies. Such workers as Nelson and Platnick (1981, p. 304), Platnick (1982), and Platnick and Rosen (1987) consider these synapomorphies as necessary and sufficient characters constituting a definition of Vertebrata. Ghiselin (1966, 1974, 1981), Hull (1976, 1983), Beatty (1982) and others suggest that it is precisely the quality of individuals that precludes characters from fulfilling the necessary and sufficient conditions necessary for a definition. (Neither 'camp' suggests that characters are unimportant.)

Below, I present the thesis that characters are neither necessary nor sufficient to recognize natural taxa. I also present the thesis that there is a necessary and sufficient condition for recognizing natural taxa, unique genealogy, and that characters can be used to recognize natural taxa so long as there is a conceptual link between the historically contingent nature of a synapomorphies and the common genealogical origin of parts of a natural taxon. I suggest that this line of reasoning will result in a criterion with which we can justify the naturalness of biological classification because it will sort out those proposed taxa that are unnatural from those that are natural.

Characters are not necessary nor sufficient

It may come as a surprise to some, but Hennig (1966) solved the problem of the nature of characters more than twenty years ago. The controversy was the same as we face today. Is morphology the 'solid base' on which we construct our hypotheses? Or, is genealogy, and all that this word implies regarding evolution, the solid base? Hennig states:

> The attempt to introduce a system based exclusively on morphological view-points or on morphological form similarities (in the broad sense) as the general reference system of biological systematics would encounter serious difficulties. Stages in metamorphosis, polymorphisms, and other semaphoronts that differ in form but obviously belong together genetically [= genealogically], would have to be grouped without regard to these genetic [genealogical] relationships (e.g., caterpillars would have to be placed in one group, and butterflies in an entirely different group). To avoid this it would be necessary to give absolute precedence to a peculiarity of the individual that is only conditionally to be recognized as a 'form character' – namely: 'the capacity under certain circumstances to alter their form in a particular way, for example in a particular cycle of metamorphosis.' – without being able to do anything comparable in the higher group categories. In discussing morphological systems, no one would think of recognizing 'the capacity to alter its form in a certain way under certain circumstances' as a morphological character (Hennig 1966, p. 23).

Some systematists (cf., Nelson 1985) have used the argument of the 'life cycle' to get at Hennig's objection. It is not important that a tetrapod have legs at all stages in its development, only that it have legs (or the embryonic precursor of legs) at some stage in its development, so that the species has legs (or its developmental equivalent) at some stage in the species' life cycle. What we are really trying to do, according to these workers, is to classify life cycles.

There is value in adopting this point of view, but it does not help solve the problem of introducing process ideas into the controversy. In suggesting this strategy, such workers are giving up on a purely morphological system. The consequence of accepting the life cycle approach is to introduce a number of process theories that are necessary to employ the approach. To understand this, let us follow Hennig's (1966) reasoning. To get around the problem of classifying caterpillars and butterflies in different groups based on the fact that both are members of a single life cycle we must embrace two ideas of process. The first is ontogeny, it is necessary if different characters are to appear and disappear during growth. The second is reproduction, it is necessary to perpetuate the life cycle itself. So far, so good, but Hennig argues that this does not solve the problem because there is nothing comparable to reproduction to justify grouping different species into higher taxa. This is an important 'but,' because it is precisely the ability to reproduce viable offspring that is missing between members of most higher taxa. Cows do not interbreed with horses. How can we place cows and horses in the same higher taxon when cows and horses have different life cycles?

There is another objection. Hennig (1966, 16–21) presents a lengthy discussion concerning the characteristics of the hierarchical system and the types of biological relationships and associated entities that display hierarchical relationships. He specifically rejects the relationship existing between semaphoronts (the form of an individual at a particular stage in its life cycle) of an individual organism as being hierarchical.[1] He also rejects the idea that the relationships existing among individuals within a species can be expressed in a hierarchy. It would be reasonable to conclude that Hennig would have considered the 'life cycle' of a species to be nonhierarchical (a common sense conclusion since cycles and hierarchies are opposites). However, Hennig did reason that the relationship among species could be expressed in a hierarchy. I conclude that if species are related in a hierarchical manner, but if their life cycles are nonhierarchical, then it would be hard to use the concept of (nonhierarchical) life cycles to justify a hierarchy of species.

In spite of this line of reasoning, not all is lost. We could hardly have the reality of homology and synapomorphy existing in multicellular organisms and the DNA products of organisms unless there was a hierarchy of life cycles. This hierarchy of life cycles corresponds to the hierarchy of species *vis-à-vis* their characters. Such workers as Platnick and Nelson are closer to the truth than one might imagine, given the above discussion. However, application of the concept of a hierarchy of life cycles depends on the acceptance of additional processes, character change (anagenesis) and speciation (cladogenesis).[2] One must acknowledge that species are historically constrained individuals and that monophyletic taxa are historical entities to apply these assumptions. Treating them as natural kinds robs them of their relationship to natural phenomena and the theories and 'laws' associated with these phenomena. As an aside, the consequences of this line of reasoning for the phylogenetic community are clear: phylogenetic hypotheses cannot be 'clado-grams' in the sense discussed by Nelson and Platnick (1981), they must be trees, with all the 'messy' connotations of trees relative to evolutionary theory.

I (Wiley 1981a) took a different approach to the 'necessary and sufficient' argument than Hennig (1966). It is not so eloquent, but it is worth a short discussion. I did not argue that the necessary and sufficient argument failed because characters appeared at different times or under different guises in the life cycle of an individual. Rather, I argued that the criterion failed because any member of a species who dies before developing the character that 'defines' the species (or higher taxon to which it belongs) could not logically be placed into that species (or higher taxon). In other words, the argument that there are necessary and sufficient characters that define taxa might have merit if all members of, say, Tetrapoda, have (and had) the capacity to develop limbs (or even embryonic limb buds that subsequently disappeared), and that all members of Tetrapoda have, did, or will, survive long enough to manifest this condition. But, many do/did/will not. Perhaps we could dismiss those cases where the individual concerned encountered an accidental death at the hands of nature or a premeditated death at the hands of a zealous biologist, by claiming the potential to develop this character. But, some individuals die natural deaths precisely because they lack (genetically) the ability to develop the characters in question. Perhaps an embryo cannot develop a tetrapod limb because it lacked the capacity to develop a notochord and was aborted. No one has argued that an aborted gastrula of tetrapod parents belongs to any other taxon than Tetrapoda. It would not, for example, be classified as a cnidarian. One might dismiss this argument as silly since one goes around classifying aborted gastrula. But, this would be no argument at all. If an aborted gastrula of tetrapod parents is a tetrapod rather than a cnidarian, then the character 'tetrapod limb present' cannot be a necessary condition for group inclusion, even if it might be a sufficient condition for group inclusion.

Finally, I approach the question of sufficiency. While it might be true that characters are not necessary for placing an organism into a natural taxon, is it not true that finding a tetrapod limb is sufficient cause to allow such a placement? Not always, it depends on whether the investigator 'knows', by parsimony arguments, whether the character in question has arisen once or more than once. For example, the weight of evidence suggests that the atlas-axis complex of birds is convergent with the atlas-axis complex of mammals. Obviously the observation that this character is found in a particular organism is not, of itself, a sufficient reason to place the organism into either Mammalia or Aves. Of course, we 'know,' by weight of other evidence, that the atlas-axis complex of birds is a different character than the atlas-axis complex of mammals. Our problem is semantic, not substantial, a fact recognized by Hennig (1966). Nevertheless, the reality of convergence and parallelisms casts doubt on the notion that characters are sufficient cause for group placement outside the context of a theory of process. This is sufficient for my argument.

What criterion defines taxa?

If characters are not necessary and sufficient to define taxa, then those who wish to use characters as if they do define taxa have a logical difficulty. No such difficulty was encountered by Hennig (1966) because he conceived of taxa as being defined by common ancestry:

> In the phylogenetic system the categories (= taxa) at all levels are determined by genetic (= genealogical) relations that exist among their subcategories (= subtaxa, ultimately species). Knowledge of these relationships is a prerequisite for constructing these categories [= taxa], but the relationships exist whether they are recognized or not. Consequently here the morphological characters have a completely different significance than in the logical and morphological systems. They are not themselves ingredients of higher categories (= supraspecific taxa) but aids used to apprehend the genetic (genealogical) criteria that lie behind them.

In other words, (1) natural higher taxa are real, they exist whether we can find them or not, (2) the basis for their existence is common genealogical connections, or common ancestry, and (3) the significance of shared similarities lies in whether or not they elucidate common ancestry relationships. What defines natural taxa is their origins not their characters. Common ancestry provides the necessary and sufficient criterion (Wiley 1981a).

Some consequences

Hennig's particular view of the logical relationship between taxa and characters is important in several respects. First, it brings into focus the logical relationships between taxa as philosophical individuals and homologies as historical contingencies. We need not worry about the logical difficulties we might encounter if we treat taxa as natural kinds while treating their characters as historical contingencies because no such logical difficulty exists.

Second, it provides the logical foundation for the relationship between evolutionary theory and biological systematics. By 'evolutionary theory' I mean the theory in its most general form, descent with modification coupled with semiconservative replication. General evolutionary theory predicts the existence of monophyletic groups, groups of species with part-whole relationships. Classifications of organisms may be considered natural classifications only to the extent that they contain, exclusively, monophyletic groups. Such classifications can be shown to be logically consistent with the known common ancestry relationships of the organisms classified (Hull 1964, Wiley 1981b). Classifications that contain groups that are not monophyletic (i.e., that contain groups that are paraphyletic or polyphyletic) are logically inconsistent with phylogeny because they destroy information concerning common ancestry relationships in direct proportion to the number of paraphyletic or polyphyletic groups they contain (Wiley 1981b). Such groupings are based on similarity, not genealogy, and thus represent classes (Ghiselin 1984). They are not even natural kinds, because there is nothing in evolutionary theory that predicts their occurrence. They are, frankly, figments of the minds of taxonomists.

The majority of taxonomists would agree with my assessment regarding polyphyletic groups, but would disagree with my assessment regarding paraphyletic groups. Paraphyletic groups are also termed grades. Typical paraphyletic groups include such 'taxa' as Reptilia and Pongidae. They are not based on genealogy, but on similarity. They are not based on descent with modification but on descent without modification. The only way to justify such groups is to claim that the 'characters' they exhibit are necessary and sufficient characters. These characters may be formulated as estimates of 'overall similarity' or adaptive distinctiveness. They may also be formulated by exclusion (for example, Reptilia is defined as a group with the amniotic egg whose members lack feathers and fur). If characters cannot define, but only diagnose taxa, there is no basis in biology or philosophy to claim that such groups function in any meaningful manner within the evolutionary paradigm.[3] They are typological constructs and classes (Ghiselin 1984). Grades are not even natural kinds because it would be a curious evolutionary theory indeed that predicted the occurrence of groups that are logically inconsistent (Hull 1964, Wiley 1981b) with the common ancestry relationships of the species classified.

Finally, let us consider the consequences of treating taxa as natural kinds. If taxa are natural kinds, then there would be no prohibition against recognizing either paraphyletic or polyphyletic groups. I have already noted the fact that convergence provides a perfectly good basis for defining powerful natural kinds such as Helium or Star. The lack of sufficient mass in a cosmic gas cloud to initiate atomic reactions is a perfectly adequate necessary and sufficient character to define Planet. Convergences, exclusion characters, and even what systematists would term primitive (symplesiomorphic) characters can define natural kinds. But, they cannot define, or even diagnose, a philosophical individual because the result would destroy the part-whole relationships that is demanded by our most general theories of evolution.

A note on the nature of evolutionary theory

There are natural processes and theories concerning natural processes. Darwinism-as-theory is not synonomous with evolution-as-process. Processes operate on matter whether we have good theories, bad theories, or no theories at all. There is a powerful tendency to confuse Darwinism (or neo-Darwinism) with the evolutionary process itself. Darwin (1859) presented two general ideas: (1) species are mutable and different species are related by common ancestry, (2) natural selection is one of the major mechanisms leading to biological change. Neither idea was new, but that does not matter in the least. The first idea was based on patterns, the second idea was a proposed mechanism. If natural selection were discarded tomorrow or twenty years from now, Darwin's major idea would not be rendered false because descent with modification would still be the best thesis to explain the patterns we observe among organisms in nature.

I do not happen to be among those who feel that Darwinism or its most recent version, neo-Darwinism are false. I do believe that neo-Darwinism is incomplete. It is incomplete because some of the patterns we observe are part-whole patterns (Vertebrata/Primates/*Homo sapiens*) while the theory is a theory of natural kinds (Allele, Population, Species, Monophyletic Group). This is hardly a criticism of neo-Darwinism *per se*. It is simply an acknowledgement of the fact that the structure of neo-Darwinism is Newtonian (Ginsburg 1981). It is a theory of natural kinds, not a theory of individuals. But, we need other theories, theories that address the individuals apart from them being examples of natural kinds (for example, Brooks and Wiley 1988). Such theories are powerful to the extent that they explain the distributions of homologies and group hierarchies in a manner analogous to the power of physical theories to explain convergences such as the atomic number of 2. There is no reason to demand of neo-Darwinism an explanation of why there are birds any more than there is a reason to demand of atomic theory an explanation of why there is a gold ring on my finger. Yet, we need theories that explain why the results of the evolutionary process should take the form of hierarchies rather than periodic tables.

Acknowledgements

Many thanks to Douglas Siegel-Causey and Darrel Frost (University of Kansas) for discussing the issues I have raised in this paper. Much of my thinking on these matters is a direct result of conversations with, and papers by, David Hull. Thanks, David, for your willingness to interact with biologists who are also interested in philosophical issues.

Notes

1. I should add that while the concept of 'life cycle' is a periodic concept, ontogeny itself is not. But the hierarchy of ontogeny is a hierarchy of cells, not a hierarchy of semaphoronts.
2. All four processes, of course, depend on the acceptance of even lower-level processes,

300

such as semi-conservative replication.
3. Indeed, as deQueiroz (1987) has argued, these groups are holdovers from preevolutionary biology and their continued appearance in biological classifications is the best evidence that systematics and taxonomy never joined the evolutionary paradigm.

References

Ax, P. (1987). *The Phylogenetic System. The Systematization of Organisms on the Basis of Their Phylogenesis*. New York: Wiley-Interscience.

Beatty, J. (1982). Classes and cladists. *Syst. Zool.* 31: 25–34.

Brooks, D.R. and Wiley, E.O. (1988). *Evolution as Entropy. Towards a Unified Theory of Biology* 2nd ed. Chicago: University of Chicago Press.

Darwin, C. (1859). *On the Origin of Species by Means of Natural Selection, or the Preservation of Favored Races in the Struggle for Life*. London: John Murray.

DeQuieroz, K. (1987). Systematics and the Darwinian Revolution. *Philo. Sci.* (in press).

Ghiselin, M.T. (1966). On psychologism on the logic of taxonomic controversies. *Syst. Zool.* 15: 207–215.

Ghiselin, M.T. (1969). *The Triumph of the Darwinian Method*. Berkeley: University of California Press.

Ghiselin, M.T. (1974). A radical solution to the species problem. *Syst. Zool.* 23: 536–544.

Ghiselin, M.T. (1981). Categories, life, and thinking. *Behavioral and Brain Sci.* 4: 269–313.

Ghiselin, M.T. (1984). 'Definition,' 'character,' and other equivocal terms. *Syst. Zool.* 33: 104–110.

Ginsberg, L.R. (1981). *Theory of Natural Selection and Population Growth*. Menlo Park, CA: Benjamin/Brown Publ. Co.

Hennig, W. (1966). *Phylogenetic Systematics*. Urbana: University of Illinois Press.

Hull, D.L. (1964). Consistency and monophyly. *Syst. Zool.* 13: 1–11.

Hull, D.L. (1976). Are species really individuals? *Syst. Zool.* 25: 174–191.

Hull, D.L. (1978). A matter of individuality. *Phil. Sci.* 45: 335–360.

Hull, D.L. (1980). Individuality and selection. *Ann. Rev. Ecol. Syst.* 11: 311–332.

Hull, D.L. (1983). Karl Popper and Plato's metaphore. *Advances in cladistics* 2: 177–189.

Kluge, A.G. (1984). The relevance of parsimony to phylogenetic inference. In: T. Duncan and T. Stuessy (eds), *Cladistics: Perspectives on the Reconstruction of Evolutionary History*, pp. 24–38. New York: Columbia University Press.

Nelson, G. (1985). Outgroups and ontogeny. *Cladistics* 1(1): 29–45.

Nelson, G. and Platnick, N.I. (1981). *Systematics and Biogeography. Cladistics and Vicariance* New York: Columbia University Press.

Platnick, N.I. (1982). Defining characters and evolutionary groups. *Syst. Zool.* 31: 282–284.

Platnick, N.I. and Rosen, D.E. (1987). Popper and evolutionary novelties. *Hist. Philo. Life Sci.* 9 (in press).

Popper, K.R. (1968). *The Logic of Scientific Discovery*. New York: Harper Torch Books.

Wiley, E.O. (1975). Karl R. Popper, systematics, and classification – a reply to Walter Bock and other evolutionary taxonomists. *Syst. Zool.* 24: 233–243.

Wiley, E.O. (1978). The evolutionary species concept reconsidered. *Syst. Zool.* 17–26.

Wiley, E.O. (1980). Is the evolutionary species fiction? – A consideration of classes, individuals, and historical entities. *Syst. Zool.* 29: 76–80.

Wiley, E.O. (1981a). *Phylogenetics. The Theory and Practice of Phylogenetic Systematics*. New York: Wiley-Interscience.

Wiley, E.O. (1981b). Convex groups and consistent classifications. *Syst. Bot.* 6: 346–358.

Williams, M.B. (1985). Species are individuals: Theoretical foundations for the claim. *Phil. Sci.* 52: 578–590.

Evolvers are Individuals: Extension of the Species as Individuals Claim

MARY B. WILLIAMS

Center for Science and Culture, University of Delaware, Newark, DE 19711, USA

In what are almost asides Ghiselin (1974, 1981), Hull (1976), Holsinger (1984), and Williams (1985) have all claimed not only that species are individuals but that populations are individuals. This larger claim has been lost in the glare of dispute over whether species are individuals. Part of the reason for this neglect is that the term 'population', although central to population biology, is considerably more amorphous than the term 'species' and it is thus considerably more difficult to dispute any claims about it. But if the individuality of species is important because of what it tells us about the way they function in biological laws (as Hull (1976, 1981, 1987) assserts), then the individuality of populations should also be understood so that we can understand how they function in biological laws.

Two kinds of arguments for the 'species are individuals' thesis have been given: One focuses on showing the relationship between species and our paradigm individuals. The other, which I will use for populations in this paper, focuses on the idea that species participate in biological processes as individuals. As Holsinger puts it:

> An individual ... is an entity that, with respect to a particular process, behaves as a whole and is independent of similar entities, i.e., with respect to the process being considered an individual is an entity that exists as a discrete unit, complete unto itself and coherent. (Holsinger 1984, p. 295).

The feature of this approach that has not received due attention is the fact that neither species nor populations (nor organisms) are individuals *simpliciter*; they are *individuals with respect to particular processes*. As explicit examples later in this paper will show, a species (or a race, or a population) may be an individual with respect to one particular selection process but not an individual with respect to a different selection process. The significant question about a particular population, or species, is not 'Is this an individual?', but rather 'With respect to what processes (if any) is this an individual?'

As Hull (1976, 1980) has pointed out, there are three different roles that individuals play in evoluton; there are individuals that are units of replication, individuals that are units of interaction, and individuals that are units of evolution.

M. Ruse (editor), What the Philosophy of Biology is. pp. 301–308.
© 1989 *Kluwer Academic Publishers, Dordrecht*

In this paper I will discuss only the third role, the unit of evolution. An entity is a unit of evolution, an evolver, if it is a chunk of the genealogical nexus and it is cohesive with respect to a particular set of selection forces; an evolver is an individual with respect to its set of selection forces, or with respect to the selection process which these forces engender. (Although Hull uses the term 'lineage' for this type of individual, I prefer 'evolver' because it does not have extraneous meanings.) The claim of this paper is that populations of many different types are evolvers, are individuals with respect to particular selection processes, and instantiate evolutionary laws.

This claim is empirical as well as philosophical. Relevant philosophical analyses have been given in Williams (1985) and Holsinger (1984), as well as in the papers of Hull and Ghiselin. This paper will concentrate on clarifying the meaning of the philosophical claim by showing its force for particular examples, and on supporting the empirical claim by displaying populations of different types together with the evidence indicating that they are evolvers.

A digression on laws, forces and cohesiveness

Since cohesiveness is a relative property, an analogy from physics may help to clarify the relationship between forces, laws, and cohesiveness. Consider the Newtonian law which gives the trajectory resulting from a force applied to an entity; one might use a sophisticated version of this law to calculate the time required for a baseball hit by a bat to reach a specified target. Now suppose that the entity in question is a banana cream pie rather than a baseball, and use a plank instead of a bat to apply the force. If, for each instant during the time the force is applied, the force is small relative to the cohesive forces holding the pie together, then the law could calculate the time required for it to hit the target. On the other hand, if the applied force is larger than these cohesive forces, the law does not apply. The banana cream pie is a Newtonian individual with respect to small applied forces, while it is a collection of molecules with respect to large applied forces. An indefinite number of such examples could be given. There is no set of Newtonian entities each of which behaves Newtonianly (i.e., is a Newtonian individual) under every possible Newtonian force; there is only a set of entities each of which behaves Newtonianly under many sets of conditions of interest to us. Similarly, there is no set of Darwinian entities each of which behaves as an evolutionary unit in every possible selection process; there is only a set of entities each of which is an evolutionary unit (an individual) with respect to many selection processes.

An example in which races are evolvers

H. melpomene and *H. erato* are two species of unpalatable butterflies; their warning coloration protects them from bird predators, who learn from experience to avoid them. Each of the species has about 30 races whose color patterns are quite distinct. In localities in which the two species coexist, the sympatric races are similar enough to fool a casual observer; sometimes the visual resemblance is so great that an expert will smell the butterfly to decide which species it belongs to. These sympatric races are Mullerian mimics of each other, originating (Sheppard *et al.*, 1985) during the cool dry periods of the Quaternary when their rain forest habitat contracted into isolated refuges. Their striking similarity is the result of a selection process in which a mutation in the locally less common species which makes it look somewhat like the locally more common species spreads through the population because of birds who have learned to avoid the more common type. Once this mutation is fixed, mutations in either species which increase the resemblance are selected for. When the rain forest habitat expands again, the mimetic races expand until they meet conspecific mimetic races formed in a different isolated refuge. Although hybridization is possible (and frequently common in hybrid zones) between the conspecific races, their color patterns differ so much that hybrids are strongly selected against; introgression of the color pattern genes does not take place. As Benson (1972) has shown, predation by birds continues to exert stabilizing selection on the races.

This is actually a set of selection scenarios, one describing the evolution of each race. To summarize:

- Trait selected: color pattern similarity of organisms of sympatric races of *H. melpomene* and *H. erato*.

- Selective force: predation by birds on sympatric organisms of the unpalatable butterflies *H. melpomene* and *H. erato*.

- Evolver: race of *H. melpomene* or *H. erato*.

The claim that this scenario describes the evolution of races is a claim that, although this selection process causes genetic change over time at every level from local population up to species (and indeed up through genus, family, etc.), the race is the entity which is cohesive with respect to this selection process. If a new allele which improves the resemblance arises in any part of the race, it will spread throughout the race because there is selection pressure in this direction on all parts of the race; it will spread no further than the race because there is a barrier to gene flow for these color pattern genes, caused by selection against them in and beyond the hybrid zone. The empirical evidence for the cohesiveness described in the preceeding sentence is extremely strong, coming from (1) a thorough knowledge of the biology of the Heliconiads (summarized in Brown, 1981), (2) experimental studies on the genetics of these races (summarized in Sheppard *et al.*, 1985), (3) experimental studies on the selection pressure (Benson 1972, and Turner 1977), and (4) from the fulfillment of theoretical predictions of Mullerian mimicry theory (see Sheppard *et al.*, 1985, pp. 438–439).

Further evidence that the evolver in this selection process is the race is given by the role that the race plays in the relevant law. To properly state this law it would be necessary to have the machinery that would be developed in an explicit detailed statement of the theory in which it is embedded (mimicry theory, which is itself embedded in natural selection theory and Mendelian genetics). But a statement of the law which is sufficient for our purposes can be given with only a single preliminary definition.

Definition: Organisms O' and O" are connected by organism predator interactions if and only if for any predator organism OP which has interacted with O', the probability of OP having interacted with O" or with an ancestor or descendant of O" is greater than zero.

Mullerian Mimicry Law:
If E1 and E2 are evolvers whose organisms
  1. are unpalatable,
  2. are structurally similar and occasionally have mutations increasing that similarity,
  3. have in common major visually hunting predators who learn to avoid prey recognizably similar to previously encountered unpalatable prey,
  4. are all connected by organism predator interactions,
then both interevolver and intraevolver polymorphism in the prey recognition characters will decrease over evolutionary time.

Note that the very same selection pressure that selects for interevolver similarity (the advantage of being recognizably similar to a known unpalatable prey) also selects for intraevolver similarity. Each sympatric pair of races of *H. erato* and *H. melpomene* is essentially monomorphic for color pattern, as is expected of sympatric evolvers under the conclusion of this law; thus, for example, the races *H. erato dignus* and *H. melpomene bellula* instantiate the law. But *H. erato* and *H. melpomene* are not instantiations of this law, since organisms of different races of *H. erato* (or of different races of *H. melpomene*) are not connected by organism predator interactions; because of this disconnection natural selection has increased the polymorphism of *H. erato* and *H. melpomene* by creating many distinct races within each species.

Thus the races of *H. erato* and *H. melpomene* are evolvers with respect to the selection pressures described in this scenario.

An example in which species are evolvers

Benson (1971) has suggested that unpalatability evolved in the Heliconiiae by kin selection on gregariously roosting adults. According to this selection scenario: an allele conferring unpalatability on organisms arose; the bird predators learned to avoid taking prey close to where they had previously found unpalatable prey; so in species in which kin groups roost together the unpalatability allele spread through the population and eventually through the species. It is assumed that there are (or were) no significant barriers to the flow of this gene throughout the species and that bird predators capable of learning this lesson exist throughout the range of the species. It is known that such bird predators exist today in the range of the various butterfly species, so that stabilizing selection pressure for unpalatability is still present. Benson presented evidence in support of the hypothesis in the form of significant correlations between the present degree of unpalatability of different species and their degree of gregariousness. The evidence for this scenario is much less than that for the mimicry scenario. However it seems clear that the species is the evolver in this case; the selective force from bird predators exists throughout the range of the species, and the barriers to gene flow whiich played a significant role in the mimicry case are assumed not to exist here.

Since *H. erato* and *H. melpomene* are two of the species which Benson suggests evolved unpalatability in this way, it is important to consider whether they were still evolvers with respect to selection for unpalatability during the period that *H. erato dignus* and *H. melpomene bellula* were evolvers with respect to selection for Mullerian mimicry. Although the Mullerian mimicry scenario assumes that unpalatability is already fixed in the evolver, there is still stabilizing selection pressure for unpalatability, as well as pressure favoring new unpalatability genes, so there is still cohesiveness due to selection pressure for unpalatability. But are the barriers to gene flow during the period discussed in the Mullerian mimicry scenario serious enough to disrupt the cohesiveness of the species with respect to selection for unpalatability? The isolation in the refugia would have been enough to disrupt cohesiveness if the predators in the different refugia found different flavors unpalatable – that is, if there were different selection pressures as well as isolation; but apparently there were not. Selection on color pattern has resulted in barriers to introgression of color pattern genes, but it has not resulted in barriers which totally prevent introgression of other genes; viable fertile hybrids are produced in the hybrid zones, and since members of the F2 generation who look like one of their grandparents are protected from the predators, there is no reason to think that introgression does not occur for new unpalatability alleles. (Brown (1981) has even suggested that limited interspecific introgression occurs in this tribe.) So the evidence indicates that these species are evolvers with respect to selection for unpalatability during the period when their races were being formed. Hence this is a case in which an evolver with respect to one selection process is a part of an evolver with respect to a different selection process.

An example in which the local population is an evolver

Smiley (1978) found that at La Selva Field Station (in north Costa Rica) *H. melpomene* is principally monophagous on *Passiflora oerstedii*, and that this difference in host plant preference is genetically inherited. He suggests that the selection pressure causing this inherited preference is predation pressure by ants, which cause high mortality on *H. melpomene* (Smiley 1985) and which are less frequent on *P. oestedii*. At the same time, Benson (1978) found that at Rincon (in south Costa Rica) *Passiflora menispermifolia* was the major host plant of *H. melpomene* even though *P. oerstedii* was present at Rincon. (In fact, Benson found no instances of *H. melpomene* on *P. oerstedii* at Rincon.) Because Smiley was not concerned with the precise extent of the selection process and did not identify his population beyond saying that it was *H. melpomene*, my identification of the evolver must be tentative. Since according to distribution maps both of these populations are in the race *H. melpomene rosina*, I conclude that the evolver is the local population.

The above conclusion would be much stronger if there were evidence that the ant-plant relationship were different in south Costa Rica, causing opposing selection pressures. So I will supplement this example with a more fully understood example of a local population evolver. Edmonds and Alstad (1978) investigated the selection process through which demes of scale insects evolved adaptation to individual pine trees. Pines defend themselves against insects by producing numerous toxic defensive compounds; there is considerable variation in defensive compounds among the trees in a pine forest, so that a population of scales which is very successful on one tree may be unsuccessful on the neighboring tree. Since scales may go through 200 generations during the life of a tree, a deme has the opportunity to evolve resistance to the compounds of a tree on which it initially can barely maintain itself. According to Edmonds and Alstad, this selection process is an important feature of the relationship between the black pineleaf scale and the ponderosa pine. The number of generations involved is relatively small, and the deme is doomed to extinction within a relatively short time (with any descendant demes forced to evolve adaptations to a different set of compounds); nevertheless, this is a distinct selection process (with a distinct set of selection pressures), and the deme is its evolver.

An evolver which is part of a species

This final example is included because it emphasizes the extent to which the selection forces, rather than gene flow boundaries, determine the boundaries of the evolver. The selection process which increased the sickle hemoglobin allele (HbS) is perhaps the most thoroughly investigated and clearly understood selection process in *Homo sapiens*. (See Livingstone (1980) for a less simplified version than will be given here.) Because persons who are heterozygous for HbS are protected against malaria, this allele has increased in African populations in which malaria is endemic. The limits of this selection process are determined not by barriers to gene flow among human beings but rather by the climatic limits to the survival of the mosquito species which transmits the germs from one person to another. Because of emigration and interbreeding the HbS allele is now found outside of the area of malaria endemism and outside of the population in which it was selected for; so it might be argued that the evolver is the whole of *Homo sapiens*. But *Homo sapiens* is not evolving in one direction under selection pressures on HbS; the part of the species in areas of malaria endemism has selection pressure toward maintaining HbS in the population, while the part of the species outside of areas of endemism has selection pressure toward eliminating HbS from the population. And in spite of recent emigration and interbreeding, the population in malarious areas is still a relatively solid chunk of a genealogical nexus; it is still the case that the parent whose HbS gene saved her from malaria passes that gene on to offspring most of whom live in malarious areas. So the evolver in this selection process is that part of the species which lives in the malarious section of Africa.

Conclusion

These examples show that populations of many different types (deme, local population, race, species, chunk of a species) are evolvers. Williams (1985) gives the theoretical foundation for the claim that evolvers are individuals; the examples given here show the way selection forces engender the cohesiveness expected in individuals. The *Heliconius* examples show that a group of organisms that is an individual with respect to one selection process can be part of an individual with respect to a different selection process; they thus clarify the notion of being an individual with respect to a particular process. The Mullerian mimicry law shows how evolvers instantiate evolutionary laws; it shows that the replicator, the interactor, and the evolver all appear in the antecedent of the law, while the conclusion is about properties of the evolver. This provides a first step in the clarification of the claim that evolutionary laws are about species (or evolvers).

References

Benson, W.W. (1971). Evidence for the evolution of unpalatability through kin selection in the *Heliconiinae* (Lepidoptera). *American Naturalist* 105: 213–226.

Benson, W.W. (1972). Natural selection for Mullerian mimicry in *Heliconius erato* in Costa Rica. *Science* 176: 936–938.

Benson, W.W. (1978). Resource partitioning in passion vine butterflies. *Evolution* 32: 493–518.

Brown, K.S. (1981). The biology of *Heliconius* and related genera. *Annual Review of Entomology* 26: 247–56.

Edmonds, G.F. and Alstad, D.N. (1978). Coevolution in insect herbivores and conifers. *Science* 199: 941–945.

Ghiselin, M. (1981). Categories, life and thinking. *Behavioral and Brain Sciences* 4: 269–283.

Ghiselin, M.T. (1974). A radical solution to the species problem. *Systematic Zoology* 23: 536–544.

Holsinger, K.E. (1984). The nature of biological species. *Philosophy of Science* 51: 293–307.

Hull, D.L. (1976). Are species really individuals? *Systematic Zoology* 25: 174–191.

Hull, D.L. (1980). Individuality and selection; pp. 311–332. In R.F. Johnston, P.W. Frank, and C.D. Michener (eds) *Annual Review of Ecology and Systematics*. Palo Alto, California: Annual Reviews Inc.

Hull, D.L. (1981). Units of evolution: a metaphysical essay. In *The Philosophy of Evolution*, pp. 23–44. Brighton: Harvester Press.

Hull, D.L. (1987). Genealogical actors in evolutionary roles, *Biology and Philosophy* 2: 168–184.

Livingstone, F.B. (1980). Natural selection and random variation in human evolution. In J.H. Mielke and M.H. Crawford (eds) *Current Developments in Anthropological Genetics*, pp. 87–109. New York: Plenum.

Sheppard, P.M., J.R.G. Turner, K.S. Brown, W.W. Benson, and M.C. Singer (1985). Genetics and the evolution of Mullerian Mimicry in *Heliconius* Butterflies. *Philosophical Transactions of the Royal Society, London Series B* 308: 433–610.

Smiley, J. (1978). Plant chemistry and the evolution of host specificity: New evidence from *Heliconius* and *Passiflora. Science* 201: 745–747.

Smiley, J.T. (1985). *Heliconius* caterpillar mortality during establishment on plants with and without attending ants. *Ecology* 66: 845–849.

Turner, J.R.G. (1977). Butterfly mimicry: the genetic evolution of an adaptation. In M.K. Hecht, W.C. Steere, and B. Wallace (eds) *Evolutionary Biology* 10. New York: Plenum.

Williams, M.B. (1985). Species are individuals: theoretical foundations for the claim. *Philosophy of Science* 52: 578–590.

A Function for Actual Examples in Philosophy of Science

DAVID L. HULL

Department of Philosophy, Northwestern University, Evanston, IL 60208, USA

In the early years of science, scientists emphasized, possibly exaggerated, the importance of actual experiments and observations in science in order to free themselves from what they took to be the sterile scholasticism of their predecessors. Their message was, instead of deciding that women have fewer ribs than men on the basis of a priori principles, count. However, thought experiments have also characterized science since its inception. The respective roles usually assigned to actual versus thought experiments is that actual experiments are used to decide truth while thoughts experiments test our conceptual boundaries, but as Kuhn (1964) has argued in his paper 'A Function for Thought Experiments,' matters of meaning and truth cannot be so clearly and conclusively distinguished. During the ongoing process of science, they are too closely intertwined. Only in retrospect, when a particular conceptual scheme is thought to be complete, can matters of truth and meaning be separated so neatly.

In this paper, I want to argue for two positions: first, the superiority of actual over imaginary examples with respect to truth in both science and philosophy of science, and second, the superiority of actual over imaginary examples even with respect to meaning in both science and philosophy of science. Except for those who reject the notion of empirical truth *tout court*, no one is likely to object to my claim that only actual counterexamples are relevant to matters of truth in science. No one is likely to reject a scientific theory just because one can imagine evidence counting against it. After all, this is supposedly one mark of a genuine scientific theory. Only actual disconfirmations are relevant to the truth of an empirical claim, and even in such circumstances, it is far from easy to decide when to reject a view in the face of apparent disconfirming evidence. There are always ample opportunities for legitimate finagling. Can speciation occur in the absence of reproductive isolation? It is hard to say, and not all the problems that arise in attempts to answer this question concern evidence.

My claim that data is also relevant to the sorts of observations that philosophers make about science is sure to be more controversial. In fact, a fairly prevalent view of philosophy of science is that it deals only with matters of meaning and never matters of fact. Philosophers provide conceptual analyses – criteria for rationality,

M. Ruse (editor), What the Philosophy of Biology is. pp. 309–321.
© 1989 *Kluwer Academic Publishers, Dordrecht*

good explanations, etc. An analysis of scientific explanation could be retained even if no scientist in the history of science ever lived up to its standards, even if no correlation could be found between the apparent 'success' of scientists and the degree to which they approached this ideal. Confronting scientific theories with data is difficult enough. Bringing data to bear on meta-scientific claims of the sort that philosophers make is even more difficult. However, I think that it can be done. Can Mendelian genetics be reduced to molecular biology? It is hard to say, and the problems that arise are not limited just to issues concerning the meaning of 'reduction'.

My assertion that actual examples are superior to imaginary examples in testing the limits of our concepts is likely to be even more controversial. After all, the grasp of any scientific theory must exceed its reach. However, as far as counterexamples are concerned, nature provides such a rich source of problem cases that rarely will a scientist have to conjure up imaginary counterexamples to add to this store. In fact, scientists are lucky to be able to rework their concepts to accommodate real phenomena without worrying too much about thought experiments. However, scientists do avail themselves of thought experiments, and a few words need to be said about the strengths and weaknesses of such imaginary examples in testing the coherence and completeness of our conceptual systems. Finally, if the claims made philosophical claims about science are normative rather than empirical, then data are irrelevant. Whatever warrant philosophical claims might have must be elsewhere. In addition to this putative warrant, all that remains are matters of meaning, and here I want to urge once again real over imaginary examples. I begin by discussing truth and meaning in science and only then turn to parallel issues in philosophy of science. Any problems that arise with respect to science are liable to be only magnified when we turn our attention to philosophical, meta-level assertions.

In the ensuing discussion I make reference to some of the papers in this anthology to illustrate my general conclusions. Some of these papers are written by people trained primarily in biology, others by people who are officially philosophers, and others by historians. In my dicussion I am interested in activities, not professions. Although I frequently refer to the people practicing these professions, to scientists, historians, and philosophers, my main concern remains the activities regardless of the professional training and affiliation of those engaging in these activities. The justification for this decision is obvious. Although those contributors who are officially scientists differ to some extent in their training and interests from those who are officially philosophers or historians, they are not shy about straying into each other's territories. In fact, the scientists who have contributed to this volume devote more time to discussing philosophical issues (such as the nature of laws) than do the philosophers, and conversely it is the philosophers who pay greater attention to the more scientific issues. And this is as it should be. In his contribution, my good friend of many years, Michael Ruse, has accurately diagnosed my general views about the relation between history of science, philosophy of science, and science. We are all engaged in the same activity, only with different though complementary training. Increased interactions between scientists, historians, and philosophers are proving productive for all those concerned, and those of us who concentrate on biology have led the way in this symbiotic relationship.

Thought experiments in science

As several authors have emphasized in this book, concepts exist in a matrix of observation statements and results of experiments, on the one hand, and laws and theories, on the other. Even the most abstruse, theoretical claims in science have some observational consequences, and conversely, even the most observational of our concepts in science are to some extent theory-laden. Neither 'raw' data nor epistemological 'givens' play any role in science. Perhaps we have sense data, but we cannot talk about them. The 'holistic' interdependence of observation statements and theories exacts a cost. It makes simple, neat analyses of both the products of scientific investigation as well as this process itself all but impossible. It is also this strong interdependence that makes thought experiments in science productive while the near absence of such interdependence frustrates a comparable use of imaginary examples in others areas of investigation. As Kitcher remarks, most of us are 'familiar with the dismaying degeneration that characterizes fields in which energy is lavished on counterexamples of no theoretical importance.'

Thought experiments to be productive must always occur in a context. They require a 'normal situation,' the more constrained and detailed the better. Thought experiments help to clarify issues so that future actual experiments can be constructed to distinguish between the most interesting alternatives. However, both actual and thought experiments carry with them a legacy of past experiments and observations. Without such a legacy, thought experiments are of no use in the production of conceptual clarity. Larry Laudan is certainly right about the effect of the rational weight of the past on science. Furthermore, scientists cannot always anticipate which constraints are going to turn out to be relevant. When scientists finally get around to running what had previously been only a thought experiment, they frequently have to go back and rethink their thought experiments because they had not envisaged certain contingencies.

For example, several of the contributors to this volume discuss the species problem. What are species? What criteria should be used to define the species category? Aborigines have their species concepts, religious fundamentalists have their species concept, alpha taxonomists theirs, the pheneticists theirs, phylogenetic cladists theirs, pattern cladists theirs, and so on. In the absence of a sufficiently specified context, too many answers are possible, and there is no way to decide among them. Just referring to species is not enough. We cannot probe our conceptual limits because there are simply too many of them and most are too amorphous. In such informal, multifarious contexts, imaginary examples do nothing but increase confusion by posing questions that we are not prepared to answer.

For example, people think that they can conceive of a centaur, and in a very superficial way, they can, but too many questions remain unanswered, and no means exist for answering them. How many pairs of lungs does a centaur have, how many hearts? How are the circulatory and pulmonary systems of these creatures connected? What happens to the food that has been digested in the human half of centaur? Does it empty into the stomach of the horse half? Of course, we are not supposed to ask such questions of mythical creatures. Legends usually tell us enough of what we need to know about centaurs and unicorns for them to play their mythical roles, but sometimes not. When I first came across stories of centaurs as a child, I wondered where baby centaurs come from because apparently there are no lady centaurs.

Only when a context is specified more narrowly can anything very definite be said about species. Because evolutionary theory is currently undergoing a period of fundamental reevaluation, even specifying species as being the things that evolve leaves numerous alternatives open. For the past couple of centuries, the two most common criteria for defining the species category have been descent via sexual reproduction (genealogy) and some sort of morphological gaps between species. More recently, increased emphasis has been placed on ecological considerations. As long as these various criteria always produced the same groupings, biologists were able to postpone opting for one criterion over the others. This situation would seem to be the ideal place to construct thought experiments or, at least, 'thought observations.' What if organisms existed which appeared in two significantly distinct morphological types but mated freely, always producing offspring of one type or the other? Conversely, what if organisms existed which exhibited no appreciable morphological differences from each other but which never mated successfully even when they had every opportunity to do so?

As it turns out, no such thought experiments are necessary because nature supplies numerous examples of both situations and many more besides. In the absence of such examples, biologists might be forced to invent them, and I am willing to bet that even the most fertile imaginations would not have been able to come up with the bizarre situations that actually occur in nature. These actual examples have additional advantages as well. When questions arise about the particulars of the examples, they can be answered, and these answers introduce issues that would not have been raised by imaginary examples tailored to test only one particular conceptual puzzle. 'Why are these two species of fish considered separate species when they mate freely whenever the occasion arises and produce fertile offspring?' Because all their offspring are female. 'But why then do not these females mate with the males of the parent species?' They do, but the original hybrid females are so much more vigorous than the organisms belonging to the parent species that they out-compete them in the struggle for life. After several generations, no males remain. Eventually all the hybrid females die as well, and extinction results.

Several of the authors in this anthology in discussing the species problem and phylogeny reconstruction refer to those ubiquitous species A, B, and C and utilize simple 'tinker toy' diagrams. Does the call for reliance on real examples preclude such devices? Not in the least, because these authors provide real examples as well, examples that can be investigated in greater detail if necessary. The diagrams and simplified examples are meant to be only illustrative, but as Kitcher warns, such simplification has its dangers: 'it is easy to draw branching diagrams and to canvass possibilities by appealing to them. But it is always worth asking how we link the organisms that the naturalist observes to the branching diagrams.' Such answers necessarily commit the author on a host of issues, issues which in other contexts and at other times might well be brought into question. Real examples allow them to be questioned, but not everything can be questioned at once. For such simplified examples to be of any use, numerous alternatives must be forestalled. The reason that Sober's tinker toy example is so decisive is that he limits himself to a narrow context – cladistic analysis. Cladists do not deny the possibility of anagenic change. They merely decline to divide a gradually evolving lineage into sequential chronospecies. But what if such species are allowed, what if outgroups can be paraphyletic, and so on? Under such circumstances, all bets are off.

In his paper on species concepts, Kitcher presents puzzles for all current attempts to define the species category and finds all proposed definitions seriously deficient. As a result, he urges that we take a 'pluralistic view of species, allowing that there are equally legitimate alternative ways of segmenting lineages – and indeed legitimate ways of dividing organisms into species that do not treat species as historical entities at all.' The problem with Kitcher's analysis is not that he introduces numerous counterexamples of no theoretical importance. His puzzles are all either real examples or else represent a class of real examples. The problem with Kitcher's discussion is that it lacks a sufficiently focused context. The 'current needs of biological research' and the species concept as 'naturalists and theoretical biologists alike use it' are not good enough.

Biologists are currently engaged in numerous, semi-independent and partially conflicting research programs as the papers in this volume amply illustrate. Although members of these research programs refer quite freely to species, it does not follow that they are referring to the same thing or even that they are attempting to converge on a univocal concept. Using the same term to refer to all these various concepts amounts to little more than equivocation. Scientists cannot entertain all alternatives. They must opt. Some biologists have opted for species as lineages. Hence, all sorts of counterexamples cease to be relevant to their species problem. Perhaps these biologists might be led eventually to give up their entire research program, but short of that, certain natural phenomena which would pose puzzles from a different perspective pose no problems for them.

Kitcher is right to expect those engaged in developing a scientifically adequate species concept to provide criteria for making 'principled' decisions, but he cannot expect such criteria to apply across all legitimate research programs currently underway in evolutionary biology, let alone all of biology. One price that we pay for scientific advance is that sometimes we must give up some of our preanalytic intuitions. Perhaps many taxospecies are not going to count as evolutionary species, perhaps organisms that reproduce asexually do not form species, perhaps some organism that reproduce sexually belong to no species whatsoever. These consequences of a particular theoretical perspective are 'puzzles' only from other theoretical perspectives or from a more global perspective, usually the hodgepodge of ill-formed, half-articulated, poorly understood conceptions of common sense and ordinary discourse. There is nothing so ill-conceived or poorly understood that someone or other has not been 'inclined' to say it.

The main message of the preceding discussion is that real counterexamples pose such serious problems for any species concept that one need not resort to imaginary thought experiments, and that the introduction of such imaginary examples are likely only to confuse the issue. Once a species concept has been devised that can accommodate all the familiar situations that occur in nature, it is time enough to introduce bug-eyed monsters. Furthermore, a species concept will be judged adequate, not from some global perspective but in the context of a particular scientific theory, e.g., some version of evolutionary theory. The puzzles that remain will be as numerous as they are irrelevant. Of course, in a different context, *some* of these puzzles may well turn out to be relevant, and one context can supplant another. If evolutionary theory or some versions of evolutionary theory are abandoned, then which examples are relevant and which irrelevant change as well. Conversely, empirical discoveries, such as the genetic heterogeneity that characterizes most species, feed back into our concepts. This feedback is only one reason why deciding empirical truth is so difficult.

Thought experiments in philosophy of science

All of the preceding has concerned the role of real versus imaginary examples in science, but the use of imaginary examples is even more prevalent in philosophical contexts, e.g., grue emeralds, clear, tasteless fluids with a chemical make-up XYZ, rabbits named 'Snow White,' and what have you. Analytic philosophers have shown especially strong partiality to imaginary examples, the sillier the better. One virtue of such examples is that they do not presuppose any technical knowledge. As a result, they are easy to conjure up and can be set out in very few words. Real examples, especially those drawn from science, require extensive knowledge and take up much more space. Goodman (1954: xix-xx) justifies his own reliance on 'commonplace and even trivial illustrations rather than more intriguing ones drawn from science' because the examples that 'attract the least attention to themselves are the least likely to divert attention from the problem or principles being explained. Once the reader has grasped a point he can make his own consequential applications. Thus although I talk of freezing radiators and the color of marbles, which are seldom discussed in books on chemistry and physics, what I am saying falls squarely within the philosophy of science.'

Sometimes philosophers of science have obscured the points they are making by illustrating them with overly technical examples. Simpler examples would do as well. In fact, sometimes the technical discussion serves to obscure the fact that the author has no philosophical point to illustrate. However, regardless of my own prejudices on the matter, one thing is clear: the commonplace, trivial examples that philosophers introduce to illustrate their principles have attracted massive attention to themselves. It is difficult to imagine how more attention could have been paid to a technical example than has been lavished on Goodman's infamous grue emeralds or Putnam's brains in a vat.

Goodman's justification for reliance on commonplace and even trivial examples depends on their being used to *illustrate* preestablished principles. But the examples used by analytic philosophers function as more than illustrations. One sort of exposition that used to be quite common in the philosophical literature was a kind of conceptual striptease. A concept is introduced (e.g., law of nature) and a general analysis is suggested. Successions of counterexamples and modifications are then introduced until either an adequate analysis is reached or else the original concept degenerates into a conceptual morass. In a single paper, the first alternative tends to prevail. However, as successive authors address the issue, the second alternative begins to predominate. One might think that a significant difference exists between the claim that all the coins in my pocket are American and Newton's law of universal gravitation, but so it seems that belief is sorely mistaken. All is one – or nothing – which amounts to the same thing.

Some of the examples that philosophers introduce into their conceptual deconstructions may serve as nothing more than illustrations, but others also function as justifications. For example, a common argument in the philosophical literature is that not all effects are functions because the sounds that hearts make as they pump blood are effects and not functions. This example and others like it not only illustrate a point but also serve to justify it. I find the basis for such justifications problematic. Because the examples are commonplace, the only foundation that seems warranted is that provided by common sense and ordinary experience, and in my book both are extremely untrustworthy. I much prefer grounding philosophical analysis of science in science, not common sense, and such grounding requires technical examples.

In science, examples function as more than illustrations; they also serve as evidence, and one thing can be said for certain about silly, science fiction examples: they cannot serve as evidence. Of course, if nothing like evidence can be brought to bear on philosophical claims, this characteristic of imaginary examples hardly counts against them. Years ago, when I was even more naive than I am now, I thought that claims about the reducibility of Mendelian genetics to molecular biology had something to do with what was going on in biology at the time (e.g., the genetic code, molecular mechanisms, reaction norms) and argued that *in point of fact* a Nagel-type reduction from Mendelian genetics to molecular biology was impossible (Hull 1972, 1973c, 1974b). Others responded by modifying and expanding upon Nagel-type reductions, attempting to show that even in the face of the empirical facts of the matter, reduction was possible. Such modifications were perfectly legitimate. They are exactly the sorts of modifications that data are supposed to elicit. If biologists can legitimately modify their gene concepts as they find out more about the genetic material, then philosophers can certainly modify their concept of reduction in the face of increased knowledge about putative cases of reduction. In this anthology, Rosenberg carries this process even further, showing the implications of more recent advances in molecular biology for any philosophical notion of reduction. As Rosenberg sees it, 'smooth' reduction will not go through in biology because selection for function is blind to structure: '*it cannot discriminate between differing structures with identical effects.*'

In science data leads scientists to modify their theories, but if we accept the thesis that even the most observational of statements are to some extent theory-laden, then no such thing as 'raw data' exist. The theories that we entertain influence how we describe natural phenomena. Parallel observations should hold for philosophical theories about science. As Lakatos (1970) has argued persuasively, no one can write history as it was without any philosophical presuppositions. Inductivism with respect to the course of science is even less justified than inductivism with respect to the genesis of Kepler's laws. If scientists cannot merely 'sum up observations,' then historians should be unable to write histories of science that record the facts and nothing but the facts.

Several of the papers in this anthology address these parallel problems in especially pointed ways. For example, Cracraft argues for a particular definition of species which combines ontological and epistemological considerations. Strange as it might sound when put so bluntly, species are observable individuals. They are observable in the sense that they are the 'smallest clusters of individual organisms sharing diagnostic character variation.' The characters in question are properties of organisms. Of course, a human being cannot literally observe character variation in species containing millions of organisms spread over thousands of square miles, let alone millions of years. A systematist observes these characters in a small percentage of organisms and *infers* patterns of variation. Cracraft also insists that species exhibit reproductive cohesion. They must be self-perpetuating. Cracraft takes this characteristic of species to be so basic that he does not bother even to argue it. One consequence of this assumption is that males, females, juvenile stages and various morphs are included in the same species even if they lack the requisite diagnostic characters.

Although Cracraft acknowledges that the contrast between species-as-observables and species as theoretical entities is far from sharp, he opts for species as belonging nearer the observable end of the continuum. At the very least, any appropriate definition of the species category should not presuppose anything about the evolutionary process. Reproduction and ontogeny are acceptable; evolution is not. Perhaps a species category can be formulated that is not laden with evolutionary theory, but it is nonetheless as theory-laden as a term can get. The main motivation for allowing certain theories to affect one's definitions but not others seems to be epistemological. For example, rarely can human beings hope to observe species evolving, speciating, and going extinct. For extant species, we can observe mating, replication, and ontogenetic development at the organismic level. We *can* observe such events at the organismic level, but in point of fact we rarely *do* – even for extant species. As Cracraft notes, most systematists study character distributions and that is all. There are simply too many species and too few systematists to do more. Species are going extinct faster than we can identify and classify them. As far as extinct forms are concerned, both organismic and species level characteristics are in the same boat. They cannot be observed.

This much being said, why the heavy emphasis on the part of scientists such as Cracraft on observability? I suspect most philosophers would argue that theoretical definitions are being confused with criteria of application, sometimes termed 'operational definitions.' Both are necessary, but criteria for application should be appended to definitions, not included as part of them. On the main, philosophers have not concerned themselves with the complex set of problems surrounding the operationalization of theoretical definitions. For the most part, we are content to observe that theoretical terms cannot be defined operationally and let it go at that. But if we are genuinely interested in science, we are mistaken to pay so little attention to matters that consume so much of scientists' time and which they take to be so important. Theoretical definitions combined with lists of techniques for operationalizing them are not good enough. Scientists feel compelled to include operational criteria in their definitions. If data is relevant to the observations that philosophers make about science, then the insistence of so many scientists on retaining certain practices that philosophers take to be so clearly mistaken cannot be dismissed lightly. If scientists were unaware of what philosophers have to say on the subject, we might conclude such mistakes are merely a function of scientists being unaware of the error of their ways. But as Cracraft clearly shows in his paper, he has read the relevant philosophical literature and still he persists. And Cracraft is not alone in the tenacity with which he resists positions widely accepted in certain philosophical quarters.

Thus, those of us who think that data about science are relevant to philosophical views about science are caught in a bind. We cannot very well use the concordance between scientific practice and our general analyses as support for our views while insisting that departures do not count against us. One plausible solution is that scientists who practice what we preach are, on the average, more successful than those who do not. Problems about what counts as success in science to one side, thus far no one has undertaken such an extensive survey. However, if such a massive effort is to be undertaken, those of us who set out general analyses had better be prepared to have the results make a difference. Perhaps we do not have to take even successful scientific practice at face value, but it must count for something. We cannot dismiss all potential disconfirming evidence in advance.[1]

As troublesome as the preceding problem may appear for meta-science, it is only a repetition of the interplay between theory and observation in science. Inductivist scientists want to keep observation reports totally free of the theories that these reports are going to be used to test; otherwise they fear that they will be caught up in circular reasoning. Philosophers are quick to point out that the sort of 'reciprocal illumination' that results from the interplay between observation and theory construction need not lead to vicious circularity, but we do not seem to appreciate how disorienting this procedure can get in particular cases. For example, if a systematist distinguishes characters in such a way that they form transformation series of primitive and derived characters, then he or she will produce nested sets of characters. The fact that these characters do nest so neatly thereby indicates that they were individuated correctly to begin with. Any characters that do not fit are not really characters. If such reasoning is legitimate in science, then it should be equally legitimate when applied to science itself, uncertainty about particular cases notwithstanding.

One reason why the inferential interplay between theory and observation is so unsettling is the fear that the theoretical component of observations statements will be so strong that we will never be able to discover when our theoretical views are mistaken. Although we can see in retrospect that in some cases our allegiance to particular theories has frustrated attempts to falsify these theories, in other cases scientists strongly committed to particular views have been forced to reject or modify them in spite of their allegiance. Observation statements are theory-laden but not so laden that they cannot refute the very theories which color them. Parallel observations hold for incommensurability. In the context of a particular semantic theory, observation statements derived from different scientific theories cannot contradict each other because certain key terms in them do not mean precisely the same thing. However, when this semantic thesis is operationalized to refer to actual scientists, it turns out to be false. Adherents of two different theories should not be able to agree on observations that would refute one theory and support the other, but they can and do. As different as William Thomson's thermodynamic theory was from Darwin's theory of evolution, the implications of these theories for the age of the earth were so different that no one could avoid the conclusion that at least one of these theories had to be seriously in error.

As is the case with so many historians, Hodge is unhappy with the use that philosophers have made of his favorite scientist. On the basis of a lifetime of philosophically informed historical research, Hodge has worked out the structure and strategy of Darwin's argumentation, not just in the *Origin of Species* but in related works as well. Philosopher after philosopher has attempted to support his or her philosophical thesis by appealing to Darwin. Hodge responds that, damn it, you have gotten Darwin wrong. One possible response to Hodge's complaint is that no description of the course of science can be totally free of general philosophical views about science. The reason that Hodge thinks that these philosophers are mistaken about the structure and strategy exhibited in Darwin's work is that he has reconstructed it on the basis of inappropriate philosophical or meta-scientific presuppositions.

This response is surely more appropriate for some historians than for others. It might be the case that the differences in general philosophical outlook between Hodge and those who want to use Darwin to illustrate (not to mention support) their general philosophical views are so subtle that the facts of the matter cannot possibly resolve these differences. Conversely the facts of the matter might be so apparent that they can be decisive. Although our meta-theories color our descriptions no less than do the more pedestrian theories constructed by scientists, sometimes even these highly theory-laden descriptions can refute the theories that laden them. For example, even if Mendelian genetics and molecular biology are reconstructed to fulfill the needs of Nagel-type reductions, the necessary correlations between Mendelian and molecular natural kinds are not forthcoming. Perhaps such a state of affairs is impossible, but it does occur. Hodge can be read as making just such a claim for Darwin and the philosophical theses that Darwin's work is supposed to exemplify. For any philosopher who claims that at least sometimes data about science can be brought to bear on at least some of the philosophical theses which philosophers of science set out about science complaints by historians such as Hodge cannot be dismissed lightly.

Conclusion

In this paper I have questioned the independence of two traditional philosophical dichotomies: between truth and meaning and between observation statements and theories. Issues of truth can cause us to change what we mean by what we say just as surely as the meanings that we assign to the terms we use influence what we take to be the case. One might be tempted to dismiss such holistic interdependence as being characteristic only of the ongoing process of knowledge acquisition. As in the case of government in the perfect Marxist state, this interdependence will wither away once total knowledge is reached. Because I think that we will never reach this ideal epistemic state, I am modest enough in my ambitions to limit my observations to the only state of affairs that will ever obtain.

Even though I think that some philosophical theses can be interpreted in such a way that data can be brought to bear on them, others are liable to be more than a little recalcitrant. For example, although I think that Rosenberg shows the clear implications that the current state of genetics and molecular biology have for Nagel-type reductions, I doubt that anything that might count as data or evidence could possibly influence the debate over instrumentalism and realism. These positions have become way too sophisticated. No matter what course the history of science takes, all sides will be able to claim victory. In any case, one promising avenue of research for those of us who are interested in science is to look for general views about science that can be tested and to test them. For example, how important are various sorts of influences on the decisions that scientists make? Do such 'external' factors as the social structure of scientific communities or society at large influence the course of science? Grene fears that such concerns might lead to the abandonment of an inadequate philosophy for none at all. I think not, but we will have to wait and see.

Kitcher expects scientists to present principled ways for drawing the distinctions that they take to be central to their areas of investigation. Although scientists might have to change these principles as time marches on, *ad hoc*, case-by-case decisions are not good enough. I agree, but these principles are not going to apply across all contexts, even across all legitimate contexts. Moving to the meta-level, philosophers should also be expected to present principled ways to draw their distinctions as well. For example, Kitcher is not a promiscuous pluralist. He argues that certain species concepts are illegitimate, in particular the species concepts urged by creationists and pheneticists. What is needed is a principled way to distinguish between legitimate and illegitimate scientific concepts. I suspect that whatever principles Kitcher devises, they will not apply across all philosophical contexts. Just as scientists must opt, so too must those of us who study science. One reason for picking one avenue of investigation over another is recent success. How much progress have phenomenologists made in their goal of reconstructing science on an adequate phenomenological foundation? As far as I can tell, not much. Have more analytically-inclined philosophers made much progress in their quite different goals? I am certainly biased in this respect, but as far as I can tell, quite a bit, and those of us who have concentrated on biology have done more than our share in contributing to this success.

Notes

1. The uneasiness that philosophers are sure to exhibit toward the possible implications of statistical studies of science for their philosophical analyses has been expressed recently by psychotherapists when the government proposed to study the efficacy of various sorts of therapies. If government money is used to pay for such therapy, then

some evidence must be provided that it does some good on some definition of 'good.' Psychotherapists responded that the results of such studies, no matter how they are conducted, cannot possibly indicate anything about the efficacy of psychotherapy. I suspect that this reflex action is more a function of the insecurity that psychotherapists feel about the efficacy of their therapies than problems in research design.

Publications of David L. Hull

Born:	June 15, 1935, Burnside, Illinois.
Educated:	Illinois Wesleyan University, AB 1960. Indiana University (Department of History and Philosophy of Science). PhD 1964.
Taught:	University of Wisconsin – Milwaukee 1964–84. Northwestern University, 1984–
Honors:	Guggenheim Fellow, 1980–81. Fellow of the American Association for the Advancement of Science, 1985.
Presidencies:	Philosophy of Science Association, 1985–86. Society of Systematic Zoology, 1984–85.

Books

1973 *Darwin and His Critics: The Reception of Darwin's Theory of Evolution by the Scientific Community*. Harvard University Press: Cambridge MA, 473 pp.; reprinted 1983, The University of Chicago Press: Chicago.

1974 *Philosophy of Biological Science*. Prentice-Hall: Englewood Cliffs NJ, 148 pp.

1988 *Science as a Process: An Evolutionary Account of the Social and Conceptual Development of Science*. Chicago: University of Chicago Press.

Papers

1964 Consistency and monophyly, *Systematic Zoology* 13: 1–11.

1965 The Effects of Essentialism on Taxonomy: Two Thousand Years of Stasis, *The British Journal for the Philosophy of Science* 15: 314–326 & 16: 1–18; reprinted in *Concepts of Species*, C.N. Slobodchikoff (ed.), University of California Press: Berkeley, 1976, pp. 46–77.

1966 Phylogenetic Numericlature, *Systematic Zoology* 15: 14–17.

1966 The Logical Structure of the Linnean Hierarchy, *Systematic Zoology* 15: 97–111, with Roger Buck.

1967 Certainty and Circularity in Evolutionary Taxonomy, *Evolution* 21: 174–189.

1968 The Conflict between Spontaneous Generation and Aristotle's Metaphysics, *Proceedings of the Seventh Inter-American Congress of Philosophy*, Les Presses de l'Université Laval: Quebec, Canada, 2: 245–250.

1968 The Operational Imperative: Sense and Nonsense in Operationism, *Systematic Zoology* 17: 438–457.

1969 What Philosophy of Biology is Not, *The Journal of the History of Biology* 2: 241–268; and *Synthese* 20: 157–184.

1970 Systemic Dynamic Social Theory, *The Sociological Quarterly* 11: 351–363.

1970 Contemporary Systematic Philosophies, *Annual Review of Ecology and Systematics* 1: 19–54; reprinted in *Topics in the Philosophy of Biology*, M. Grene and E. Mendelsohn (eds), D. Reidel: Dordrecht-Holland, 1976, pp. 396–440, and *Conceptual Issues in Evolutionary Biology*, E. Sober (ed.), MIT Press: Cambridge MA., 1984, pp. 567–602.

1972 Darwin and Nineteenth Century Philosophies of Science, *The Foundations of*

Scientific Method: The Nineteenth Century, R. Giere and R. Westfall (eds), Indiana University Press: Bloomington IN, pp. 115–132.

1972 Reduction in Genetics – Biology or Philosophy? *Philosophy of Science* 39: 491–499.

1973 Reduction in Genetics – Doing the Impossible, *Proceedings of the IVth International Congress of Logic, Methodology and Philosophy of Science*, P. Suppes (ed.), North-Holland Publishing Company, pp. 619–635.

1974 Darwinism and Historiography, *The Comparative Reception of Darwinism*, T. Glick (ed.), University of Texas Press: Austin TX, pp. 388–402; reprinted by the University of Chicago Press, 1988.

1975 Central Subjects and Historical Narratives, *History and Theory* 14: 253–274.

1976 Are Species Really Individuals? *Systematic Zoology* 25: 174–191.

1976 Informal Aspects of Theory Reduction, *Boston Studies in the Philosophy of Science: PSA 1974*, R.S. Cohen and A. Michalos (eds), D. Reidel: Dordrecht-Holland, pp. 653–670; reprinted in *Conceptual Issues in Evolutionary biology*, E. Sober (ed.), MIT Press: Cambridge MA, pp. 462–476.

1977 The Ontological Status of Species as Evolutionary Units, *Foundational Problems in the Species Sciences*, R. Butts and J. Hintikka (eds), D. Reidel Publishing Company: Dordrecht-Holland, pp. 91–102.

1978 Génétique et réductionisme, *La Recherche* 9: 220–227.

1978 Altruism in Science: A Sociobiological Model of Cooperative Behaviour among Scientists, *Animal Behaviour* 26: 658–697.

1978 A Matter of Individuality, *Philosophy of Science* 45: 335–360; reprinted in *Conceptual Issues in Evolutionary Biology*, E. Sober (ed.), MIT Press: Cambridge MA, pp. 623–645.

1978 Planck's Principle, *Science* 202: 717–723.

1978 Sociobiology: Scientific Bandwagon or Traveling Medicine Show? *Society* 15: 50–59; also in *Sociobiology and Human Nature*, M.S. Gregory, A. Silvers and D. Sutch (eds), 1979, Jossey-Bass, Inc.: San Francisco, pp. 136–163.

1978 The Sociobiology of Sociobiology, *New Scientist* 79: 862–856; an unabridged version reprinted as, Sociobiology: Another New Synthesis, in *Sociobiology: Beyond Nature-Nurture?* W. Barlow and J. Silverberg (eds), 1980, Westview Press: Boulder CO, pp. 77–96.

1979 In Defense of Presentism, *History and Theory* 18: 1–15.

1979 The Limits of Cladism, *Systematic Zoology* 28: 414–438.

1979 Philosophy of Biology, *Current Researches in Philosophy of Science*, P.D. Asquith and H.E. Kyburg (eds), Philosophy of Science Association: East Lansing MI, pp. 421–435.

1979 Teaching at a State University, *Teaching Philosophy* 2: 119–121.

1980 Individuality and Selection, *Annual Review of Ecology and Systematics* 11: 311–332.

1981 Units of Evolution: A Metaphysical Essay, *The Philosophy of Evolution*, U.J. Jensen and R. Harré (eds), The Harvester Press: Brighton, England pp. 23–44; reprinted in *Genes, Organisms, Populations*, R.N. Brandon and R.M. Burian (eds), 1984, The MIT Press: Cambridge MA, pp. 142–160.

1981 Historical Narratives and Integrating Explanation, *Pragmatism and Purpose: Essays Presented to Thomas Goudge*, L.W. Summer, J.G. Slater, and F. Wilson (eds), University of Toronto Press: Toronto, Canada, pp. 172–188.

1981 Reduction in Genetics, *The Journal of Medicine and Biology* 6: 125–143; the first half of paper in Fløistad (1982).

1981 The Herd as Means, *PSA 1980*, Vol. 2, P.D. Asquith and R.N. Giere (eds), Philosophy of Science Association: East Lansing MI, pp. 73–92.

1981 The Principles of Biological Classification: The Use and Abuse of Philosophy, *PSA 1978*, P.D. Asquith and I. Hacking (eds), Philosophy of Science Association: East Lansing MI, pp. 130–153.

1982 The Naked Meme, *Learning, Development, and Culture*, H.C. Plotkin (ed.), John Wiley: London, pp. 273–317.

1982 Biology and Philosophy, *Contemporary Philosophy: A New Survey*, G. Fløistad (ed.), Martinus Nijhoff: The Hague, pp. 281–316.

1983 Darwin and the Nature of Science, *Evolution from Molecules to Men*, D.S. Bendall (ed.), Cambridge University Press: Cambridge, pp. 63–80.

1983 Popper and Plato's Metaphor, *Advances in Cladistics*, Vol. 2, N. Platnick and V. Funk (eds), Columbia University Press: New York, pp. 177–189.

1983 Thirty-One Years of Systematic Zoology, *Systematic Zoology* 32: 315–342.

1983 Exemplars and Scientific Change, *PSA 1982*, Vol. 2, P.D. Asquith and T. Nickles (eds), Philosophy of Science Association: East Lansing MI, pp. 479–503.

1984 Cladistic Theory: Hypotheses that Blur and Grow. In *Cladistic Perspectives on the Reconstruction of Evolutionary History*, T. Duncan and T. Stuessy (eds), New York: Columbia University Press, pp. 5–23.

1984 Lamarck among the Anglos. Introduction to Lamarck's *Zoological Philosophy*, Chicago: University of Chicago Press, pp. xi–lxvi.

1984 Historical Entities and Historical Narratives. *Minds, Machines and Evolution*, C. Hookway (ed.), Cambridge: Cambridge University Press, p. 17–42.

1985 Bias and Commitment in Science: Phenetics and Cladistics. *Annals of Science* 42: 319–338.

1985 Darwinism as a Historical Entity: A Historiographic Proposal. *The Darwinian Heritage*, D. Kohn (ed.), Princeton: Princeton University Press, pp. 773–812.

1985 Openness and Secrecy in Science: Their Origins and Limitations. *Science Technology, and Human Values* 10: 4–13.

1985 Linné as an Aristotelian. *Contemporary Perspectives on Linnaeus*, J. Weinstock (ed.), Lanham, MD: University Press of America, pp. 37–54.

1986 Conceptual Evolution and the Eye of the Octopus. *Logic, Methodology and Philosophy of Science*, Vol. VII, R.B. Marcus, G.J.W. Dorn and P. Weingartner (eds), Amsterdam: North-Holland, pp. 643–665.

1986 Les fondements épistémologiques de la classification biological, *L'Ordre et la Diversité du Vivant*, Nouvelle Encyclopédie des Sciences et des Techniques, Ourage Coordonné par Pascal Tassy, Fayard: Fondation Diderot, pp. 161–203.

1987 Genealogical Actors in Ecological Plays, *Biology & Philosophy* 1: 44–60.

Notes and replies

1967 Definitions of Taxa: A Reply, *Systematic Zoology* 16: 349, with Roger Buck.

1968 The Syntax of Numericlature, *Systematic Zoology* 17: 472–474.

1969 The Natural System and the Species Problem, *Systematic Biology*, C.G. Sibly (ed.), Publication 1692, National Academy of Science: Washington, D.C., pp. 56–61.

1969 Reply to Gregg, *Systematic Zoology* 18: 354–357, with Roger Buck.

1969 Contemporary Logic and Evolutionary Taxonomy: A Reply to Gregg, *Systematic Zoology* 18: 347–354, with Paul Snyder.

1969 Reply to Randal and Scott, *Systematic Zoology* 18: 468–469.

1970 Morphospecies and Biospecies: A Reply to Ruse, *The British Journal for the Philosophy of Science* 21: 280–282.

1973 A Belated Reply to Grüner, *Mind* 72: 437–438.

1973 Contemporary Systematic Philosophies: Introduction, *Systematic Zoology* 22: 337.

1974 Are the 'Members' of Biological Species 'Similar'? *The British Journal for the Philosophy of Science* 25: 332–334.

1977 History, Philosophy and Sociology of Science, *Forum 1977*, University of Wisconsin-Milwaukee, Milwaukee WI, pp. 19–23.

1979 Discussion: Reduction in Genetics, *Philosophy of Science* 46: 316–320.

1979 Universality and Species Specificity, *The Behavioral and Brain Sciences* 2: 38–39.

1979 Philosophical Issues in Systematics, *Systematic Zoology* 28: 520.

1981 Discussion: Kitts and Kitts and Caplan on Species, *Philosophy of Science* 48:

141–152.
1981 The Essence of Sociobiology, *The Behavioral and Brain Sciences* 4: 342–343.
1983 Comments on Beatty, *Nature Animated* M. Ruse (ed.), D. Reidel: Dordrecht-Holland, pp. 101–108.
1984 Can Kripke Alone Save Essentialism? *Systematic Zoology* 33: 110–112.

Book reviews

1969 Giovanni Blandino, *Theories of the Nature of Life* (1969), *Systematic Zoology* 18: 238–240.
1969 Michael Ghiselin, *The Triumph of the Darwinian Method* (1969), *Systematic Zoology* 18: 447– 450.
1972 William B. Provine, *The Origin of Theoretical Population Genetics* (1971), *Systematic Zoology* 21: 132–134.
1973 Stephen Toulmin, *Human Understanding* (1972), *Science* 182: 1121–1124.
1973 Michael Simon, *The Matter of Life* (1971), *Second Order* 2: 100–108.
1974 Michael Ghiselin, *The Economy of Nature and the Evolution of Sex* (1974), *Systematic Zoology* 23: 560–562.
1975 Carl Hempel, *Philosophy of Natural Science* (1965), Thomas Kuhn, *The Structure of Scientific Revolutions* (1970), Dudley Shapere, *Galileo: A Philosophical Study* (1974), *Systematic Zoology* 24: 394–401.
1975 F.J. Ayala and T. Dobzhansky (eds), *Studies in the Philosophy of Science* (1974), *Nature* 257: 429
1976 W.W. Wagar, *Good Tidings: The Belief in Progress from Darwin to Marcuse* (1972) *Theory and Society* 3: 146–149.
1976 R. Harré (ed.), *Problems of Scientific Revolution: Progress and Obstacles to Progress in the Sciences* (1975), *Philosophy of the Social Sciences* 6: 375–380.
1976 H.H. Krebs and J. Shelley (eds), *The Creative Process in Science and Medicine* (1975), *The Quarterly Review of Biology* 51: 290–291.
1976 René Dubois, *Beast or Angel? Choices that Make Us Human* (1974), *The Bulletin of the Atomic Scientists* 32: 46–47.
1976 G.E. Russett, *Darwin in America: The Intellectual Response* (1976), *Nature* 265: 144.
1976 Peter Singer, *Animal Liberation* (1975), *Science* 192: 679–680.
1977 Michael Ruse, *Philosophy of Biology* (1973), *The British Journal for the Philosophy of Science* 28: 181–194.
1977 J. Burchfield, *Lord Kelvin and the Age of the Earth* (1975), *Systematic Zoology* 26: 99–100.
1977 P.H. Barrett (ed.), *The Collected Papers of Charles Darwin* (1977), *Philosophy of Science* 44: 662–663.
1978 M.A. Finocchiaro, *History of Science as Explanation* (1973), *Clio* 7: 335–338.
1978 G. Holton, *The Scientific Imagination: Case Studies* (1978), *American Scientist* 66: 638.
1978 W. Coleman and C. Limoges (eds), *Studies in History of Biology*, Vol. 1 (1977), *The Quarterly Review of Biology* 53: 45.
1978 R.W. Burkhardt, Jr. *The Spirit of System* (1977), *Systematic Zoology* 27: 248–250.
1978 G. Hardin, *The Limits of Altruism* (1977), *The Bulletin of the Atomic Scientists* 34: 53–54.
1978 T.H. Clutton-Brock and P.H. Harvey (eds), *Readings in Sociobiology* (1978), *The Quarterly Review of Biology* 53: 441.
1979 L. Laudan, *Progress and its Problems* (1977), *Philosophy of the Social Sciences* 9: 457–465.
1979 W. Coleman and C. Limoges (eds), *Studies in History of Biology*, Vol. II (1978), *The Quarterly Review of Biology* 54: 169–170.
1979 J.D. Paradis, *T.H. Huxley: Man's Place in Nature* (1978), *Victorian Studies* 54: 66.

1979 R.S. Westfall, *The Construction of Modern Science* (1971, 1979), *The Quarterly Review of Biology* 54: 66.

1979 A. Flew, *A Rational Animal and Other Philosophical Essays on the Nature of Man* (1978), *Isis* 70: 278.

1979 W. Coleman, *Biology in the Nineteenth Century: Problems of Form, Function, and Transformation* (1971, 1977), *The Quarterly Review of Biology* 54: 67–68.

1980 M. Ruse, *The Darwinian Revolution: Science Red in Tooth and Claw* (1979), *Nature* 284: 670–671.

1980 E.O. Wilson, *On Human Nature* (1979), *Environmental Ethics* 71: 656–657.

1980 E. Nagel, *Teleology Revisited and Other Essays in the Philosophy and History of Science* (1979), *Isis* 71: 656–657.

1980 J. Gaston, *The Reward System in British and American Science* (1978), *Philosophy of Science* 47: 160–161.

1981 J. Cracraft and N. Eldredge (eds), *Phylogenetic Analysis and Paleontology* (1979), *Paliobiology* 6: 131–136.

1980 *Synthese* (1980), Vol. 43, No's 1 & 2, *Systematic Zoology* 29: 408–412.

1980 W. Coleman and C. Limoges (eds), *Studies in History of Biology* Vol. III (1979), *The Quarterly Review of Biology* 55: 170–171.

1981 M. Midgley, *Beast and Man: The Roots of Human Nature* (1978), *Philosophical Review* 90: 307–310.

1981 H.F. Judson, *The Search for Solutions* (1980), *The Quarterly Review of Biology* 56: 322.

1982 J. Greene, *Science, Ideology, and World View* (1981), *The Quarterly Review of Biology* 57: 171–172.

1982 A. Kelly, *The Descent of Darwin* (1981) and D.F. Bratchell, *The Impact of Darwin* (1981), *4S Newsletter* 7: 59–62.

1982 D. Ospovot, *The Development of Darwin's Theory* (1981), *Nature* 259: 719.

1983 J. Browne, *The Secular Ark* (1983), *Nature* 303: 551.

1983 P.J. Bowler, *The Eclipse of Darwin* (1983), *Nature* 306: 174–175.

1983 P. Kitcher *Abusing Science* (1982), *The Quarterly Review of Biology* 306: 174–175.

1983 J.G. Murphy *Evolution, Morality and the Meaning of Life* (1982), *The Quarterly Review of Biology* 58: 407–408.

1983 M. Ruse *Darwinism Defended* (1981), *Isis* 74: 106–107.

1984 M. Grene (ed.), *Dimensions of Darwinism* (1981), *Science* 223: 923–924.

1984 E.O. Wilson, *Biophilia* (1984), *Nature* 312: 205.

1984 J.L. Brooks, *Just Before the Origin* (1984) and H. Clements, *Alfred Russel Wallace* (1983), *Nature* 308: 798–799.

1984 A. Desmond, *Archetypes and Ancestors* (1982), *Paleobiology* 10: 384–388.

1985 R.W. Clark, *The Survival of Charles Darwin* (1984), *Nature* 314: 679–680.

1985 F. Burkhardt and S. Smith (eds), *The Correspondence of Charles Darwin*, vol. I. (1985), *The Times Higher Education Supplement*, no. 661, p. 19.

1986 R. Levins and R. Lewontin, *The Dialectical Biologist* (1985), *Nature* 320: 23–24.

1986 L.R. Godfrey, *What Darwin Began* (1985), *American Scientist* 74: 316.

1986 P. Kitcher, *Vaulting Ambition* (1985), *Isis* 77: 356–357.

1986 D. Depew and B. Weber (eds), *Evolution at a Crossroads* (1985), *Isis* 77: 128–129.

1986 A. Rosenberg, *The Structure of Biological Science* (1985), *Cladistics* 2: 100–104.

1986 E. Gellner, *Relativism and the Social Sciences* (1985), *Ethics* 96: 898–899.

1986 A. Rosenberg, *The Structure of Biological Science* (1985), *Ethics* 96: 899.

1986 D. Raup, *The Nemesis Affair* (1986), *Field Museum of Natural History Bulletin* 57: 6–8.

1987 S.M. Friedman, S. Dunwoody, and C.L. Rogers (eds), *Scientists and Journalists* (1986), *Science* 235: 93–94.

1987 H. Himsworth, *Scientific Knowledge and Philosophic Thought* (1986), *History and Philosophy of the Life Sciences* 9: 384–385.

1987 R. Dawkins, *The Blind Watchmaker* (1986), *Quarterly Review of Biology* 6: 289–292.

Authors' Index

Subject Index

Nijhoff International Philosophy Series

1. Rotenstreich, N.: Philosophy, History and Politics. Studies in Contemporary English Philosophy of History. 1976. ISBN 90-247-1743-4.
2. Srzednicki, J.T.J.: Elements of Social and Political Philosophy. 1976. ISBN 90-247-1744-2.
3. Tatarkiewicz, W.: Analysis of Happiness. 1976. ISBN 90-247-1807-4.
4. Twardowski, K.: On the Content and Object of Presentations. A Psychological Investigation Translated and with an Introduction by R. Grossman. 1977. ISBN 90-247-1926-7.
5. Tatarkiewicz, W.: A History of Six Ideas. An Essay in Aesthetics. 1980. ISBN 90-247-2233-0.
6. Noonan, H.W.: Objects and Identity. An Examination of the Relative Identity Thesis and Its Consequences. 1980. ISBN 90-247-2292-6.
7. Crocker, L.: Positive Liberty. An Essay in Normative Political Philosophy. 1980. ISBN 90-247-2291-8.
8. Brentano, F.: The Theory of Categories. 1981. ISBN 90-247-2302-7.
9. Marciszewski, W. (ed.): Dictionary of Logic as Applied in the Study of Language. Concepts, Methods, Theories. 1981. ISBN 90-247-2123-7.
10. Ruzsa, I.: Modal Logic with Descriptions. 1981. ISBN 90-247-2473-2.
11. Hoffman, P.: The Anatomy of Idealism. Passivity and Activity in Kant, Hegel and Marx. 1982. ISBN 90-247-2708-1.
12. Gram, M.S.: Direct Realism. A Study of Perception. 1983. ISBN 90-247-2870-3.
13. Srzednicki, J.T.J., Rickey, V.F. and Czelakowski, J. (eds.): Leśniewski's Systems. Ontology and Mereology. 1984. ISBN 90-247-2879-7.
14. Smith, J.W.: Reductionism and Cultural Being. A Philosophical Critique of Sociobiological Reductionism and Physicalist Scientific Unificationism. 1984. ISBN 90-247-2884-3.
15. Zumbach, C.: The Transcendent Science. Kant's Conception of Biological Methodology. 1984. ISBN 90-247-2904-1.
16. Notturno, M.A.: Objectivity, Rationality and the Third Realm. Justification and the Grounds of Psychologism. A Study of Frege and Popper. 1985. ISBN 90-247-2956-4.
17. Dilman, I. (ed.): Philosophy and Life. Essays on John Wisdom. 1984. ISBN 90-247-2996-3.
18. Russell, J.J.: Analysis and Dialectic. Studies in the Logic of Foundation Problems. 1984. ISBN 90-247-2990-4.
19. Currie, G. and Musgrave, A. (eds.): Popper and the Human Sciences. 1985. ISBN 90-247-2998-X.
20. Broad, C.D.: Ethics. Edited by C. Lewy. 1985. ISBN 90-247-3088-0.
21. Seargent, D.A.J.: Plurality and Continuity. An Essay in G.F. Stout's Theory of Universals. 1985. ISBN 90-247-3185-2.
22. Atwell, J.E.: Ends and Principles in Kant's Moral Thought. 1986. ISBN 90-247-3167-4.
23. Agassi, J. and Jarvie, I.Ch. (eds.): Rationality. The Critical View. 1987. ISBN 90-247-3275-1.
24. Srzednicki, J.T.J. and Stachniak, Z. (eds.): S. Leśniewski's Lecture Notes in Logic. 1988. ISBN 90-247-3416-9.
25. Taylor, B.M. (ed.): Michael Dummett. Contributions to Philosophy. 1987. ISBN 90-247-3463-0.
26. Bar-On, A.Z.: The Categories and Principle of Coherence. Whitehead's Theory of Categories in Historical Perspective. 1987. ISBN 90-247-3478-9.
27. Dziemidok, B. and McCormick, P. (eds.): On the Aesthetics of Roman Ingarden. Interpretations and Assessments. 1989. ISBN 0-7923-0071-8
28. Srzednicki, J.T.J. (ed.): Stephan Körner. Philosophical Analysis and Reconstruction. 1987 ISBN 90-247-3543-2.
29. Brentano, F.: On the Existence of God. Lectures given at the Universities of Würzburg and Vienna (1868–1891). 1987. ISBN 90-247-3538-6.
31. Pawlowski, T.: Aesthetic Values. 1989. ISBN 0-7923-0418-7.
32. Ruse, M. (ed.): What the Philosophy of Biology Is. Essays Dedicated to David Hull. 1989 ISBN 90-247-3778-8.
33. Young, J.: Willing and Unwilling: A Study in the Philosophy of Arthur Schopenhauer. 1987 ISBN 90-247-3556-4.
37. Winterbourne, A.: The Ideal and the Real. 1988. ISBN 90-247-3774-5.
38. Szaniawski, K. (ed.): The Vienna Circle and the Lvov-Warsaw School. 1989. ISBN 90-247-3798-2.
39. Priest, G.: In Contradiction. A Study of the Transconsistent. 1987. ISBN 90-247-3630-7.